Maintenance of plant genetic resources –

towards an understanding of seed deterioration

POST-DOCTORAL THESIS

Maintenance of plant genetic resources – towards an understanding of seed deterioration

PD Dr. sc. agr. Manuela Nagel

University of Hohenheim

Faculty of Agricultural Sciences

October 2018

INDEX OF ABBREVIATIONS

a_w	Water activity
AA	Accelerated seed ageing
ABA	Abscisic acid
ABI	*ABSCISIC ACID INSENSITIVE*
AFR	Ascorbic free radical
AFRR	Ascorbic free radical reductase
AP2	*APETALA2*
AP2/EREBP	APETALA2/ethylene-responsive element (ERE)-binding proteins
APR	Adenosine 5'-phosphosulphate reductase
APX	Ascorbate peroxidase
AS	Ambient storage at 20 °C and 50 % RH
ASC	Ascorbate
C_H	Seed viability equation, species-specific temperature constant
C_Q	Seed viability equation, species-specific temperature constant
C_W	Seed viability equation, species-specific moisture constant
CO_2	Carbon dioxide
Canola	Canadian oil, low acid oilseed rape
CAT	Catalase
CBD	Convention on Biological Diversity
CD	Controlled seed deterioration
CENARGEN	Embrapa Recursos Genéticos e Biotecnologia
CGIAR	Consultative Group on International Agricultural Research
CIMMYT	Centro Internacional de Mejoramiento de Maíz y Trigo
CS	Cold storage at -18/0 °C
Cys	Cystein
DArT	Diversity Array Technology markers
DH	Double-haploid
DHA	Dehydroascorbate
DHAR	Dehydroascorbate reductase
DMA/DMTA	Dynamical mechanical (thermal) analysis

DOG	*DELAY OF GERMINATION*
DOI	Digital Object Identifiers
DREB	Dehydration responsive element binding proteins
DSC	Differential scanning calorimetry
DW	Dry weight
γ-ECS	γ-glutamylcysteine synthase
$E_{GSSG/2GSH}$	Glutathione half-cell reduction potential
EPPO	Elevated partial pressure of oxygen
EPR	Electron paramagnetic resonance
ERE	Ethylene-responsive element
ERF	Ethylene response factor
EST	Expressed sequence tag marker
FAO	Food and Agriculture Organization of the United Nations
FT	*FLOWERING TIME*
FUS	*FUSCA*
FW	Fresh weight
GA	Gibberellic acid
GAAS1	*Germination Ability After Storage1*
Glu	Glutamate
Gly	Glycine
GPX	Glutathione peroxidase
GR	Glutathione reductase
GS	Glutathione synthetase
GSH	Reduced glutathione
GSSG	Oxidized glutathione disulphide
GST	Glutathione S-transferase
GWAS	Genome-wide association mapping
H_2	Hydrogen
$HO^{\bullet -}$	Hydroxide radical
$HO_2^{\bullet}$	Hydroperoxyl radical

H_2O	Water
H_2O_2	Hydrogen peroxide
H_3O^+	Hydronium ion
$H_5O_2^+$	Zundel ion
$H_9O_4^+$	Eigen ion
HDO	Semi heavy water (one hydrogen atom is replaced by deuterium)
HGGT	Homogentisate geranylgeranyl transferase
HPT1	Homogentisate phytyltransferase or vitamine E (VTE2)
HSF	Heat shock factor
HSP	Heat shock protein
IBPGR	International Board for Plant Genetic Resources
ICARDA	International Centre for Agricultural Research in the Dry Areas
ICGR-CAAS	Institute of Crop Germplasm Resources - Chinese Academy of Agricultural Sciences
ICRISAT	International Crops Research Institute for the Semi-Arid Tropics
IPK	Leibniz Institute of Plant Genetics and Crop Plant Research
IRRI	International Rice Research Institute
ISTA	International Seed Testing Association
K_E	Seed viability equation, species-specific moisture constant
K_i	Seed viability equation, initial viability of the seed lot at p=0 days
LD	Linkage disequilibrium
LEA	Late embryogenic abundant
LEC	*LEAFY COTYLEDON*
LMW	Low-molecular weight
m	Seed viability equation, seed equilibrium MC
MC	Moisture content
MDA	Malondialdehyde
MTA	Marker-trait association
N_2	Nitrogen
NBPGR	National Bureau of Plant Genetic Resources
NIAS	National Institute of Agrobiological Sciences

NPGS	National Plant Germplasm System
O_2	Oxygen
$^1O_2$	Singlet oxygen
$O_2^{\bullet -}$	Superoxide radical
OWB	Oregon Wolfe Barley mapping population
p	Seed viability equation, survival period
P50	Half-viability period
PGRC	Plant Gene Resources of Canada
PUFA	Polyunsaturated fatty acids
QTL	Quantitative trait locus
RFO	Raffinose family oligosaccharides
RH	Relative humidity
Rht	*REDUCED HEIGHT*
ROS	Reactive oxygen species
SAT	Serine acetyltransferase
SGSV	Svalbard Global Seed Vault
SOD	sodium dismutase
T3	Tocotrienols
T_g	Glass transition temperature
t	Seed viability equation, storage temperature
TF	Transcription factor
TOC	Tocopherols
v	Seed viability equation, viability after any storage period
VIR	N.I. Vavilov all-Russian Scientific Research Institute of Plant Industry
VTE	Vitamin E
VTE1	Tocopherol cyclase
VTE3	VTE3 methyltransferase
ψ	water potential
WC	Water content
Zeo1	*ZEOCRITON1*

1 INTRODUCTION

Agriculture depends on seeds, and the plant's ability to grow across a wide range of environmental conditions. During the last 10,000 years, plant genetic resources have evolved and may harbour allele combinations required for crop improvements in future. To safeguard valuable crop wild relatives, locally adapted landraces and varieties; *ex situ* gene banks were established at the beginning of the 20th century. Most accessions are maintained as seeds under low moisture contents and sub-zero temperatures. However, as any other material on Earth, seeds undergo ageing and loose viability during prolonged storage and, consequently, require frequent rejuvenation.

Orthodox, desiccation tolerant seeds are able to withstand extensive losses in water content and to survive adverse environmental conditions over long terms. The major factors determining seed viability and deterioration processes are the genotype, the environmental conditions during seed development and the storage conditions including relative humidity, temperature, gas composition and pressure. Most components, if not all, affect the water activity in the seed tissue and, consequently, biochemical and thermodynamic mechanisms.

The aim of the post-doctoral thesis is to elucidate differences in physiological, biochemical and genetic mechanisms of seed germination and deterioration in dependency of the seed storage conditions. Fundamental processes are compared in seeds of wheat, barley and oilseed rape genetic resources subjected to long-term ambient and cold storage and artificial ageing conditions using elevated temperatures, increased water contents and in some cases increased oxygen levels and atmospheric pressures. The relevance of artificial ageing for seed preservation and the relationship between seed germination, seed dormancy and seed longevity and the water content is discussed in conjunction with the genetic background and seed's morphology and chemical composition.

2 PROLOG

2.1 Maintenance of plant genetic resources in *ex situ* gene banks

2.1.1 The importance to collect plant genetic resources

Plant genetic resources are considered as a strategic resource at the heart of sustainable crop production (FAO, 2014). They include the sum of genes, gene combinations or genotypes which are available for the genetic improvement of crops (Gepts, 2006). Since the beginning of agriculture, selection of plants and seeds during sowing, growing, harvest and storage gave a rise of locally adapted varieties, so-called landraces, that reveal a specific variation of morphological and yield characteristics and quality traits (de Carvalho et al., 2013). At the mid of 19th century, the rediscovery of Gregor Mendel's work and the introduction of breeding schemes led to the development of high-yielding and more stress-tolerant varieties at the cost of local landraces (Damania, 2008).

At the end of the 19th century, the vanishing of plant genetic resources had already been recognized which motivated great plant explorers, i.e. Frank N. Meyer, Nicolai I. Vavilov, to initiate collection missions (Damania, 2008; Hummer and Hancock, 2015). At this time, plant genetic resources were generally considered as 'global public good' (Halewood et al., 2018). The legal transfer and exchange of plant material facilitated the establishment of eight international plant resources centres majorly located in the industrialized countries. Due to foundation of the 'Consultative Group on International Agricultural Research' (CGIAR) in 1972 and of the 'International Board for Plant Genetic Resources' (IBPGR) later renamed to 'Bioversity International' in 1975, the number of long-term genetic conservation centres increased to 33 in 1983 (Damania, 2008). The last survey of the 'Food and Agriculture Organization of the United Nations' (FAO) revealed more than 1,750 gene banks, nowadays (FAO, 2010). The *ex situ* approach, where plant genetic resources are maintained out of their natural environments, is supplemented by the *in situ* conservation. Here, species are protected in their natural habitats. In 2010, the FAO estimated a protected area of about 17.5 million km^2 (FAO, 2010). Due to the focus of this work on seed deterioration, *in situ* conservation is not discussed further.

When industrialized countries claimed the international recognition of the 'intellectual property protection for living material', a debate around commercialisation and protection of genetic resources was pushed forward by developing countries (Halewood et al., 2018). Therefore, the 'Convention on Biological Diversity' (CBD) was established in 1993 (CBD, 2018a) and adopted the 'Global Strategy for Plant Conservation'. The strategy includes, i.e. target 8 that aims to keep 75 % of threatened plant species in *ex situ* collections and to make use of 20 % for restoration programs. Target 9 describes the goal to conserve 70 % of the genetic diversity of crops and major socio-economically valuable plant species (CBD, 2018b). The continuing discussions about the recognition of sovereign rights to regulate access to genetic resources and benefit sharing agreements resulted in the development of the 'Nagoya Protocol on Access to Genetic Resources and the Fair and Equitable Sharing of Benefits Arising from their Utilization' (Nagoya Protocol) which came into force in 2014.

Meanwhile, in 2004, the 'International Treaty on Plant Genetic Resources for Food and Agricultur' created a multilateral system of access and benefit-sharing for contracting parties and international organisations. Furthermore, access to the genetic diversity of 64 crops (Annex I) for the purposes of conservation, research, training and plant breeding was enabled and commercial users are obligated to make financial payments to an international benefit-sharing fund (Halewood et al., 2018). The first payment to the 'Treaty's Benefit-sharing Fund' was made by Nunhems Netherland in July 2018 (Nunhems Netherlands, 2018).

2.1.2 Long-term storage and utilization of *ex situ* collections

Since the 16th century, botanical gardens have collected and preserved a variation of more than 80,000 plant species in about 2,500 collections. *Ex situ* gene banks have been established since the mid of the 20th century. They preserve more than 7.4 million accessions in about 1,750 facilities and focus on the maintenance of genetic diversity of crop species and their wild relatives (FAO, 2010). About 45 % of the accessions are cereals, i.e. wheat (*Triticum aestivum* L.), rice (*Oryza sativa* L.), barley (*Hordeum vulgare* L.), maize (*Zea mays* L.) and *Sorghum* (L.) Moench, followed by food legumes (15 %), forages (9 %) and vegetables (7 %). A quarter of the material are landraces and wild species, whereby major parts originated in North and South America, Europe and South Asia. It is estimated that only 1.9 to 2.2 million of the 7.4 million accessions are unique (FAO, 2010). Among the 10 largest centres for plant conservation are those held by the CGIAR, i.e. the 'Centro Internacional de Mejoramiento de Maíz y Trigo' (CIMMYT) and the 'International Centre for Agricultural Research in the Dry Areas' (ICARDA) (Table 1). In 2008, the first and only global germplasm conservation facility, the 'Svalbard

Global Seed Vault' (SGSV) built by the Norwegian government and operated by the 'Global Crop Diversity Trust' and NordGen (FAO, 2010) opened and houses about 1,061,909 accessions currently (SGSV, 2018).

Table 1. The ten biggest *ex situ* gene banks storing crop genetic resources. Based on FAO (2010) and updated in * June 2018 based on IPK (2018) and personal communication with Chinese gene bank manager and *2 September 2018 based on SGSV (2018).

	Gene bank	Location	Genus	Species	Accessions	SGSV
1	NPGS	USA	2,128	11,815	508,994	30,868
2	ICGR-CAAS	China	-	2,386*	470,295*	-
3	NBPGR	India	723	1,495	366,333	-
4	VIR	Russia	256	2,025	322,238	945
5	NIAS	Japan	341	1,409	243,463	
6	CIMMYT	Mexico	12	48	173,571	80,492
7	IPK	Germany	756	3,127*	150,751*	48,655*
8	ICARDA	Lebanon*	86	570	154,000*	62,834
9	ICRISAT	India	16	180	118,882	20,003
10	IRRI	Philippines	11	23	109,161	4,008
	SGSV	Norway	>664	5,979*2	1,061,909*2	-

CIMMYT, Centro Internacional de Mejoramiento de Maíz y Trigo (Mexico); ICARDA, International Centre for Agricultural Research in the Dry Areas (Lebanon); ICGR-CAAS, Institute of Crop Germplasm Resources, Chinese Academy of Agricultural Sciences; ICRISAT, International Crops Research Institute for the Semi-Arid Tropics (India); IPK, Leibniz Institute of Plant Genetics and Crop Plant Research (Germany); IRRI, International Rice Research Institute (Philippines); NBPGR, National Bureau of Plant Genetic Resources (India); NIAS, National Institute of Agrobiological Sciences (Japan) NPGS, National Plant Germplasm System (United States of America); SGSV, Svalbard Global Seed Vault; VIR, N.I. Vavilov all-Russian Scientific Research Institute of Plant Industry (Russian Federation)

The international 'Genebank Standards for Plant Genetic Resources for Food and Agriculture' provide the guidelines the maintenance of plant genetic diversity. In 2014, the standards were renewed and aspects on the legal status, physical security, identity of accessions, monitoring of viability and genetic integrity were updated. In general, it was agreed to dry orthodox seeds between 5 °C and 20 °C and 10 % and 25 % relative humidity (RH) and initial germination should exceed 85 % for most cultivated crop. Seeds in active collections are used for distribution and are held between 5 °C and 10 °C and 15 % RH for about 30 years. Seeds in long-term base collections should maintain high seed quality for more than 30 years. Therefore, seeds are packed in airtight containers and stored between -20 °C and -15 °C at different sites. Some repositories can only provide short-term storage conditions. Here, the minimum requirements are that high seed quality should be preserved at stable temperatures of maximum 25 °C for up to eight years. However, at any case, an active viability monitoring program should ensure that viability is tested before it drops below 85 % of the initial viability (FAO, 2014).

To capture the genetic variability of gene bank collections, molecular marker assisted germplasm curation offers a powerful tool for germplasm managers, basic researchers, and plant breeders. In the past, molecular information generated in studies to investigate the diversity, domestication, evolution and phylogeny of plant genetic resources have not sufficiently used (Kilian and Graner, 2012; McCouch et al., 2012). Therefore, FAIR principles of Findability, Accessibility, Interoperability and Reusability (Wilkinson et al., 2016) and an improved meta data management system is acquired to facilitate data integration and the potential for data sharing (Halewood et al., 2018). To promote the networking of high quality data repositories, the 'Global Information System' of the 'International Treaty on Plant Genetic Resources for Food and Agriculture' plans to link existing information systems by using 'Digital Object Identifiers' (DOIs) (Alercia et al., 2018). The major goals are to take advantage of the ongoing revolution in the exploration, manipulation and synthesis of biological systems, to increase the efficiency and effectiveness of conservation, trait discovery and utilization of plant genetic resources in gene banks (Halewood et al., 2018).

However, the security status of the collections should be prioritized and range ahead utilization and exploration of the genetic diversity. The need for a backup system of gene banks have been exemplified by ICARDA. Due to the civil war in Syria, the institute was forced to leave the collections in Aleppo and move to the Lebanon. Fortunately, most accessions were backed up in Svalbard and in total, 38,073 seed samples have been delivered to Morocco and Lebanon since 2015. The ancient varieties, majorly seeds of wheat, barley, lentil (*Lens culinaris* Medikus), chickpea (*Cicer arietinum* L.), wild cereals and pulses were used to re-establish an active collection (Crop Trust, 2015) which comprises about 154,000 accessions currently (ICARDA, 2016). In general, many *ex situ* collections are still maintained under suboptimal conditions which have negative consequences on the viability status (FAO, 2010). Low seed viability involves the necessary of rejuvenation. Seed regeneration in gene banks is mostly limited to small field plots which increase the potential to create population bottlenecks and to loose genetic diversity (Parzies et al., 2000). Therefore, the maintenance of a high quality seed collection and a sufficient control of seed deterioration should be prioritized for gene bank collections.

2.1.3 Wheat genetic resources

Wheat (*Triticum aestivum* L. and *Triticum durum* Desf.) is among the 'big three' cereal crops and adapted to a wide range of temperate environments (Shewry, 2009). After maize and rice, 750 million tonnes of wheat seeds are produced annually on more than 220 million ha (www.fao.org/faostat). In world's gene banks, 856,000 wheat accessions are maintained whereby about 13 % of the accessions are stored at CIMMYT in Mexico, 7 % at the 'National Plant Germplasm System' (NPGS) in the USA, 5 % at the 'Institute of Crop Germplasm Resources - Chinese Academy of Agricultural Sciences' (ICGR-CAAS) in China and 4 % at the 'National Bureau of Plant Genetic Resources' (NBPGR) in India in 2010 (FAO, 2010). In Gatersleben, at the 'Leibniz-Institute of Plant Genetics and Crop Plant Research' (IPK), 28,206 accessions have been preserved in 2018 (IPK, 2018).

The great genetic diversity of bread wheat (*Triticum aestivum* L.) has been developed as part of the 'Neolithic Revolution' over the past 10,000 years. During domestication, the tetraploid cultivated emmer (AABB) hybridized with unrelated wild grass *Aegilops tauschii* Coss. (DD) and formed the allohexaploid wheat (2n=6x=42) with three sub-genomes, A, B, D. The A genomes of tetra- and hexaploid wheat are closely related to the A genomes of wild and cultivated einkorn and the B genomes are derived from the Sitopsis section of *Aegilops*. Here, *Aegilops speltoides* Tausch is the closest species (Shewry, 2009). Just recently, the 'International Wheat Genome Sequencing Consortium' achieved to annotate a reference sequence in the form of 21 chromosome-like sequence assemblies that is giving an access to 107,891 high-confidence genes (IWGSC, 2018).

The wheat caryopsis (Figure 1), for simplicity termed seed, comprises a large starchy endosperm, the embryo and the pericarp fused with the seed coat (Xiong et al., 2013). The endosperm cells contain 60 % to 70 % starch and 10 % to 15 % storage protein which form the gluten protein fraction. The unique composition of wheat seeds determines its functional properties among milling efficiency, bread making and nutritional value. During domestication, wheat changed from a hulled to the free-threshing naked form (Shewry, 2009).

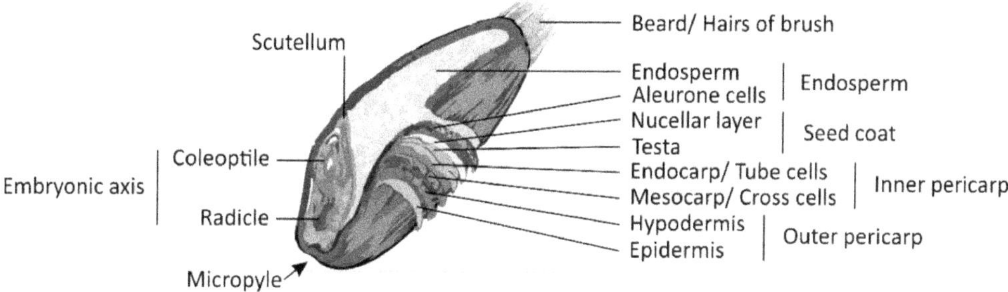

Figure 1. Wheat caryopsis including major structures. Based on Rathjen et al. (2009).

2.1.4 Barley genetic resources

Barley (*Hordeum vulgare* L.) has been important since the dawn of agriculture and was among the earliest domesticated crops (Mayer et al., 2012). Barley can be grown over a wide range of environmental conditions, from 70°N in Norway to 46°S in Chile (Grando and Gormez Macpherson, 2005). About 141 million tonnes of seeds are produced annually for food, feed and the malting industry (www.fao.org/faostat). The genetic variation is presented by 466,531 accessions, whereby 23 % of the material is classified as landraces. About 9 % of the collections is hold by 'Plant Gene Resources of Canada' (PGRC), 6 % by the NPGS in the USA, 6 % by the 'Embrapa Recursos Genéticos e Biotecnologia' (CENARGEN) in Brazil, 6 % by ICARDA. About 23,600 accessions, 5 % of the total collection, are preserved at IPK in Germany (IPK, 2018).

Barley is a diploid (2n=2x=14), inbreeding plant that has considered as a model plant for cereal genetics. *Hordeum vulgare* ssp. *spontaneum* is the immediate ancestor (Pankin and von Korff, 2017). During domestication, a great diversity in morphological forms evolved ranging between spring, winter, 6-row, 2-row, awned, awnless, hooded, hulled and naked, feed, malting and food type barleys (Fan, 2017). In 2012, the International Barley Genome Sequencing Consortium presented an ordered physical, genetic and functional sequence resource including a physical map of 4.98 Gb and 26,159 high confidence genes (Mayer et al., 2012). This achievement is supplemented by map-based reference sequence of the barley genome including the first comprehensively ordered assembly of the pericentromeric regions of a Triticeae genome (Mascher et al., 2017) and the first barley Pan-genome which will be published shortly.

Comparable with wheat, barley produces a caryopsis (Figure 2), for simplicity termed seed. Interestingly, only barley varieties show a strong hull-caryopsis adhesion in which the hull (outer lemma and inner palea) is firmly adherent to the pericarp epidermis at maturity. Few accessions are of a free-threshing variant called naked (hulless) barley (Taketa et al., 2008). About 70 % of the barley seed consists of starch, whereby amylose content can vary between 0 % and 100 %. The protein content of the starch ranges between 0.07 % and 0.3 % and the amount of free and bound lipids are estimated to between 0.05 and 0.85 %. Due to functional components such as tocochromanols and β-glucan, naked barley seeds are preferred for human food (Fan, 2017).

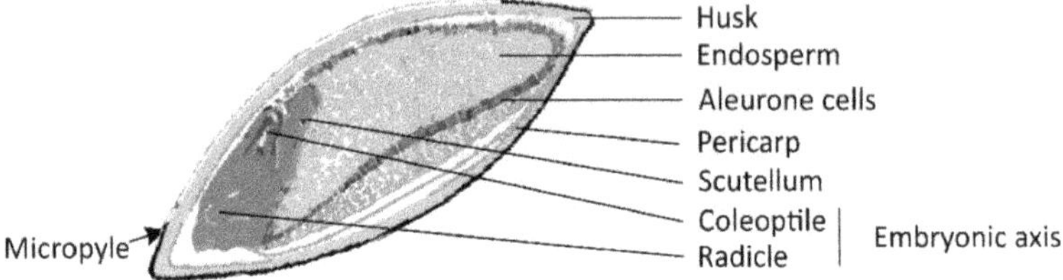

Figure 2. Barley caryopsis including major structures. Based on Fox (2010).

2.1.5 Oil seed rape genetic resources

About 16 % of the global oil production, produced by 69 million tonnes of seeds, is contributed by oilseed rape (*Brassica napus* L.) (www.fao.org/faostat). Only 25,566 oilseed rape accessions are kept in gene banks, whereby the ICGR-CAAS in China and the NBPGR in India maintain 16 % and 10 % of the collection, respectively (FAO, 2010). The IPK preserves 1,205 accessions at the satellite collections North on the Island Poel, Germany (IPK, 2018).

The oilseed rape genome (AACC, 4n=2x=38) comprises about 1.2 Gb and has been formed about 7,500 years ago (Chalhoub et al., 2014). The short domestication history is assumed to be the reason for the limited genetic variability. Oilseed rape derived from a small number of hybridization events between the diploid progenitors *Brassica rapa* L. and *Brassica oleracea* L. which contributed the A and C genomes, respectively (Bancroft et al., 2011). Since the Middle Ages, oil seed rape cultivation have began in Europe and led to the selection of valuable agronomic traits such as those for oil biosynthesis, glucosinolate contents, disease resistance and flowering time (Chalhoub et al., 2014). Canadian oil, low acid (Canola) also known as low erucic acid-type oil seed rape, derived from the selection of plants having an improved seed composition. Those seeds are characterized by high yield, low erucic acid and glucosinolate content and are predominantly used for culinary purpose (Wang et al., 2017). High erucic acid-types are used in the industry (Woodfield et al., 2017). Recent releases of the oilseed rape genome (Chalhoub et al., 2014; Schmutzer et al., 2015; Bayer et al., 2017) enabled first predictions of gene content. However, due to problems of mis-annotations, it is still a challenging task (Bayer et al., 2017).

The oilseed rape embryo comprises a central embryonic axis embraced by two cotyledons (Figure 3). The whole embryo is encased by a liquid endosperm, a cellular aleurone layer, and a seed coat. When the embryo imbibes and germination is initiated, the two cotyledons fold in toward the embryonic axis, one staying outermost while the other becomes restricted to the inner part of the seed. During seed development seeds accumulate about 45 % oil by dry weight (DW) which consists of 62 % oleic acid (18:1), 22 % linoleic acid (18:2) and 10 % α-linolenic acid (18:3) (Woodfield et al., 2017). In addition, the inner integuments produce significant amounts of flavonoids assuming a function as protective substance (Moise et al., 2005).

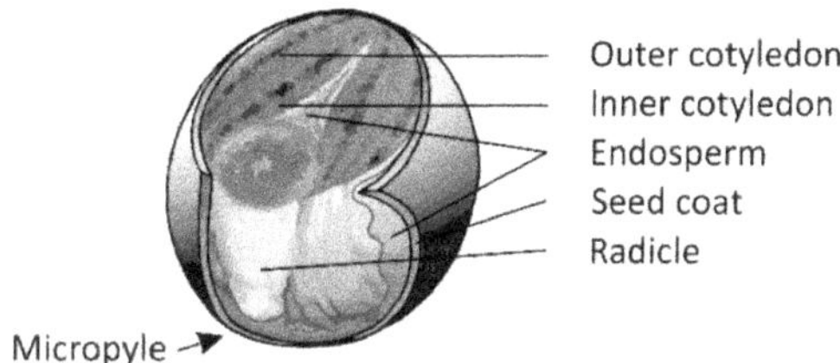

Figure 3. Oilseed rape seed including major structures. Based on Munz et al. (2017).

2.2 Survival in the dry state

Life on Earth fundamentally depends on water. Over 1 billion years ago, plant terrestrialization was only possible by the transition of plants to live on lands including adaptation to high-stress environments (Vries and Archibald, 2018). Land plants, formally called embryophytes (Adl et al., 2005), have developed important features to survive fluctuating environments (Delwiche and Cooper, 2015). Most important ones were the acquisition of desiccation tolerance in parallel with the acquisition of dormancy. Dormant organisms can reduce metabolic functions to a minimum under unfavourable conditions and resume their activities when the conditions become supportive to life (Lubzens et al., 2010; Costa et al., 2016). The outstanding features of desiccation tolerance are the ability to survive loss of more than 90 % of cellular water

(≤0.1 g H_2O g^{-1} DW) (Leprince et al., 2017), the rapidity of rehydration and recovery, the tolerance against multiple cycles of dehydration and rehydration, the stability of the dry tissue and its longevity in the dry state (Gaff and Oliver, 2013).

Vegetative desiccation tolerance has been first developed in bryophytes (Oliver et al., 2000) and is widespread in less complex clades such as algae and lichens; but rare in larger and more complex groups of vascular land plants (Oliver et al., 2000; Hilhorst et al., 2018). It is assumed that genes evolved for cellular protection and repair had been recruited for other tolerance mechanisms. However, genes were conserved in seeds and re-evolved multiple times in the history of flowering plants (angiosperms) (Oliver et al., 2000). About 135 angiosperm species display a certain degree of desiccation tolerance in the vegetative tissue. Most commonly, their reproductive organs, seed embryos and pollen (Gaff and Oliver, 2013), are desiccation tolerant and seeds of most of them are considered as orthodox (Table 2). About 95 % of the spermatophytes, angiosperms and gymnosperms, produce orthodox seeds with the ability to survive long-term storage. In contrast, about 3 % have 'recalcitrant' seeds. These seeds are desiccation- and chilling sensitive and have short life spans. About 1 % is categorized as 'intermediate' of variable degrees of desiccation and cold tolerances (Kew, 2018).

Table 2. The numbers of angiosperm and gymnosperm species with desiccation tolerant or desiccation-sensitive seeds. Seed types are categorised according to the Royal Botanic Gardens Kew Seed Information Database (Kew, 2018).

Seed type	Number of species	% of the total species
Orthodox	18,858	76.2
Presumably orthodox	4,587	18.5
Total orthodox seeds	**23,445**	**94.7**
Intermediate	73	0.3
Presumably intermediate	77	0.3
Total intermediate seeds	**150**	**0.6**
Recalcitrant	345	1.4
Presumably recalcitrant	275	1.1
Total recalcitrant seeds	**620**	**2.5**
Uncertain	540	2.2

The extraordinary stability of orthodox seeds to resist prolonged storage have been fascinated researchers over centuries. Around 370 BC, Theophrastus of Lesbos, the father of Botany, recognized the importance of the dry state for seeds and that seeds differ in their germinability according to the place in which they are stored (Leprince and Buitink, 2015). In the 19th century, in parallel the American William Beal and the Austrian Friedrich Haberlandt initiated long-term seed storage experiments (Steiner and Ruckenbauer, 1995; Telewski and Zeevaart, 2002). By accidence, Friedrich Haberlandt died but the crop seeds stored at ambient temperatures and ultra-dry conditions [<5 % moisture content (MC)] were discovered and germinated up to 90 % after 110 years (Steiner and Ruckenbauer, 1995). In contrast, William Beal mixed seeds of 23 different species with moist sand, buried and monitored them for selected intervals. The last excavation revealed that most seeds lost germination after 40 years of storage. However, seeds of *Verbascum* have survived 120 years of burial (Telewski and Zeevaart, 2002).

Nowadays, radio-carbon dating can estimate precisely the age of tissues and has elucidated that seeds may even survive thousands of years. Most prominent, the seeds of *Nelumbo nucifera* Gaertn. found in north-eastern China (Shen-Miller et al., 1995) and the seeds of a date palm (*Phoenix dactylifera* L.) found near the Dead Sea (Sallon et al., 2008) grow to fully developed plants after 1,300 years and 2,000 years of storage, respectively. However, even more spectacular but disputed (Zazula et al., 2009), *Silene stenophylla* Ledeb. plants were regenerated from tissues of 30,000 years old fruits (Yashina et al., 2002; Yashina et al., 2012) which would be, if true, the most impressive example of extreme tissue survival.

The first systematic studies on detailed deterioration processes were conducted by Roberts (1961) who developed the artificial ageing approaches and the seed viability equation. He postulated that a simple mathematical relationship between temperature and MC can be used to predict the expected life of the seed; or alternatively, predict various combinations of storage conditions necessary to achieve a required period of viability.

The ability of seeds to survive in the dry state is a result of molecular processes during maturation drying and involves the acquisition of desiccation tolerance followed by the acquisition of seed longevity (Leprince et al., 2017).

2.3 The establishment of seed longevity during seed maturation

Seed maturation is considered as the final period of seed development that is essential for the synthesis of storage compounds, i.e. starch, storage proteins, oil, and is associated with a reorganization of the metabolism (Angelovici et al., 2010). A schematic and simplified overview covering physiological and molecular processes across different species is given in Figure 4.

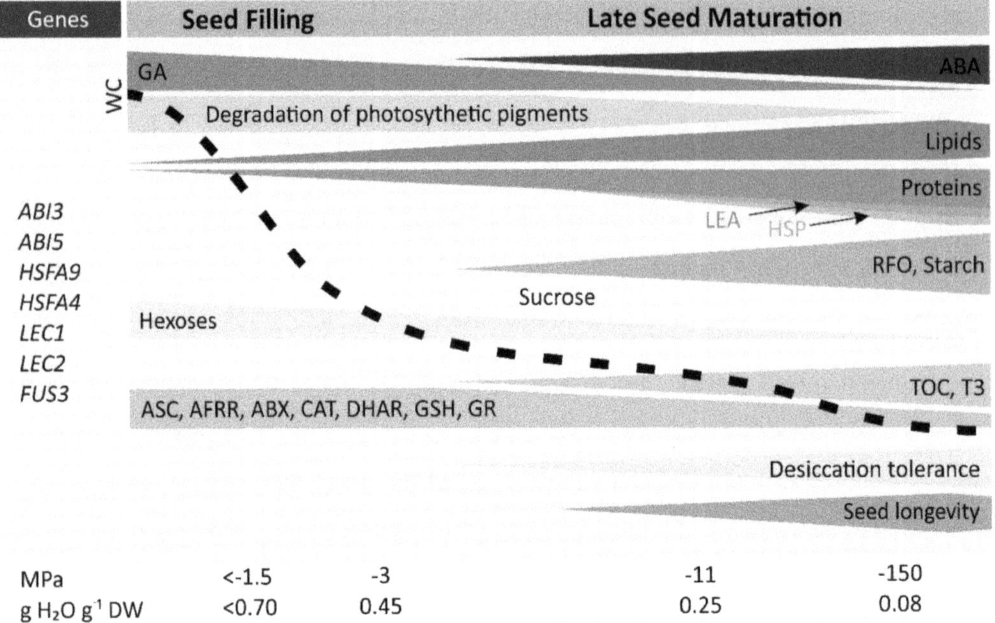

Figure 4: Metabolic and physiological changes during seed maturation. Schematic overview comprises data of transcriptional regulation (Holdsworth et al., 2008), storage reserves in Arabidopsis (*Arabidopsis thaliana* L.) (Baud et al., 2002), in faba bean (*Vicia faba* L.) (Borisjuk et al., 2002), in durum wheat (*Triticum durum* Desf.) (De Gara et al., 2003) and in bread wheat (*Triticum aestivum L.)* (Lehner et al., 2006) including antioxidants, hormone signalling in wheat (Kong et al., 2015) and physiological response in *Medicago truncatula* (Righetti et al., 2015). MPa shows water potential (Costa et al., 2017) and g H_2O g^{-1} DW corresponds to the water content (WC) of the curve above shown during maturation. ABA, abscisic acid; AFRR, ascorbic free radical reductase; ABI; abscic acid insensitive; APX, ascorbate peroxidase; ASC, ascorbate; CAT, catalase, DHAR, dehydroascorbate reductase; FUS; Fusca; GA, gibberellic acid; GR, glutathione reductase; GSH, glutathione; HSF, heat shock factor; HSP, heat shock protein; LEA, late embryogenesis abundant; LEC, leafy cotyledon; RFO, raffinose family oligosaccharides; TOC, tocopherols; T3, tocotrienols.

Late seed maturation is considered as the end of seed filling, when the embryo and/or endosperm ceases to expand. It varies widely between species, ranging from 78 % of the total developmental time in rice to 10 % in red cabbage (Leprince et al., 2017). Most seeds have already obtained their germination capacity before they acquire desiccation tolerance and enter the later maturation phase. The reverse is true for wheat and barley and is assumed to be related to the presence of seed dormancy (Probert et al., 2007). Dormant seeds are intact, viable seeds but incapable to complete germination under favourable environmental conditions (Baskin and Baskin, 2004; Finch-Savage and Leubner-Metzger, 2006). During the late maturation phase, seed longevity increases gradually to 30- till 50- fold (Leprince et al., 2017).

Seed longevity is at its maximum during final drying when water content (WC) decreases below 0.1 g H_2O g^{-1} DW and the cytoplasm transforms into a glass (Ballesteros and Walters, 2011).

The acquisition of desiccation tolerance and seed longevity is related with a plethora on molecular events which are controlled by a network of transcriptional regulators. In Arabidopsis, four major gene, namely *FUSCA3* (*FUS3*), *ABSCISIC ACID INSENSITIVE 3* (*ABI3*), *LEAFY COTYLEDON 1* (*LEC1*) and *LEC2* have been identified which severely affect desiccation tolerance, seed longevity and/or dormancy (Gutierrez et al., 2007; Holdsworth et al., 2008). *ABI3*, *FUS3* and *LEC2* encode related plant-specific transcription factors (TFs). Deficiencies in *ABI3* or *LEC1* lead to lower protein denaturation temperature (Wolkers et al., 1998). Plants in which *ABI3* is knocked down show the absence of chlorophyll degradation and a reduced sensitivity to abscisic acid (ABA) during maturation (Delmas et al., 2013). *LEC1* is required for normal development during embryogenesis and *lec1* mutant plants have defects in the cotyledon identity (Holdsworth et al., 2008). *FUS3* controls embryo-derived dormancy and determination of cotyledon epidermis cell identity (Tiedemann et al., 2008).

Changes in the transcriptome have downstream effects on the hormone signalling. Gibberellins (GA) are required in the earlier phases of seed development (Singh et al., 2002) and were shown to be regulated by *LEC2* and *FUS3*. In later phases, the synthesis is down-regulated by a cross-talk of ABA, ethylene, jasmonic acid and/or auxin (Curaba et al., 2004; Sreenivasulu et al., 2006; Kong et al., 2015). Thereby, an increase of ABA is necessary to acquire desiccation tolerance and dormancy (Costa et al., 2015) and to mediate responses to environmental stresses (Weber et al., 2005). ABA inhibits cell-cycle progression, the storage product synthesis (Sreenivasulu et al., 2006), embryo growth and germination. Thereby, the ABA/GA ratio has turned out as a key regulator of both maturation and germination processes (Santos-Mendoza et al., 2008).

The most visible sign of late seed maturation is the degradation of chlorophyll. In Arabidopsis (*Arabidopsis thaliana* (L.) Heynh.), the presence of a higher chlorophyll content in dry mature seeds of the double mutant *delay of germination1-1 (dog1-1) abi3-1* was related with a significant reduction of seed longevity (Dekkers et al., 2016). Seed longevity was also significantly negatively correlated with chlorophyll fluorescence in rice (Hay et al., 2015) indicating that the loss of chlorophyll is required for an extended storability for most species. Leprince et al.

(2017) hypothesized that chlorophyll breakdown is required for the accumulation of the antioxidant tocopherol. In leaves, chlorophyll serves as a substrate for phytol which can be phosphorylated to phytyl-phosphate and phytyl-diphosphate that is required for the tocopherol synthesis controlled via *VITAMIN E6* (*VTE6)* (vom Dorp et al., 2015).

The late maturation phase is mostly associated with changes in the content and compositions of soluble sugars (Weschke et al., 2003). Sucrose is the major product of photosynthesis, transported to the seed and used to synthesize arabinose, galactose or raffinose family oligosaccharides (RFO). RFOs are mono-, di-, tri- and tetra-galactosides of sucrose including raffinose, stachyose, verbascose and ajugose, respectively (Obendorf and Gorecki, 2012). About 60 % to 80 % of total soluble sugars are sucrose or RFOs (Leprince et al., 2017) and may contribute to the glass formation by stabilizing membranes during desiccation drying (Angelovici et al., 2010). Sucrose is also assumed to act as signal molecule triggering storage associated processes (Koch, 2004). A low ratio of mono- to oligosaccharides rather than total content was suggested to control the acquisition of desiccation tolerance (Wolkers et al., 1998). However, their functional role in relation to seed longevity is not well understood (Buitink and Leprince, 2008).

A gradual increase of the expression of heat shock proteins (HSPs) and late embryogenic abundant (LEA) proteins is essential in the late maturation phase (Verdier et al., 2013). HSPs were the first molecular chaperones studied. They protect proteins from unfolding-induced aggregation and may assist refolding into the native conformation and minimize the impact of environmental variations on the proteome (Jacob et al., 2017). Chaperones are induced by heat shock factors (HSFs) which belong to a large and highly divers family of TFs. Most increase their abundance in response to high temperatures and other stresses such as *HEAT SHOCK FACTOR A3* (*HSFA3)* during chilling (Re et al., 2017). Co-expression of *HSFA9* and *HSFA4*, that are only expressed in seeds, contribute to seed longevity and desiccation tolerance in different species (Almoguera et al., 2015). Although LEA proteins do not show classical chaperone function, i.e. prevention of heat aggregation of enzymes, some of them preserve the activity of enzymes after desiccation and stabilize membranes during freezing and desiccation. About 51 LEA proteins categorized in nine groups are described for Arabidopsis (Tolleter et al., 2010; Hincha and Thalhammer, 2012) and their abundance coincides with the final decrease in WC (Candat et al., 2014). In *Medicago truncatula* Gaertn., seeds showing a knock-out for the TF

Mtabi5 had less LEA polypeptides, accumulated less RFOs and were significantly impaired in seed longevity and dormancy (Zinsmeister et al., 2016).

To ensure survival in the dry state, seeds activate regulatory defence pathways to set up protective processes against oxidative stress. During maturation drying, the changes in WC facilitate a reorganization of the protective compounds (Santos-Mendoza et al., 2008). In wheat, the metabolic changes lead to reduction of enzyme activities, i.e. ascorbic free radical reductase (AFRR), dehydroascorbate reductase (DHAR), ascorbate peroxidase (APX), glutathione reductase (GR), catalase (CAT), that are either responsible for the recycling of the redox couples dehydroascorbate (DHA)/ascorbate (ASC) and glutathione disulphide (GSSG)/glutathione (GSH) and/or detoxification of hydrogen peroxide (H_2O_2). At full maturity, except for a low amount of DHA, ASC vanished completely whereas GSSG/GSH reduced slightly but shifted to reducing conditions (De Gara et al., 2003). The lipophilic antioxidants tocotrienols (T3) and tocopherols (TOC), both commonly termed tocochromanols, accumulated during desiccation drying until 80 % of the final DW was reached in barley (Falk et al., 2004).

In conclusion, the establishment of desiccation tolerance and seed longevity has some common aspects between the species. However, great variations in seed composition and environmental impacts lead to differences in acquisition time including consequences on mass maturity and natural seed dispersal (Probert et al., 2007).

2.4 The physics of water and water activity in seeds

After hydrogen (H_2), water (H_2O) is the second most common molecule in the Universe. It covers two-third of our planet and is involved in all biological processes. Liquid water functions as a solvent, a solute, a reactant, a catalyst and a biomolecule, structuring proteins, nucleic acid and gels (Ball, 2008).

Water simply consists of two hydrogen atoms attached to single oxygen atom and forms a transparent, odourless, tasteless and ubiquitous substance. Purified liquid water consists of a mixture of molecules and ions, including H_2O, semi heavy water (one hydrogen atom is replaced by deuterium) HDO ($\approx 10^{-2}$ %), hydronium and hydroxide ions (H_3O^+ and $HO^{\bullet -}$ $\approx 10^{-6}$ %), H_2O_2 ($\approx 10^{-7}$ %), carbon dioxide (CO_2 $\approx 10^{-4}$ %), oxygen (O_2 $\approx 10^{-4}$ %) and nitrogen (N_2 $\approx 10^{-3}$ %) (Needham, 2008). Water is smaller and lighter than other common atmospheric molecules, such as O_2 and N_2, which might explain the complexity of its action. In an ideal tetrahedral structure, water can possess four hydrogen bonds: two accepting and two donating (Chaplin, 2011; Chen et al., 2017) (Figure 5).

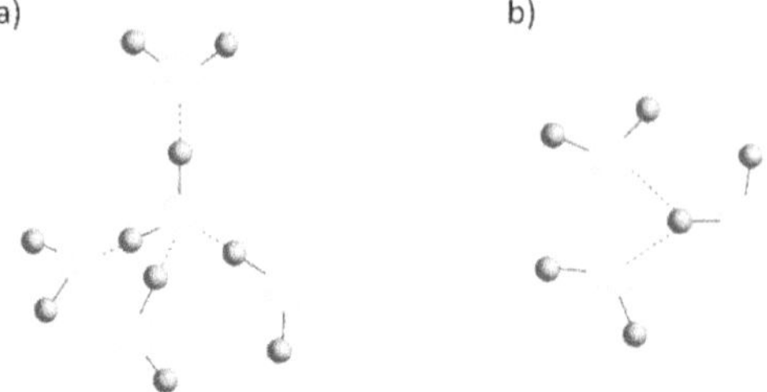

Figure 5. Water molecules shown in a) an ideally tetrahedral coordination geometry and b) with bifurcated bonds. In the latter one hydrogen bonds are shared between two water molecules. Based on Chaplin (2018).

The differences in atom mass (oxygen is 16 times heavier than hydrogen) facilitate molecular rotation and movements, i.e. symmetric and asymmetric stretches, librations and bends (Chaplin, 2011). The opposite charges, negatively charged oxygen atoms and positively charged hydrogen atoms, cause that the water molecules attract each other. When the O-H bond from one molecule points at a nearby oxygen atom in a straight line (O-H·O), the so-called hydrogen bonding is formed. However, the molecules are very unstable. Thermal fluctuations break and reform hydrogen bonds and distort the tetrahedral structures in liquid water. The average lifetime of hydrogen bonds is about 1 ps. The hydrogen atoms are constantly exchanging between water molecules. Therefore, only 80 % of the molecule electrons are considered to be bond (Chaplin, 2018). In addition, the hydrogen bond network is produced by a delicate competition among 1) covalent bonds, 2) the hydrogen bonds and 3) interactions of Van der Waals forces and generate a variety of anomalous properties (Chen et al., 2017). Chaplin (2018) counted 74 anomalous properties, among those, water expands on freezing. The modelling of this complex interplay has been a highly challenging task (Chen et al., 2017).

In actively growing plants, the relative WC, equivalent to MC, is considered between 85 % and 100 % (Gaff and Oliver, 2013). Dependent on the purpose, cells differ in the WC. However, the intracellular environment is extremely crowded. Estimates show that the concentration of biological macromolecules, i.e. proteins, nucleic acids, ribonucleoproteins, polysaccharides, inside cells is between 80 and 400 mg ml^{-1}. This corresponds to a volume occupancy of 5 % to 40 % (Kuznetsova et al., 2014). The remaining water might be altered by solutes or attracted by hydrophilic interfaces. Hydrophobic interactions of proteins are assumed to be the driving force for protein folding (Baldwin, 2014; Ball, 2017). In addition, hydrogen bonded chains are considered as "water wires" and translocate the protonic charge via H_3O^+, Zundel ($H_5O_2^+$) or Eigen ($H_9O_4^+$) cation into and through proteins (Swanson et al., 2007). The molecular crowding and interactions restrict the diffusion rates and the molecular motions which alter the state and dynamics of water in the cytoplasm (Thoke et al., 2018). Around 10 % to 25 % of the water molecules in cells have a slower orientational dynamics, so-called slow waters, and experience longer relaxation times (Persson and Halle, 2008; Ball, 2017). Furthermore, the average distance of macromolecules in cells is around 1 nm which corresponds to about 4 layers of water molecules which can hardly considered as free water (Ball, 2017). To estimate the effects of water in systems that restrict the water availability, different methods of quantification have been introduced.

The WC on a wet or fresh weight (FW) basis, also referred to as MC, has been adopted by the International Seed Testing Association (ISTA, 2018). It describes the percentage mass fraction of water of the total tissue mass (Equation 1).

$$WC\left[\%\,\mathrm{FW}\right]=\frac{(FW-DW)}{FW}\cdot 100 \qquad (1)$$

During dehydration the WC (% FW) does not necessarily reflect the exact extent of water loss because the change of the WC is the change of the reciprocal of tissue FW. For dehydration measurements of plant tissues, the WC on DW basis (Equation 2) have often been preferred. The change of WC is proportional to the loss of water in the tissue (Sun, 2002).

$$WC\left[\mathrm{g\,H_2O\,g^{-1}\,DW}\right]=\frac{(FW-DW)}{DW}\cdot 100 \qquad (2)$$

For the survival of organisms under water stress, the effective available water is more important than the WC. Therefore, the water potential (ψ) and the water activity are commonly measured parameters. The water potential describes the potential energy of water per unit mass and combines hygrostatic (ψ_P), osmotic (ψ_π) and gravitational (ψ_h) components. At higher water potentials, all binding sites are saturated and solvent water is available for biochemical activities (Sun, 2002; Bewley et al., 2013). The term 'water activity' (a_w) describes the equilibrium amount of water available for hydration of materials and thermodynamic processes (Equation 3). It is related to the equilibrium RH of the surrounding system.

$$RH[\%] = a_w \cdot 100$$

It has been shown to be relevant for specific processes occurring during dehydration/desiccation and rehydration/seed imbibition. At a_w=0, water is either absent or completely bond to solutes and surfaces and not available for other interactions. An a_w of 1 indicates that free water is available. It can be determined by the hygrometric instrument methods which measure the equilibrium RH and, in seed science, the relation between WC and a_w is often shown as sorption isotherms (Figure 6). The a_w is dependent on the species, or even genotype, and usually increases with temperature or pressure (Sun, 2002; Chaplin, 2018).

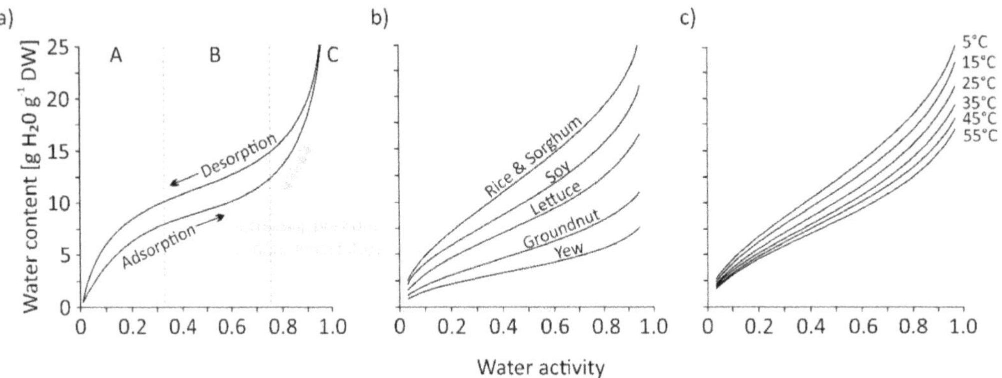

Figure 6. Seed moisture sorption isotherms. The relationship between water content and water activity is shown in a) for desorption (drying) and adsorption (imbibition) in conjunction with binding sites whereby A refers to strongly bound monolayers, B to less strongly bound water layers and capillary adsorbed water and C to solvent and free water. b) Isotherms of seeds from different species distinguish, and are dependent on the seed composition and c) on the temperature changes. Based on Chaplin (2018) and Bewley et al. (2013).

Water is essential for plant growth and development and it functions as a plasticizer for proteins and membranes (Chaplin, 2018). At temperatures below 0 °C, water crystallizes and expands which may disrupt most plant cells. During freezing, all hydrogen bonds to the four neighbouring water molecules are occupied and can form eighteen different crystalline phases (Bartels-Rausch et al., 2012). Thereby, water is very bad glass former. In nature a water glass is formed by vapour deposition on extremely cold surfaces. Artificially, it is achieved by compression of ordinary ice until it amorphized. The glass transition temperature is considered to be between -153 and -113 °C (Angell, 2008). However, plants, especially orthodox seeds, have evolved different protection mechanisms to cope with dehydration and freezing, among the property to form intercellular glasses in combination with other solutes.

2.5 The glass formation as requirement for seed desiccation and storability

A glass is defined as amorphous, non-crystalline solid and is marked by a random, disordered molecular structure. Dehydration or rapid cooling of a solution or melt, respectively, retains the molecular disorder and allows the formation of a solid-like but disordered, non-crystalline glass (Figure 7) (Roos, 2010).

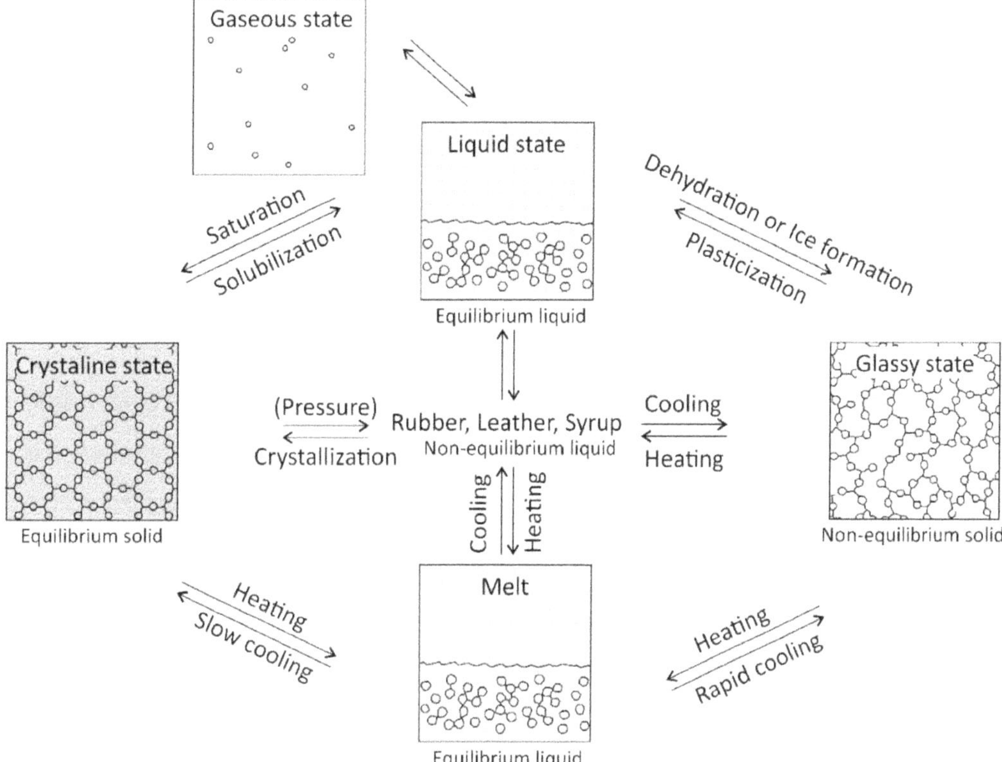

Figure 7. Effects of changes of temperatures and water contents on the physical states of (bio)polymers. Rapid cooling and dehydration is required to form the glassy state. Based on Roos (2010).

The glass state is a model adapted from polymer science and has been used to study the stability and mechanical properties of biopolymers in food. At the glassy state, molecules are frozen to their positions and the viscosity increases extremely up to flow rates of 10^{-14} m s^{-1}. Molecular movements are only limited to rotations and vibrations (Buitink and Leprince, 2004; Sperling, 2005). An increase of temperature can "melt" the structure. Water functions as a 'plasticizer' and changes the fluidity in dependency on WC and temperature (Walters et al., 2010). When a glass is heated or rehydrated, the mobility of molecules increase, the plasticizing solutes loosen intermolecular constraints, enlarge pores and allow molecules to move a bit more until they enter a liquid-like state (Sperling, 2005). The WC–temperature relationship at which an amorphous solid forms or melts is called the glass transition, the second phase transition (Sperling, 2005; Walters, 2015). The most common technique to measure glass-to-liquid transitions is the differential scanning calorimetry (DSC) which detects changes in the heat capacity during cooling and heating. The glass transition temperature (T_g), which indicates when fluid becomes viscous, is known as α-relaxation. Several other relaxations can appear and are named from higher to lower temperatures as β, γ, δ-relaxation (Walters et al., 2010). These various relaxations can be measured by dynamical mechanical (thermal) analysis (DMA/DMTA) (Sperling, 2005; Roos, 2010).

The glass concept entered seed science in the late 80ies when Williams and Leopold (1989) hypothesized that seeds may survive desiccation by the formation of a glassy state. By using DSC and electron paramagnetic resonance (EPR), Buitink et al. (1999) demonstrated that during maturation drying, below 0.8 g H_2O g^{-1} DW, cells start to shrink, solution concentration increases and the cytoplasm becomes viscous (Figure 8). The further removal of water restricts the mobility and at around 0.1 g H_2O g^{-1} DW the cytoplasm vitrifies and enters the glassy state (Buitink et al., 1999). The amorphous state decreases the possibility of detrimental processes, such as Maillard reactions (Karmas et al., 1992), and prevents conformational changes of proteins (Chang et al., 1996).

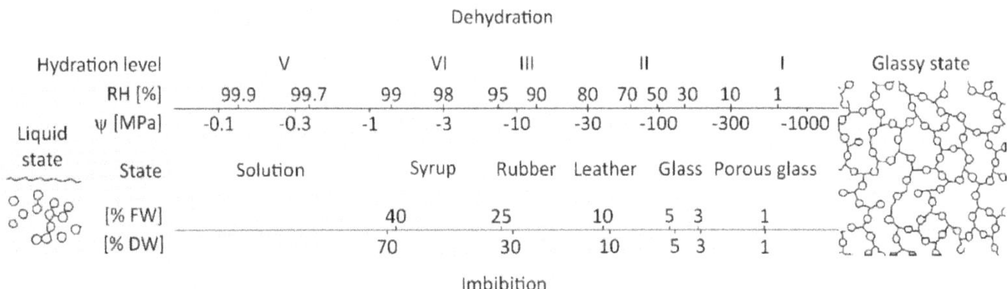

Figure 8. The physical states of water associated to the seed hydration levels (I-V). Both parameters are shown in parallel with corresponding relative humidities (RH), water potentials (ψ), water contents as % of fresh (FW) and dry weight (DW). Based on Walters (2015) and Bewley et al. (2013).

The accumulations of sucrose and oligosaccharides are key components for cytoplasmic vitrification (Sun and Leopold, 1997). T_g increases in the order of sucrose < trehalose < raffinose < stachyose which is associated with an increasing molecular weight, length of hydrogen bonding and a decrease in the average strength of the hydrogen bonding. Monosaccharides exhibit a stronger hydrogen bonding and can be packed more tightly, whereas oligosaccharides have greater degrees of freedom to rearrange hydrogen bonds during temperature changes (Wolkers et al., 2004). Hincha et al. (2003) could show that RFOs are able to protect liposomes against leakage and membranes against fusion with increasing degree of polymerization. In accordance, fructans and glucans of different chain lengths interact with phospholipid head groups and stabilize liposomes (Hincha et al., 2002). The direct hydrogen bonding interactions between sugars and phospholipid head groups reduce the liquid-to-crystalline phase transition temperature of the liposomes and prevent crystallization (Hincha et al., 2003). However, sugar glasses are not a comprehensive model to estimate the complex interactions in seeds. Wolkers et al. (2004) demonstrated that the molecular density of dry seeds is more comparable to protein-sugar glasses. Furthermore, oligosaccharides are important to set up the amorphous structure during maturation drying but they do not have specific effects on seed storability (Bentsink et al., 2000).

Seed longevity is a function of seed WC and temperature which, both, determine the molecular mobility and the kinetics of chemical and physical reactions in the cytoplasm (Buitink et al., 2000; Walters et al., 2010). Changes in the endothermal properties during glass transition have often been associated with physical ageing of glasses (Sperling, 2005). The DMA visualizes that various combinations of WC and temperature affect α-, β-, γ-relaxations differently. These differences are assumed to be based on different types of molecular motions that are restricted to rotations and vibrations. Physical ageing is related to processes which are ongoing between α- and β-relaxation when molecular motions occur only in the short side chains. However, some ageing processes, e.g. loss of crispness in cereal products, cannot be measured and might be the result of reorganisation events in the glass (Champion et al., 2000).

2.6 Seed viability loss during storage

2.6.1 Prediction and factors of seed viability loss

E. H. Roberts was among the first who investigated systematically factors of seed viability loss. By exposure of crop seeds to different WC and temperatures, he realized that the loss of seed germination followed a sigmoidal pattern (Roberts, 1960) that could be mathematically described as negative cumulative normal distribution (Roberts and Abdalla, 1968). He hypothesized that the expected life of the seeds can be predicted by a combination of storage conditions (Equation 4) and made use of the probit transformation that adapts a cumulative normal distribution to a linear form by converting the percentages axis to a probability scale (Roberts, 1972).

$$v = K_i - p / 10^{K_E - C_W \log m - C_H t - C_Q t^2} \quad (4)$$

The so-called seed viability equation can be used to calculate either the seed survival period (p) or the viability after any storage period (v) based on seed MC in equilibrium (m), the storage temperature (t), the initial viability of the seed lot at p=0 days (K_i) and species-specific temperature (C_H and C_Q) and moisture constants (K_E and C_W) (Ellis and Roberts, 1980). Increasing seed MC and temperature and pro-longed storage lead to the acceleration of seed viability loss (Figure 9) and to the development of 'artificial ageing' approaches (Bewley et al., 2013). Based on the relationship between field emergence and storage potential, seed testing associations have recognized the potential as seed vigour tests (Hampton and TeKrony, 1995). The ISTA applies the so-called 'Accelerated Ageing' (AA) test to soybean (*Glycine max* Merr.) and the 'Controlled deterioration' (CD) test to Brassica species (ISTA, 2018).

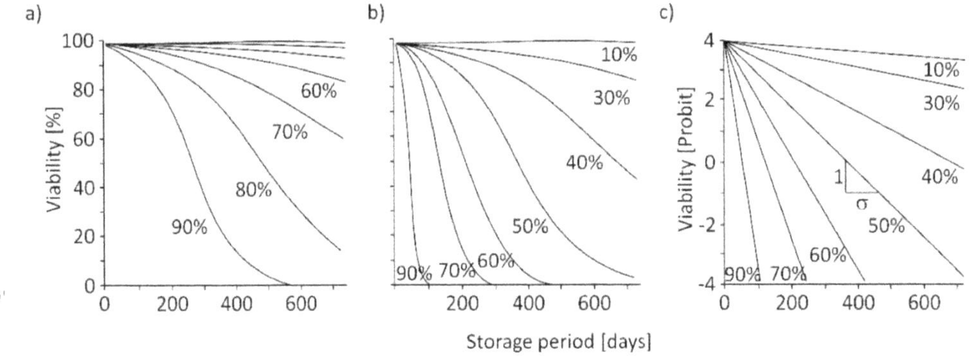

Figure 9. Seed viability curves for soybean (*Glycine max* Merr.). Seeds were stored at different seed moisture contents at a) 10 °C and b) 30 °C. In c) viability curves is plotted to a probit scale where 1 σ^{-1} represents the slope. Based on Bewley et al. (2013) and Kew (2018).

The AA exposes the seed to a short period of high temperature, mostly between 40 °C and 45 °C, and high RH (>95 %), commonly above free available water. During ageing seeds equilibrate and viability is lost within a short period. To avoid these fluctuations of seed MC, seeds undergoing controlled deterioration (CD) are equilibrated to a target MC before high temperature is applied to seeds packed in sealed aluminium foil bags. Usually the initial seed MC is measured by the oven method and a defined water volume added to reach a calculated target MC (ISTA, 2018). To improve the comparability between genotypes, Hay et al. (2008) used unsaturated lithium chloride solution for to adjust the RH during CD. Seeds were equilibrated at 47 % RH and 20 °C and are transferred to 60 % RH and 45 °C.

The atmosphere can modulate seed storability and increased concentrations of N_2 or vacuum storage can improve seed viability (Specht and Börner, 1998). Therefore, Groot et al. (2012) hypothesized that a higher O_2 concentration or O_2 partial pressure increases the deterioration rate. Here, seeds dried at 20 °C and 35 % RH are exposed to an elevated partial pressure of oxygen (EPPO) at 18 MPa (atmospheric pressure is about 0.1 MPa). Under these conditions, lettuce (*Lactuca sativa* L.) and barley decreased viability within two and seven weeks, respectively, but damages were comparable with seeds stored under long-term storage conditions. However, all approaches that apply elevated seed MC, temperature or oxygen pressure lead to a rapid seed viability loss in comparison to, i.e. wheat seeds, that have been stored under gene bank cold storage (CS) conditions at -18 °C (Walters et al., 2005) or ambient storage (AS) at 20 °C and 50 % RH (Nagel and Börner, 2010). Hence, artificial approaches allow the investigation of seed viability loss mechanisms within a limited time frame (Figure 10).

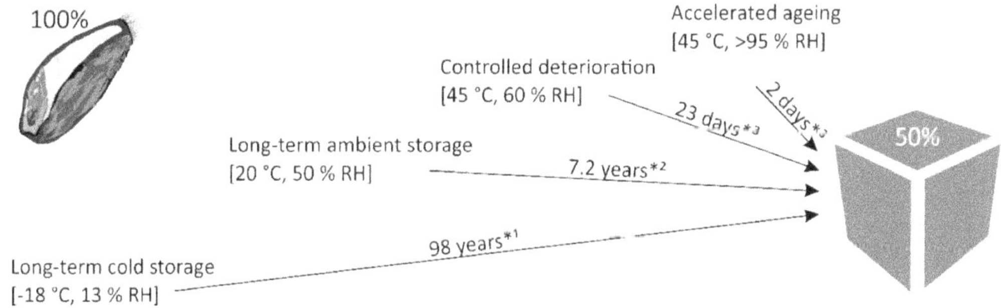

Figure 10. Wheat storage periods until seed viability decreased to 50 %. Based on *[1] Walters et al. (2005), *[2] Nagel and Börner (2010) and *[3] the Royal Botanic Gardens Kew Seed Information Database (Kew, 2018). RH, relative humidity.

2.6.2 Biochemical mechanisms of seed viability loss

Higher plants produce O_2 via photosynthesis. In immature green seeds, the photosynthetic active surface (embryo or pericarp) produce oxygen whereby concentrations vary in dependency on the species and the location. In all seeds, the O_2 concentration drops dramatically towards the interior regions (Rolletschek et al., 2004; Borisjuk and Rolletschek, 2009). However, photosynthesis and cellular respiration require high rates of electron and energy transfer and lead inevitably to the formation of reactive oxygen species (ROS). These oxygen-containing molecules exhibit a higher chemical reactivity than O_2 (Figure 11). The major forms are singlet oxygen($^1O_2$), superoxide anion ($O_2^{\bullet-}$), H_2O_2 and $HO^{\bullet-}$ (Waszczak et al., 2018). During photosynthesis, within the thylakoid membranes of the chloroplasts, the first excited electronic state of molecular oxygen ($^1O_2$) is generated and energy transferred to the ground state molecular oxygen ($^3O_2$) (Krieger-Liszkay et al., 2008). Other production sites of $^1O_2$ can be the mitochondrial electron transport chain and plant-pathogen interaction where $^1O_2$–generating phototoxins are used to kill pathogens (Waszczak et al., 2018). $^1O_2$ has an extremely short lifetime and is scavenged by antioxidative molecules such as carotenoids, tocochromanols and membrane lipids (Mène-Saffrané et al., 2009; Triantaphylidès and Havaux, 2009). $^1O_2$ can be also dismutated by sodium dismutase (SOD) into H_2O_2 on the stromal side of the thylakoid membrane. In addition, H_2O_2 is produced in peroxisomes. Though H_2O_2 has a larger dipole moment than H_2O and cannot diffuse through membranes, it is quickly transferred through the plasma membrane and from the chloroplast to the cytosol (Mubarakshina et al., 2010).

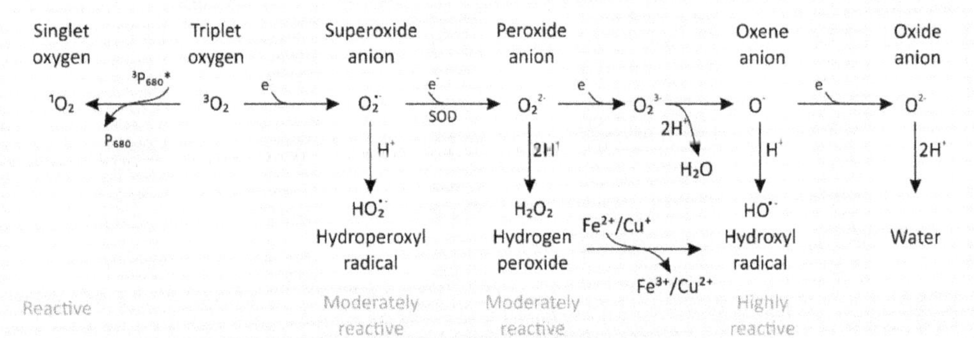

Figure 11. Generation of reactive oxygen species (ROS). Electron and energy transfer or sequential univalent reduction of ground state triplet oxygen is shown. Based on Waszczak et al. (2018) and Apel and Hirt (2004). ^{3}P680*, triplet state of the primary electron donor of photosystem II; SOD, sodium dismutase.

When the production of ROS becomes uncontrolled, the oxidative stress can provoke reversible and irreversible damages to membranes, DNA, RNA and proteins (Choudhury et al., 2017). ROS can induce post-translational modifications of proteins. Thereby, sulfonylation, carbonylation, glutathionylation and S-nitrosylation affect the protein conformation and can alter the protein/enzyme activity. These structural changes can be also understood as signalling pathways that alter transcriptions and translations (Miller et al., 2008; Choudhury et al., 2017) and may result in activation of processes during imbibition and germination.

Seed vigour loss appears frequently in parallel with an increased seed leachate conductivity (Thornton et al., 1990; Demir et al., 2008) indicating cell membranes are disrupted during prolonged storage. Cell membranes are important to control the physiological state of the organelles that maintain redox potential and pH. The $HO^{\bullet -}$ and hydroperoxyl ($HO_2^{\bullet -}$) radicals affect predominantly lipids. $HO_2^{\bullet -}$ are short-lived and react unspecific with all biomolecules which are in a distance of few nanometres. When $O_2^{\bullet -}$ is protonated, the very strong oxidant $HO_2^{\bullet -}$ is formed and can yield in H_2O_2 (Ayala et al., 2014). Saturated fatty acids are more resistant to oxidative modification. Major targets of radical attacks are glycolipids, phospholipids and polyunsaturated fatty acids (PUFAs) that contain carbon-carbon double bonds and form phospholipid carbon centred radicals. These react immediately with molecular oxygen and form peroxyl radicals ($LOO^{\bullet}$) which are highly reactive and can form further carbon- or oxygen – centred radicals such as $L^{\bullet}$ (alkyl), $OL^{\bullet}$ (epoxy alkyl), $HOL^{\bullet}$ (hydroxyl-alkyl), $LO^{\bullet}$ (alkoxyl) and $OLOO^{\bullet}$ (epoxy-peroxyl). The continuous propagation of radicals is terminated either when the concentration of radicals is sufficiently high to support radical-radical coupling or by lipophilic antioxidants such β-carotene or α-tocopherol. Another pathway of lipid oxidation would be the electrophilic attack by certain non-radical oxidants due to susceptibility of unsaturated phospholipids for two electrons. However, the profile of oxidized lipids is influenced by the oxidant species, the type of linkage of the fatty acids to the glycerol backbone and fatty acyl chain (Reis and Spickett, 2012).

Another source of oxidants is the browning reaction (Jeevan Kumar et al., 2015). The presence of reducing monosaccharides, such as glucose, fructose and maltose, is adverse during prolonged storage. They foster the non-enzymatic reactions with amino acids, also known as Maillard reaction (Evershed et al., 1997; Narayana Murthy and Sun, 2000). Thereby, high-molecular weight heteropolymeric products, some of them brown products called melanoidins, are formed during ageing processes (Evershed et al., 1997; Narayana Murthy and Sun, 2000). The by-products of the reactions are a complex mixture of low-molecular weight (LMW) volatile compounds and can be linked to degradative reaction of organic matter (Evershed et al., 1997).

2.7 Seed protection mechanisms during quiescence

To cope with stresses appearing during seed desiccation, quiescence and germination, seeds have established a variety of protection mechanisms. LEA proteins, HSPs and soluble sugars play important roles to set up desiccation tolerance and seed longevity. Oxidative stress is handled by an antioxidant machinery that include enzymatic and non-enzymatic LMW antioxidants. In seeds, SOD, CAT, APX, DHAR, glutatione S-transferase (GST), glutathione peroxidase (GPX) have been shown to be associated with longevity and germination (Lee et al., 2010). They can be found in all subcellular compartments, where i.e. electron transport chains facilitate the production of $O_2^{\bullet -}$. However, the relative contribution to ROS detoxification is dependent on the enzyme concentration, substrate affinity and reaction rate (Mittler et al., 2004).

Sodium dismutases (SODs) can be categorized by their metal cofactors into copper/zinc (Cu/Zn-SOD), manganese (Mn-SOD) and iron (Fe-SOD). The Cu/Zn SODs are located in the chloroplast, the cytosol, and possibly the extracellular space, the Mn-SODs in the mitochondrion and the peroxisome and the Fe-SODs in the chloroplast. SODs catalyses the dismutation of one $O_2^{\bullet -}$ to H_2O_2 and another O_2. Via the metal catalysed Haber-Weiss-type reaction, it also removes $O_2^{\bullet -}$ and decreases the risk of $HO^{\bullet -}$ formation (Alscher et al., 2002; Mittler et al., 2004). Dependent on the type of environmental stress, e.g. light, radiation, different SOD proteins are expressed (Kliebenstein et al., 1998).

The H_2O_2 produced by SODs can be broken down by catalases (CAT) into H_2O and O_2. The rate of dismutation for one CAT molecule is extremely high and the maximal turnover rate is estimated to be 16,000,000 to 44,000,000 s^{-1} per folded tetrameric heme containing enzyme (Heck et al., 2010). CATs are considered among the most important enzymatic antioxidants that prevent seed viability loss during seed deterioration (Kibinza et al., 2011), protect the photosystems and control plant growth and development (Queval et al., 2007).

An alternative pathway to catalyse the reduction of H_2O_2 or other hydroperoxides is used by glutathione peroxidase (GPX). Glutathione or thioredoxin functions as electron donor and produce H_2O or alcohols (Gill and Tuteja, 2010). The GPX family consists of multiple isoenzymes whereby plant GPX contain cysteine instead of selenocysteine in their active side. The thioredoxin peroxidase function of some GPXs indicates that they are involved in the redox homeostasis by maintaining the thiol/disulphide or NADPH/$NADP^+$ balance (Bela et al., 2015).

Glutatione S-transferases (GST) are highly abundant proteins encoded by a large gene family (He et al., 2016a). They catalyse the conjugation of electrophilic xenobiotic substrates with reduced glutathione. GST proteins are involved in several development processes among are herbicide detoxification, signal transduction and protection against heavy metals. Based on amino acid sequence similarity, plant GSTs can be grouped into eight classes (Dixon et al., 2011). Thereby, lambda and DHAR GSTs, both containing cysteine instead of serine, are considered to be less involved in the detoxification of xenobiotics and more in the redox and thiol transfer reactions (Dixon et al., 2002).

APX, DHAR, GR are the key enzymes of the ascorbate-glutathione cycle, also termed Halliwell-Asada cycle (Figure 12), and are important for the regulation, modulation and maintenance of the redox homeostasis (Kranner et al., 2010b; Foyer and Noctor, 2011; Szarka et al., 2012; Couto et al., 2016). APX, also a heme containing peroxidase, has a high affinity to H_2O_2 and catalyses the oxidation of ASC. This molecule forms an ascorbic free radical (AFR) that is reduced by DHA or a NAD(P)H-dependent AFR-reductase (AFRR). DHA, the final product of ASC oxidation, is then reduced to ASC by DHAR, an enzyme using GSH to form GSSG (De Gara et al., 1997; Gill and Tuteja, 2010). In the latter reaction, GR, a flavo-protein oxidoreductase located mainly in the chloroplast and mitochondria (Edwards et al., 1990), regenerates GSH by using the reducing power of NADPH (De Gara et al., 1997; Gill et al., 2013).

LMW antioxidants such as tocochromanols, ASC and glutathione are mainly involved in ROS scavenging and redox regulation, in addition to several other physiological functions. Glutathione is a tripeptide, γ-L-glutamyl-L-cysteinylglycine, and present in all mammalian tissues (Lu, 2013). In plants, glutathione and the functionally homologous thiols, i.e. homoglutathione in legumes (Colville et al., 2015), are the principal LMW thiols in most cells. Reduced glutathione (GSH) is synthesized from its constituent amino acids, i.e. cysteine (Cys), glutamate (Glu) and glycine (Gly), by two ATP-dependent steps involving γ-glutamylcysteine synthase (γ-ECS) and glutathione synthetase (GS) (Pasternak et al., 2008). The synthesis is controlled by the genes *GSH1* and *GSH2*. Thereby, *GSH1* encodes the first enzyme γ-glutamate cysteine ligase that is responsible for the synthesis of γ-ECS (Cairns et al., 2006) and located in the plastids. The second gene *GSH2* encodes GS and is located in both, plastids and cytosol. Most limiting factors for glutathione accumulation are the γ-ECS activity and Cys availability (Noctor et al., 2012). In turn, accumulation can be triggered by oxidative stress that causes an increased abundance of transcripts encoding adenosine 5'-phosphosulphate reductase (APR) and serine acetyltransferase (SAT) which are important for cysteine synthesis (Queval et al., 2009).

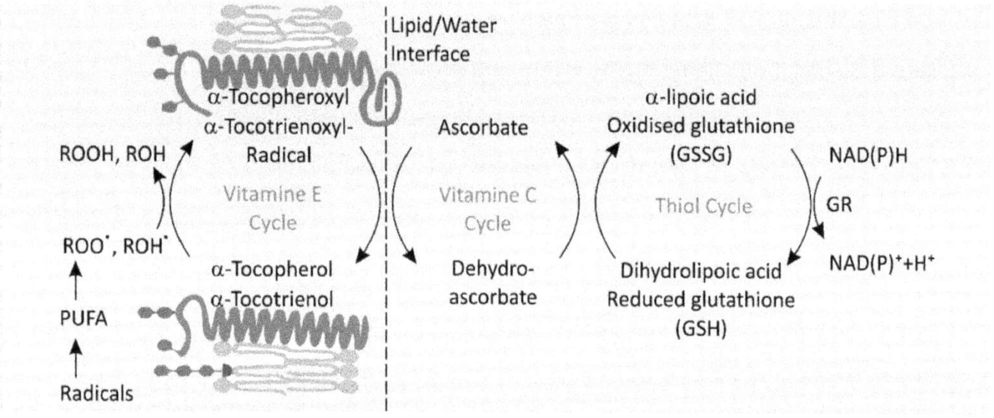

Figure 12. Tocopherol-ascorbate-glutathione cycle and the connected radical scavenging system. GR, glutathione reductase; GSSG, glutathione disulphide; PUFA, poly unsaturated fatty acid. Based on Noctor et al. (2012) and Foyer and Noctor (2005).

In seeds, glutathione has been widely discussed as LMW antioxidant and its contribution to the intracellular redox environment (Kranner et al., 2010b). Glutathione can scavenge ROS directly, whereby two moles GSH form one mole oxidized glutathione (GSSG, glutathione disulphide) via Cys disulphide bonds (Potters et al., 2002; Noctor et al., 2012). During seed maturation, the glutathione pool increases consistently but shifts to oxidizing conditions at full maturity (De Gara et al., 1997). Verdier et al. (2013) could show that the two enzymes, GPX and GST, involved in detoxification pathways, are accumulated during maturation drying. In contrast, the absence of seed maturation and embryo lethality can be provoked by complete lack of GSH in *gsh1* Arabidopsis T-DNA insertion mutants (Cairns et al., 2006). In the dry state, GSH cannot be regenerated and GSSG accumulates and/or reacts with thiol groups of proteins forming protein-SSG mixed disulphides, which protect proteins against desiccation-induced oxidative injuries (Kranner and Grill, 1996). Consequently, the glutathione pool decreases

(Lehner et al., 2008; Nagel et al., 2015). The ratio between GSH and GSSG determines the half-cell reduction potential of glutathione $E_{GSSG/2GSH}$ which has been considered as viability marker (Schafer and Buettner, 2001). A shift to between -180 and -160 mV indicate a viability reduction by 50 % (Kranner et al., 2006). During seed germination, Cys concentration increases and GSH is accumulated (Gerna et al., 2017). Thereby, GSH is assumed to be an essential transport metabolite as it is one of the major forms of organic sulphur (Herschbach and Rennenberg, 1994; Kopriva and Rennenberg, 2004). Changes in Cys and GSH leading to more reducing conditions might be related to the mobilisation of storage proteins during early seedling growth (Gerna et al., 2017).

ASC is synthesized in plant cells and one of the most abundant, water soluble antioxidants. It is mostly located in chloroplasts of photosynthetic active cells and meristems. ASC scavenges $O_2^{\bullet-}$ and $HO^{\bullet-}$ directly or it regenerates α-tocopherol by donating electrons to tocopheroxyl radical which are further transferred through the ASC-GSH cycle involving DHA and APX (Gill and Tuteja, 2010). Dry, photosynthetic inactive seeds do not maintain neither ASC nor APX and show only low levels of DHA. At a later stage, during imbibition, ASC accumulate in parallel with APX and GR and indicate a complete resumption of the metabolism (De Gara et al., 1997).

Tocochromanols are a group of the lipophilic antioxidants and comprise tocopherols tocotrienols and tocomonoenols. They differ structurally by the presence of three trans-double bonds in the hydrocarbon tail of tocotrienols and the four isomers α, β, γ and δ (Munné-Bosch and Alegre, 2002). Their hydrophobic isoprenoid tail is anchored into membrane bilayers, whereas the more hydrophilic hydroxychroman ring sits nearer the surface (Smirnoff, 2010) which, both, lead to membrane structure-stabilizing functions (Falk and Munné-Bosch, 2010). Tocochromanols are synthesized exclusively by photosynthetic organisms involving homogentisate phytyltransferase (HPT1, VTE2), tocopherol cyclase (VTE1), VTE3 methyltransferase and in monocots (homogentisate geranylgeranyl transferase, HGGT) (Dormann, 2007). In seeds of most dicots and embryos of monocots, α-, or γ-tocopherol are predominant (Falk and Munné-Bosch, 2010; Matthaus et al., 2016). Tocotrienols are mainly accumulated in the seed endosperm of monocots (Horvath et al., 2006; Moreau et al., 2007). Tocochromanols protect lipids in photosynthetic membranes and limit the oxidation of membrane and storage lipids during maintenance, germination and seedling growth (Leprince et al., 1990; Sattler et al., 2004; Mène-Saffrané et al., 2010). They interact with polyunsaturated acyl groups and protect

PUFAs from lipid peroxidation by scavenging lipid peroxy radicals (Smirnoff, 2010), quenching ROS produced e.g., by photosystem II and during membrane lipid peroxidation and reacting chemically with $^1O_2$ (Krieger-Liszkay et al., 2008). Mutants deficient in tocopherols and tocotrienols have reduced germination after artificial seed ageing and are assumed to have important roles of seed protection during storage (Chen et al., 2016).

References of 'Prolog' and 'Epilog' are combined and attached at the end of the thesis.

2.8 Aim of the study

Orthodox (desiccation tolerant) seeds are used to preserve the crop genetic diversity in seed-banks. Though secure storage, seeds loose viability and the deterioration rate depends on the genetic make-up, the growth and storage environment. To protect the genetic integrity of crop species in seedbanks, the study aims to understand biochemical and genetic mechanisms of seed deterioration in conjunction with storage environment, genotypic variance and seed dormancy in endospermic seeds of various chemical compositions. Therefore, seeds of wheat, barley and oil seed rape are exposed to different storage conditions and artificial ageing treatments and analysed by techniques involving HPLC, lipidomic and proteomic as well as quantitative trait locus (QTL) and genome-wide association (GWAS) mapping (Figure 13). Based on ten peer-review papers and three manuscripts the following questions are discussed.

a) Are artificial ageing treatments an appropriate instrument to conclude on molecular processes appearing in long-term storage?
b) How does genetic variability and growth conditions affect seed longevity?
c) Can biochemical or genetic markers be applied to a wide range of genotypes/species to predict correctly seed viability and storage behaviours?

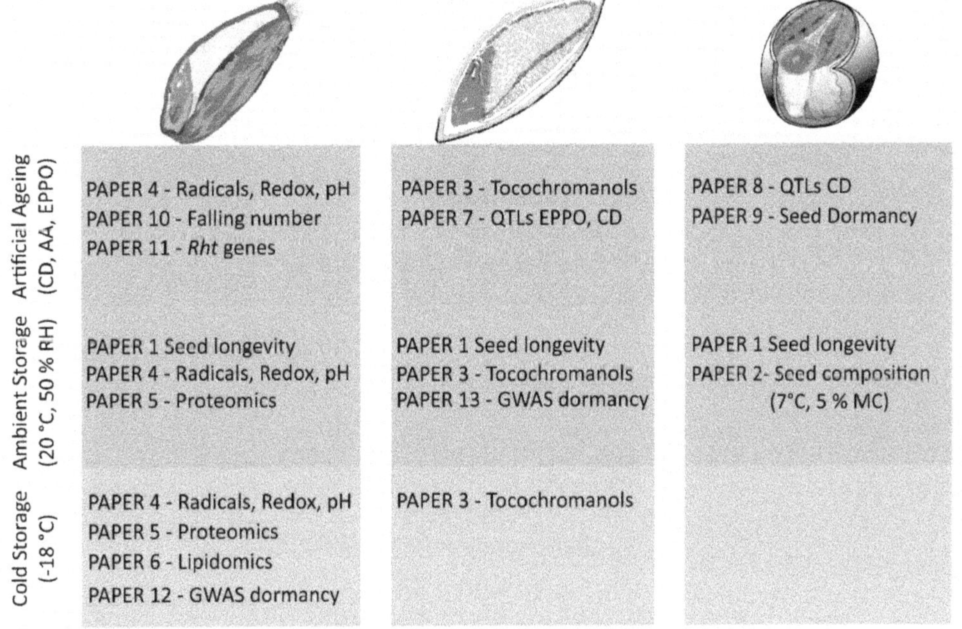

Figure 13. Overview of peer-reviewed papers and manuscripts discussed in the 'Epilog'.

3.1 PAPER 1:

The longevity of crop seeds stored under ambient conditions

by

Manuela Nagel and Andreas Börner

Published in

Seed Science Research (2010) 20, 1-12

https://doi.org/10.1017/S0960258509990213

Seed Science Research (2010) **20**, 1–12

doi:10.1017/S0960258509990213

The longevity of crop seeds stored under ambient conditions

Manuela Nagel and Andreas Börner*
Leibniz-Institut für Pflanzengenetik und Kulturpflanzenforschung (IPK), Corensstrasse 3, D-06466 Gatersleben, Germany

(Received 24 June 2008; accepted after revision 31 July 2009 – First published online 19 November 2009)

Abstract

The ability of crop seeds to retain their viability over extended periods of uncontrolled temperature and/or relative humidity conditions has not been widely investigated, although this is an important issue for genebank management. We report here the response of 18 crop species to storage for up to 26 years at 20.3 ± 2.3°C and 50.5 ± 6.3% relative humidity. Germination rates decreased in a sigmoid fashion, but the curve parameters were species characteristic. Pea, common bean and maize seeds retained their viability over the longest period (23, 21 and 19 years, respectively). In contrast, chive seeds survived for only 5 years and lettuce for 7 years. In addition to this interspecific variability, there were also indices for intraspecific variability, particularly in bean and chive seeds, just as in collard, lupin, poppy, wheat and maize seeds. A significant correlation was obtained between germination performance in the laboratory and seedling emergence following autumn sowing. Seeds in which oil was the major seed storage component were more short lived, whereas carbohydrates or proteins did not show an effect on seed longevity.

Keywords: ambient storage, conservation, field establishment, genebank, genetic diversity, longevity, seeds

Introduction

Agriculture depends on the ability of seeds to survive until the next growing season under ambient conditions. This may have been a factor for which our modern crops were chosen for domestication, thousands of years ago. Normally it would not be have been important that those seeds should survive over decades, but in the case of natural disasters seeds would only be useful if they were still able to germinate after the incident. Priestley *et al.* (1985) have shown that different crops are able to maintain 50% of their germination even in different storage locations (e.g. Australia, Russia) at least for 2 years. But most of this investigated material showed a longer half-viability period. Plants with high agricultural utilization but which could not produce long-living seeds (e.g. onion), have never obtained the same agronomical importance as wheat, rice and maize. The agronomical trait 'seed longevity' was not considered highly relevant until the beginning of the 20th century. At that time the Russian botanist and ecologist Nicolai I. Vavilov was one of the first to recognize that our plant genetic resources are in particular danger and need to be conserved for the future (Maxted *et al.*, 1997). These efforts were followed by the establishment of modern genebanks around the world (Linington and Pritchard, 2001), which led to an increasing interest in the storage behaviour of seeds. Haferkamp *et al.* (1953) showed that germination of open-stored seeds decreased with increasing seed age, but many seeds, including barley, wheat, oat, alfalfa, pea and maize, remained viable after storage for as long as 31–33 years. Moreover, different discoveries have demonstrated that under certain conditions seeds can survive for hundreds of years.

A famous example of extreme longevity is provided by the seeds of a date palm (*Phoenix dactylifera* L.). These seeds were discovered beneath a Heriodin fortress in Israel in 1965 and, after four decades in storage, one seed germinated. The radiocarbon dating of the remaining seeds revealed an age of 2000 years (Sallon *et al.*, 2008). Furthermore, a deliberate long-term cereal seed-storage experiment, involving wheat, barley, rye and oats, was initiated in November 1877 by F. Haberland at the Universität für Bodenkultur in Vienna. The seeds were maintained at 10–15°C, and after 90 years of storage, 3.5% of the wheat seeds were still viable (Ruckenbauer, 1971). After a further 20 years, the remaining vials were opened, and while the rye seeds no longer germinated, 81% and 90% of the oat and barley seeds, respectively,

*Correspondence
Email: boerner@ipk-gatersleben.de

were still viable (Steiner and Ruckenbauer, 1995). In addition, the seeds from five weed species (*Sinapis arvensis* L., *Vaccaria hispanica* (Mill.) Rauschert, *Agrostemma githago* L., *Sinapis alba* L. and *Lolium temulentum* L.) present in the oat sample as impurities, were also able to germinate.

However, the seeds of the above-mentioned discoveries were all found and stored under ultra-dry conditions; even the samples of Haferkamp *et al.* (1953) were stored in a dry continental climate zone. Is it possible to have comparable longevities in a temperate zone? Seed companies, breeders, farmers and many seed banks store their material under open storage conditions. How long can this material survive in terms of high germination (>80%)? The following study on genebank accessions addresses these questions. The plant material was planted, harvested and stored in central Germany, where the 10-year average relative humidity is 80.4 ± 1.5% and the average temperature is 9.8 ± 0.5°C. The seeds belonged to the *ex situ* genebank for agricultural and horticultural crop plants in Gatersleben, which houses some 150,000 accessions, covering over 3000 species within 890 genera. In general, orthodox seeds are conserved at −15°C, while vegetatively reproduced plant materials are maintained by *in vitro* culture and cryo-conservation. Seed storage accounts for about 90% of the stored material (Börner, 2006). In addition to the long-term cold-storage facilities, reserve seeds are routinely stored in paper bags under room-temperature conditions. These seeds are provided for the enquiries of different users during the first years after harvest.

Our aim was to assess the longevity of this reserve seed. In comparison with standard papers by Walters *et al.* (2005), Priestley *et al.* (1985), Roberts (1972) and Ellis and Roberts (1980), we investigated the seed from 18 crop species, which had been stored under ambient conditions for 6–26 years. We addressed the following questions: (1) How long can seeds survive in ambient but uncontrolled storage? (2) How much variation exists within species? (3) Does crop year make a difference? (4) Do laboratory tests of longevity correspond to field emergence? (5) How long can the seeds be stored, maintaining germination over 80%?

Materials and methods

Plant material

Seeds were obtained from various plantings performed between 1980 and 2005. The 18 test species included starchy-seeded cereals: oat (*Avena sativa* L.), barley (*Hordeum vulgare* L.), bread wheat (*Triticum aestivum* L.), rye (*Secale cereale* L.), triticale (×*Triticosecale* Wittm.) and maize (*Zea mays* L.); proteinaceous legumes: lupin (*Lupinus albus* L.), common bean (*Phaseolus vulgaris* L.), pea (*Pisum sativum* L.) and common vetch (*Vicia sativa* L.); oilseeds: collard (*Brassica oleracea* L.), sunflower (*Helianthus annuus* L.), linseed (*Linum usitatissimum* L.) and poppy (*Papaver somniferum* L.); and miscellaneous species, including chives (*Allium schoenoprasum* L.), cucumber (*Cucumis sativus* L.), carrot (*Daucus carota* L.) and lettuce (*Lactuca sativa* L). All seeds were kept in paper bags in a room where the temperature was maintained at 20.3 ± 2.3°C, and the relative humidity (RH) at 50.5 ± 6.3%. Three to five different accessions per crop species were chosen randomly from almost every harvest year between 1980 and 2005, giving a total of 30–148 accessions per crop, ranging from 1 to 26 years in age (Table 1). The species were classified into cereals, legumes, oil crops and miscellaneous crops.

Seed germination

All samples were germinated in 2006, using 50 seeds for each accession, according to ISTA rules (ISTA, 2005). Depending on their size, the seeds were placed either over or between moistened filter papers on a Jacobsen apparatus, which provided a temperature regime of 25 ± 2°C during the day and 23 ± 2°C during the night. Germination percentage was based on the number of seedlings with a normal appearance. A parallel field experiment tested seed material derived from a subset of accessions (Table 1). Seeds were sown in September 2006 in the form of two replicated rows of 25 seeds each. The sowing depth was adjusted to seed size, with small seeds being placed at a depth of 1–2 cm, and large ones at 3–5 cm. The number of emerged seedlings was counted after 2 and 4 weeks.

Longevity calculations

The age of each seed sample was taken as the number of years that had elapsed since their harvest. Based on the germination of the five accessions per reproduction cycle, a probit analysis was performed. A transformation of cumulative, normally distributed germinabilities converts the expected sigmoid germination curve to a linear one which can be described by equation (1), where ν is probit percentage viability at storage time p and its initial probit percentage viability K_i, whereas σ is the standard deviation of the distribution of deaths in time (Ellis and Roberts, 1980).

$$\nu = K_i - \left(\frac{1}{\sigma}\right) \cdot p \quad (1)$$

In addition, the half-viability period (P50) was calculated in equation (2) by setting $\nu = 0$ (equal to

Table 1. Numbers of harvest years (HY) and accessions (Acc.), averaged amount (% of dry mass) of carbohydrates (CH), proteins and oil, and 1000 kernel weight (g) taken from data by Earle and Jones (1962), Jones and Earle (1966) and Sinclair and DeWit (1975) with respect to the species and crop group investigated

Crop	Species	Germination in lab		Field emergence		CH (%)	Protein (%)	Oil (%)	Wt/1000 (g)
		No. of HY	Total Acc.	No. of HY	Total Acc.				
Cereals									
	Avena sativa L.	20	96	5	24	77.0	13.0	5.0	–
	Hordeum vulgare L.	16	79	4	20	80.0	6.0	1.0	–
	Secale cereale L.	12	59	4	19	82.0	14.0	2.0	–
	× *Triticosecale* Wittm.	10	39	5	13	–	–	–	–
	Triticum aestivum L.	15	75	4	20	82.0	14.0	2.0	–
	Zea mays L.	20	92	4	17	84.0	10.0	5.0	–
Legumes									
	Lupinus albus L.	15	67	4	18	–	36.9	9.4	276.7
	Phaseolus vulgaris L.	25	120	5	25	70.0	28.1	1.1	297.3
	Pisum sativum L.	26	129	5	25	68.0	24.6	1.3	151.6
	Vicia sativa L.	25	80	4	17	–	29.2	0.9	48.8
Oil crops									
	Brassica oleracea L.	13	62	4	18	–	28.2	37.9	3.2
	Helianthus annuus L.	8	37	4	17	48.0	27.9	43.4	47.2
	Linum usitatissimum L.	9	50	4	20	–	26.1	40.2	6.2
	Papaver somniferum L.	6	30	3	15	–	–	–	–
Miscellaneous crops									
	Allium schoenoprasum L.	6	30	3	15	–	–	–	–
	Cucumis sativus L.	13	52	4	18	–	22.9	32.1	8.3
	Daucus carota L.	7	33	3	14	–	25.6	21.2	1.2
	Lactuca sativa L.	9	35	3	15	–	29.3	37.7	0.7

50% germination) and transforming p to P50:

$$P50 = K_i \cdot \sigma \qquad (2)$$

In some cases, extrapolation was required and conducted on the basis of the slope ($-1/\sigma$) and the intercept (K_i) of the seed survival curve, as given in equation (1). Probit parameters and uncertainty of P50 values are given as confidence interval in Table 2.

Statistical analysis

Based on the three to five accessions per harvest year and species, arithmetic mean and standard deviation (SD) were calculated. Intraspecific variation was evaluated on the basis of these SDs. Correlations between absolute longevities, which define the storage period until the last year of seed germination, and P50 values, and also between germinations and field emergence, were calculated using the software program SPSS 17.0 (SPSS Inc., Chicago, Illinois, USA, 2008). Correlations are expressed by coefficient of determination (R^2). Relationships between absolute longevities, P50 values (Table 3), chemical composition and 1000 kernel weight (Table 1) are based on data by Priestley *et al.* (1985), Walters *et al.* (2005), Earle and Jones (1962), Jones and Earle (1966) and Sinclair and DeWit (1975). P50 values (Table 3) of barley, wheat, maize, bean, pea, sunflower, flax and lettuce seeds were also calculated using Ellis's viability equation (Ellis and Roberts, 1980). Coefficients were reported by Liu *et al.* (2008), the storage temperature was assigned as 20.3°C (average storage temperature) and water content was calculated by using average storage RH (50.5%), average temperature (20.3°C) and oil contents provided by Liu *et al.* (2008). The initial germination was assigned as intercept based on the longevity calculation (Table 2).

Results

Viability after 1–26 years of storage was assessed among the different crop species. Most crops showed high germination when germinated within 2 years post harvest, but germination of most species declined to nearly 0% after 20 years, except for barley, maize, lupin, common bean and pea. Final germination was not available for lupin, poppy, linseed, cucumber and carrot. A correlation between germination and field establishment was considered significant at R^2 of 0.60 ($P < 0.001$).

Cereals

Maize was the most robust of the cereals. After 12 years of storage, germination was still over 80%

Table 2. Calculated longevity parameters (intercept, slope) in accordance with the longevity equation $\nu = K_i - (1/\sigma)p$ including the time for the level of germination to fall to 50% (P50). [ν = probit viability; K_i = initial germination (probit); $-1/\sigma$ = slope; p = storage time (years); SE = standard error]

Crop	Species	Probit parameters			95% Confidence limits		
		Intercept (K_i)	Slope ($-1/\sigma$)	SE (slope)	Lower bound	Upper bound	P50 (years)
Cereals							
	Avena sativa L.	1.7	−0.22	0.006	7.2	8.3	7.8
	Hordeum vulgare L.	1.3	−0.15	0.004	8.2	10.2	9.2
	Secale cereale L.	2.0	−0.35	0.008	5.4	6.4	5.9
	× *Triticosecale* Wittm.	1.6	−0.27	0.008	5.4	6.5	5.9
	Triticum aestivum L.	1.5	−0.21	0.004	6.3	8.0	7.2
	Zea mays L.	2.3	0.19	0.003	11.4	13.2	12.3
	Ø						8.1
Legumes							
	Lupinus albus L.	1.6	−0.12	0.004	12.0	15.7	13.5
	Phaseolus vulgaris L.	1.3	−0.14	0.002	8.6	10.9	9.8
	Pisum sativum L.	3.0	−0.22	0.003	12.8	15.0	13.9
	Vicia sativa L.	3.5	−0.32	0.006	10.3	11.4	10.8
	Ø						12.0
Oil crops							
	Brassica oleracea L.	1.9	−0.26	0.006	6.6	8.1	7.3
	Helianthus annuus L.	1.4	−0.40	0.012	3.8	4.9	4.3
	Linum usitatissimum L.	2.7	−0.26	0.006	9.6	11.4	10.4
	Papaver somniferum L.	1.6	−0.19	0.015	6.6	25.8	8.2
	Ø						7.6
Miscellaneous crops							
	Allium schoenoprasum L.	1.2	−0.60	0.022	2.0	2.4	2.0
	Cucumis sativus L.	3.0	−0.20	0.007	13.5	17.3	14.9
	Daucus carota L.	1.1	−0.18	0.011	4.8	8.5	6.1
	Lactuca sativa L.	5.8	−1.27	0.041	4.2	5.0	4.6
	Ø						6.9

Ø, crop average.

(82.8 ± 12.6%) but it decreased to 55.2% with an increasing variability (SD: ±47.2%) after 13 years, and germination only ceased after year 20 of storage (Fig. 1). In the field the highest variability of emergence (28.4 ± 18.2%) could be detected in year 14, whereas the correlation between germination and the field emergence was 0.73 (R^2; $P < 0.001$). On the basis of the single lab germination, the half-viability period (P50) was computed and resulted in 12.3 years in the case of maize. In barley, P50 was 9.2 years but the germination percentage remained above 80% only during the first 4 years of storage (4 years: 80.4 ± 6.8%), falling thereafter to 44.1 ± 30.7%, including the highest variability within this species, by year 7, and to zero by year 17. In the field, year 13 of the four tested years presented the highest variability (34.8 ± 27.6%) and field emergence correlated highly with lab germination ($R^2 = 0.75$; $P < 0.001$). Oat showed the same results in this relation ($R^2 = 0.76$; $P < 0.001$). The mean germination of oat remained close to 80% (79.2 ± 9.0%) after 4 years of storage, but fell to 52.8% by year 6, showing the highest standard deviation (±29.7%), just as in the field emergence (67.2 ± 25.2%). After 16 years, oat germination dropped to zero and the half-viability period was calculated to be 7.8 years. In bread wheat, the behaviour was noticeably variety-dependent, falling to 4.4% by year 12, and to zero by year 16. Its SD increased between 2 years (77.6 ± 10.2%) and 9 years (44.8 ± 32.3%) and decreased after 10 years (42.0 ± 28.9%), which was also detectable in the field (3 years: 86.4 ± 5.5%; 9 years: 24.8 ± 19.8%; 13 years: 0.8 ± 1.0%). There was a strong correlation between germination and field emergence ($R^2 = 0.82$; $P < 0.001$). The half-viability was reached after 7.2 years. Rye seed viability declined rapidly after 5 years of storage (65.6 ± 19.3%), reached the highest SD in the lab (29.5 ± 32.6%) and in the field (38.5 ± 44.4%) after 8 years, falling to zero by year 13. The correlation (R^2) between both traits was 0.54 ($P < 0.01$). Triticale behaved in a similar fashion, decreasing from 89.6 ± 3.8% in year 2, 68.0 ± 5.6% in year 5 and 17.0 ± 24.0% in year 6 to 14.0 ± 5.7% in year 11. However, the correlation of lab germination and field emergence was much better ($R^2 = 0.89$; $P < 0.001$). P50 of both species was reached after 5.9 years. Ranking the cereals along their P50 values, the species with the longest-living seeds was maize followed by

Table 3. Comparison of seed longevity values (in years) for each species as represented by absolute longevities, analysed P50 results and P50 values by Ellis's viability equations, Priestley *et al.* (1985) and Walters *et al.* (2005)

Crop	Species	Absolute longevity	P50	Ellis's viability equations	Priestley *et al.* (1985)	Walters *et al.* (2005)
Cereals						
	Avena sativa L.	15	7.8	–	13.0	117
	Hordeum vulgare L.	16	9.2	1.4	7.2	84
	Secale cereale L.	12	5.9	–	4.5	36
	× *Triticosecale* Wittm.	–	5.9	–	–	–
	Triticum aestivum L.	15	7.2	0.9	7.6	54
	Zea mays L.	19	12.3	2.3	9.6	49
	Ø	15.4	8.1	1.5	8.4	68.0
Legumes						
	Lupinus albus L.	–	13.5	–	–	–
	Phaseolus vulgaris L.	21	9.8	5.1	16.0	31
	Pisum sativum L.	23	13.9	14.8	15.9	97
	Vicia sativa L.	18	10.8	–	7.3	71
	Ø	20.7	12.0	10.0	13.1	66.3
Oil crops						
	Brassica oleracea L.	11	7.3	–	7.2	23
	Helianthus annuus L.	8	4.7	0.9	5.4	50
	Linum usitatissimum L.	–	10.4	3.5	8.7	53
	Papaver somniferum L.	–	8.2	–	7.3	62
	Ø	9.5	7.7	2.1	7.1	47.0
Miscellaneous crops						
	Allium schoenoprasum L.	5	2.0	–	–	24
	Cucumis sativus L.	–	14.9	–	4.9	87
	Daucus carota L.	–	6.1	–	6.6	30
	Lactuca sativa L.	7	4.6	3.2	6.4	23
	Ø	6.0	6.9	3.2	6.0	41.0

Ø, crop average.

barley, oat and wheat. Rye and triticale were the species with the shortest-living seeds.

Legumes

The longest-living seeds among the legumes were those of pea, in which >80% germination was maintained for 11 years (11 years: 82.0 ± 16.2) (Fig. 2). From this time point on, the accessions behaved less consistently, and by year 16 the germination dropped to 43.6%, the SD increased to 40.4% but the overall mean germination decreased, reaching zero by year 24. In the field, the germination decreased from year 1 (96.0 ± 3.7%) to year 12 (58.8 ± 40.5%) and showed an increasing SD. But older seeds, by year 16, revealed a decreasing SD (16.0 ± 14.7%). The similar germination behaviour in the lab and in the field led to a significant high correlation ($R^2 = 0.71$; $P < 0.001$). Furthermore, probit analysis showed that pea seeds performed the best, with P50 = 13.9 years, followed by lupin with P50 = 13.5 years. Lupin germination percentage varied between 80 and 100% over the first 7 years (7 years: 83.6 ± 10.4%), dropping thereafter in some accessions, but being maintained at a high value in others (10 years: 44.0 ± 32.7%). The oldest available seed stock was 17 years (39.6 ± 28.9%), so the storage time before germinability was completely lost could not be established. The germination variability of genotypes in the field was high over all tested harvest years (SD: ±12.4 to ±23.5) but the correlation with lab germination appeared to be lower ($R^2 = 0.33$; $P < 0.01$). Common vetch accessions maintained their germinability at ~95% for 7 years (7 years: 94.8 ± 4.8%), thereafter behaving much like pea. The variability of germination between the genotypes rose in year 11 (48.0 ± 30.0%) and dropped until year 18 (0.55 ± 1.3%). Only the behaviour in the field changed such that the highest SD could be detected in year 4 (79.2 ± 33.4%) and correlation between field and lab behaviour was low ($R^2 = 0.16$). Probit analysis revealed that the half-viability period of common vetch was 10.8 years. For common bean, 70–90% lab germination was maintained over 9 years (9 years: 76.8 ± 26.6%), after which it fell gradually and reached its highest SD in year 12 (lab: 32.4 ± 39.8%; field: 21.6 ± 40.7%), whereas germination ceased by year 22. Germination and field emergence were highly correlated ($R^2 = 0.77$; $P < 0.001$). On the basis of the lower initial germination, P50 of common bean

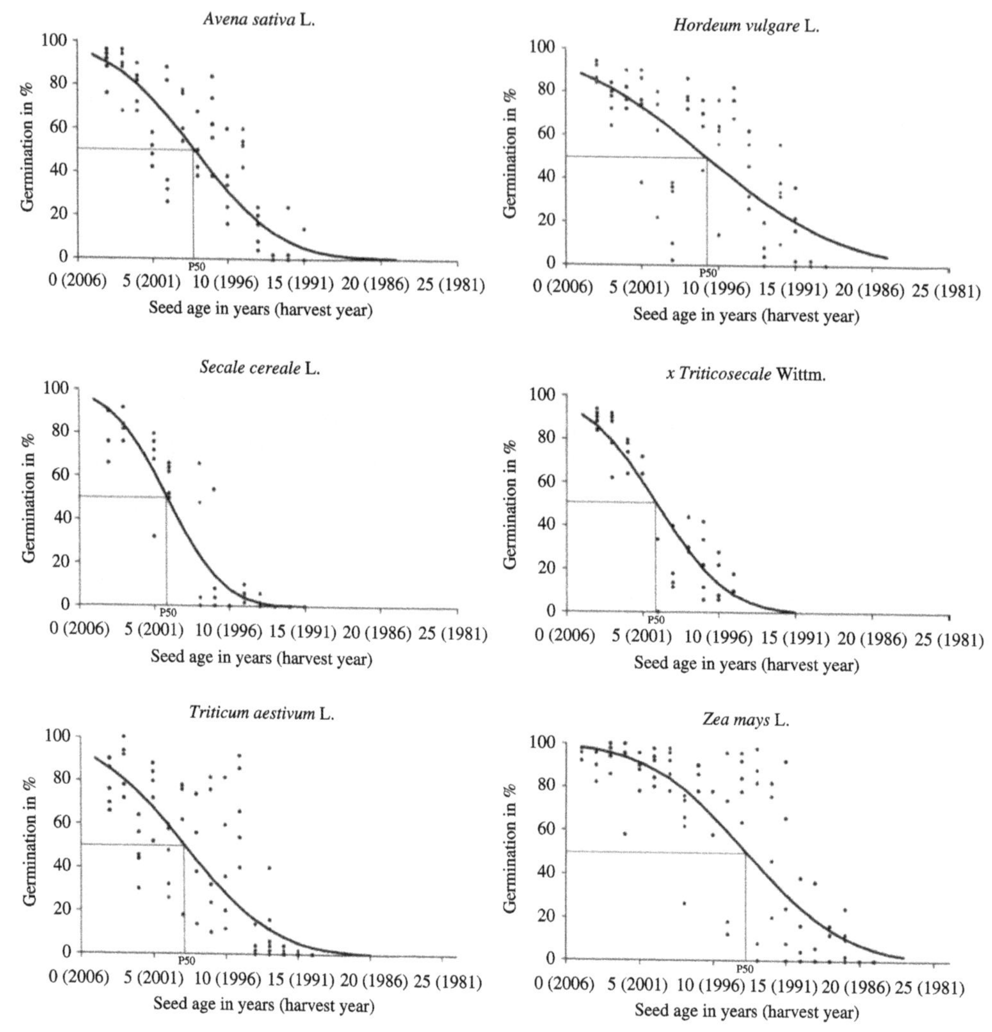

Figure 1. Percentage germination of six cereal species including the calculated probit curve as a function of storage time (harvest year). •, Germination (%) per accession; —, calculated probit curve; P50, half-viability period.

was calculated to be 9.8 years. Corresponding to germination, the half-viability period of the different legume seeds could be ranked with pea seeds as longest-living, followed by lupin, common vetch and common bean seeds.

Oil crops

Linseed retained an average germination of around 90% during the first 8 years of storage (Fig. 3) (8 years, lab: 90.4 ± 3.8%; field: 84.9 ± 13.9%), which dropped to ~10% by years 13 (lab: 5.2 ± 11.6%) and 16 (lab: 10.8 ± 8.6%). After this storage period the germination in the field was better (16 years: 27.2 ± 17.5%) but corresponded very well with the lab germination ($R^2 = 0.88$; $P < 0.001$). No seed older than this was available for testing. The half-viability period P50 was 10.4 years, which made linseed the best performing oil crop. Poppy seeds followed with a P50 of 8.2 years but included a high uncertainty between 6.6 and 25.8 years due to missing germination results in older harvest years. The mean germination of poppy seeds fell from 92.4 ± 7.3% to 62.0 ± 44.3% between years 4 and 9, and the highest variability of genotypes was reached in year 9. Older seeds than these were not available for testing. In the field, poppy seeds performed worse than in the lab, resulting in a correlation (R^2) of 0.63 ($P < 0.001$). Collard seeds maintained a high germination rate for 5 years

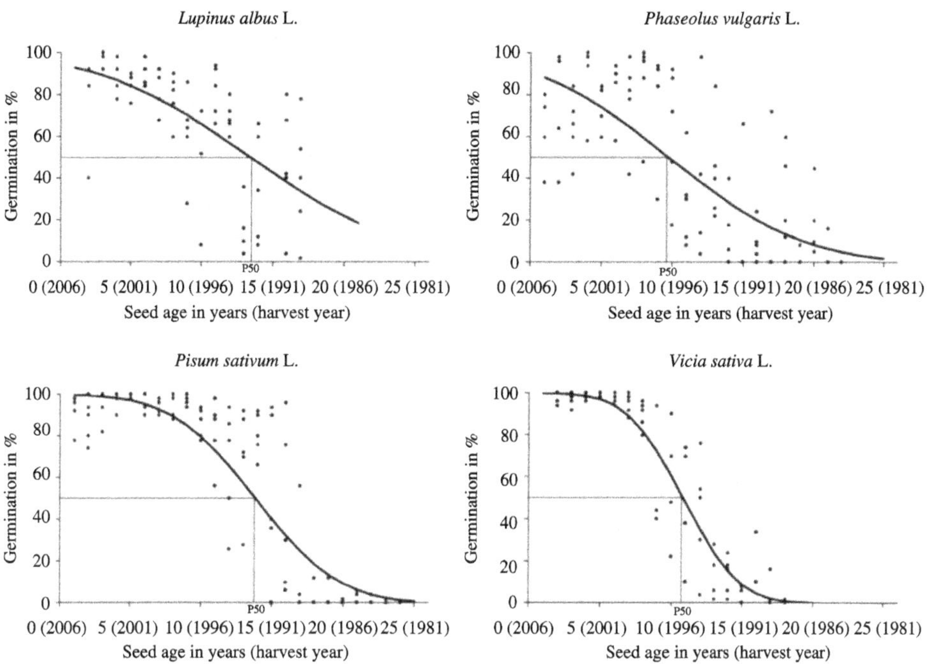

Figure 2. Percentage germination of four legume species including the calculated probit curve as a function of storage time (harvest year). •, Germination (%) per accession; —, calculated probit curve; P50, half-viability period.

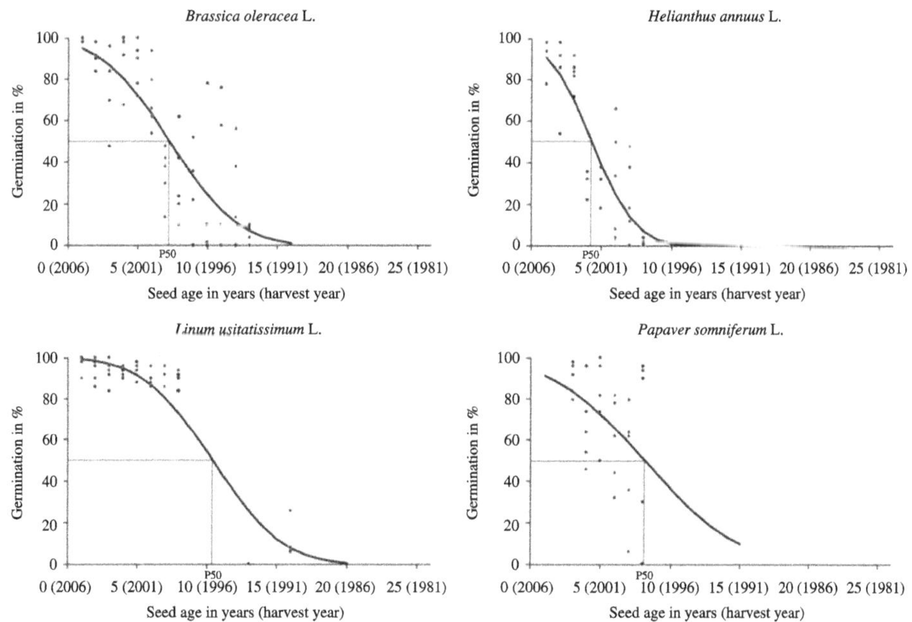

Figure 3. Percentage germination of four oil crop species including the calculated probit curve as a function of storage time (harvest year). •, Germination (%) per accession; —, calculated probit curve; P50, half-viability period.

(lab: 86.8 ± 11.5%; field: 84.8 ± 14.4%), the standard deviation increased in 11 years (28.8 ± 35.7%) thereafter decreasing to 4.4 ± 4.6% by year 13. The appearance of the seedlings in the lab was similar to those in the field and was highly correlated ($R^2 = 0.87$; $P < 0.001$). On the basis of these lab germination results, the half-viability period of collard was 7.3 years followed by sunflower with P50 = 4.3 years. For the latter crop large differences between accessions were noted but the mean germination rate fell rapidly from year 1 (90.0 ± 10.6%) to 32.4 ± 26.7% in year 6, reaching zero by year 10. In the field the performance of sunflower seeds was only 77.0 ± 18.4% germination after 1 year of storage, and after 7 years 49.2 ± 28.8% germinated, but the results corresponded well with lab germination ($R^2 = 0.73$, $P < 0.01$). The P50 values of the oilseeds were inferior to those of the legume seeds and could be ranked from long-living linseed to poppy and collard seeds. Sunflower seeds appeared to be short-living in this experiment.

Miscellaneous crops

Cucumber seed germination was maintained between 70 and 100% over 12 years of storage (12 years: 70.4 ± 42.7%) (Fig. 4). In year 13 the variation of genotypes rose to 50.4 ± 44.1%, falling to 49.3 ± 42.8% by year 16 and ceasing in year 17. For further tests in field performance no seeds were available. On the basis of the lab germinations P50 was 14.9 years with a slightly higher uncertainty between 13.5 and 17.3 years. In contrast, carrot seed half-viability was more than half shorter and calculated to be 6.1 years. Carrot seed germination declined slowly from 75.2 ± 7.8% (2 years) to 15.5 ± 28.4% by year 9 and reached highest standard deviation by year 5 (46.4 ± 31.2%). The same behaviour was detectable in the field: germination dropped from year 3 (64.8 ± 6.6%) to year 9 (8.5 ± 17.0%) and reached highest SD in year 6 (49.2 ± 18.5%). The results in the field correlated with the lab germination at $R^2 = 0.61$ ($P < 0.01$). A considerably shorter survival curve was presented by lettuce, falling from 98.0 ± 2.5% (2 years) to 26.0 ± 32.0% by year 5 and 0% by year 8. In the field it already reached 64.8 ± 6.6% in year 3 and declined to 0.8 ± 1.1% in year 6, thus a high correlation was found between both experiments ($R^2 = 0.79$, $P < 0.001$). The least storable seed of all the test species were those of chives, for which germination started at 62.4 ± 22.9% (1 year), the SD was increased in year 3 (23.6 ± 29.6%) and reached zero by year 6. In the case of the field experiment, germination was detected only in the first storage year (24.4 ± 10.1%) and the correlation resulted in $R^2 = 0.63$ ($P < 0.001$). Lettuce and chive seeds behaved in a similar fashion and their survival curves resulted in a half-viability period of 4.6 years and 1.9 years, respectively.

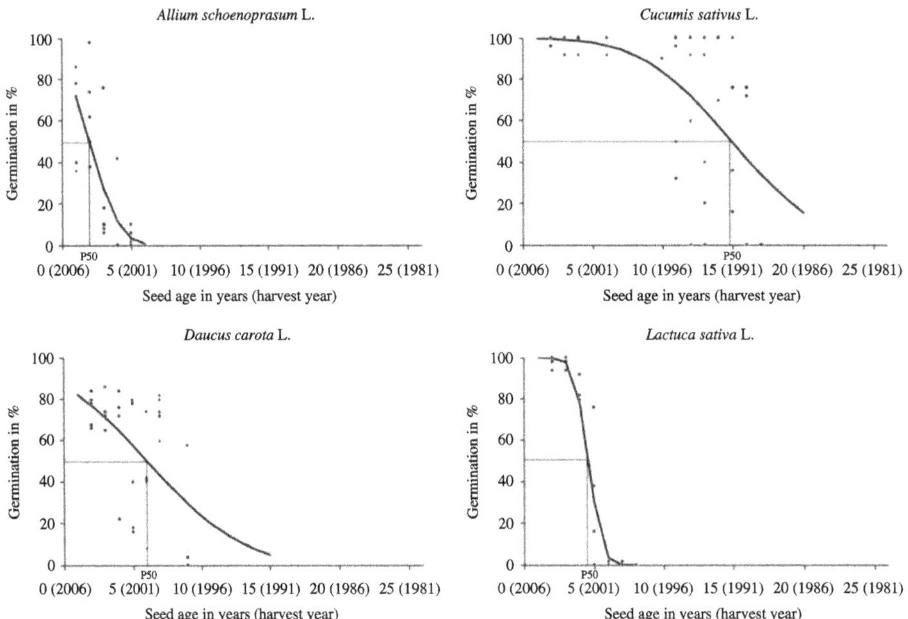

Figure 4. Percentage germination of four miscellaneous species including the calculated probit curves as a function of storage time (harvest year). •, Germination (%) per accession; —, calculated probit curve; P50, half-viability period.

Finally, the longest-living miscellaneous crop seeds were cucumber followed by carrot and lettuce. The lowest P50 value was associated with chive seeds.

Discussion

We have determined longevity for seeds from 18 crop species, which were stored under ambient conditions (average temperature: 20.3 ± 2.3°C; average RH: 50.5 ± 6.3%) for up to 26 years. Viability remained relatively high for at least the first 2 years after harvest but declined to 0 within 5–23 years, giving an average P50 among all species of 8.6 ± 3.6 years.

As has been documented by Roberts (1972), germination decreased over storage time in a sigmoid fashion whereby the parameters of these curves seemed to be species specific. Despite the fact that the probit analysis method is sensitive to deviation with regard to the shape of the survival curve (Moore and Roos, 1982), the correlation between P50 and absolute longevities resulted in a high R^2 (0.90; $P < 0.001$). A higher uncertainty was only seen for those species (lupin, poppy, cucumber, carrot) that did not have available seeds of older harvest years (Table 2). When germination results of a whole survival curve were available, the half-viability period P50 proved to be a valuable parameter for evaluation of seed longevity and may give information about seed storage behaviour under ambient conditions.

On the basis of P50 values, seeds of maize, lupin, common vetch, pea, linseed and cucumber (P50 > 10 years) tended to be relatively long-living, whereas those of sunflower, chives and lettuce (P50 < 5 years) were not. According to the results of Walters *et al.* (2005) seeds of maize, common vetch, pea and lettuce were equally classified. In contrast, sunflower and linseed seeds were defined as medium, cucumber seeds as medium long, whereas chive and lupin seeds were not included. On the whole, the hypothesis that seed deterioration behaviour is species characteristic is supported by significant correlations among different seed ageing experiments (Tables 3 and 4). In relation to this fact, the absolute longevity correlated significantly with results by Priestley *et al.* (1985), whereas P50 correlated with data by Walters *et al.* (2005) (Table 4). The viability equation by Ellis and Roberts (1980) also showed a significant R^2 (0.51, $P < 0.05$) but only by using a ranked correlation with P50. Due to open storage conditions, averaged temperatures and RH, Ellis's viability equations underestimated almost all calculated species, except pea. However, the equation is thought to predict seed longevity under controlled conditions in a range of temperatures from −13°C up to 90°C. The validity of predicting, by extrapolation, longevities in the region of millennia for dry, cold-stored seeds is often criticized and overestimations are assumed (Pritchard and Dickie, 2004). Viability equations would provide helpful tools for genebank managers but overestimation would end in a disaster of dead seeds. Underestimations, as shown here, are also of low value and involve higher management cost and higher risk for the material, considering intermixture, natural disasters and genetic drift. Nevertheless, referring to the coefficients of determination, they do not explain the whole variance. Therefore further factors of influence, e.g. harvest year, processing and storage conditions, are conceivable.

Besides an interspecific variation in seed longevity, a varying behaviour between different genotypes was indicated by the standard deviations. Over all species the mean SD was 9.9 ± 8.1% in the second year, increasing to 21.6 ± 11.4% in the year around the species-characteristic P50 value and attaining the highest SD value (32.1 ± 9.4%) after 9.3 ± 3.4 years. Chive and common bean seeds, particularly, showed an SD above 20% over the half-survival period. In the case of common bean seeds, unpublished data by

Table 4. Correlations – (a) Pearson, (b) Spearman, ranked correlation – between P50, absolute longevities, Ellis's viability equations and literature data. Longevity values are given in Table 3. Correlation coefficients (R^2) are given for each linear regression, with the number of species considered given in parentheses

	Absolute longevity	Ellis viability equation	Priestley *et al.* (1985)	Walters *et al.* (2005)
(a)				
P50	0.902 (12)***	0.420 (8)	0.152 (15)	0.336 (16)*
Absolute longevity		0.378 (7)	0.586 (11)**	0.279 (12)
Ellis viability equations			0.618 (8)*	0.256 (8)
Priestley *et al.* (1985)				0.105 (15)
(b)				
P50	0.895 (12)***	0.510 (8)*	0.235 (15)	0.364 (16)*
Absolute longevity		0.413 (7)	0.646 (11)**	0.299 (12)
Ellis viability equations			0.581 (8)*	0.000 (8)
Priestley *et al.* (1985)				0.082 (15)

* Indicates a significant trend at $P < 0.05$; **, a significant trend at $P < 0.01$; ***, a significant trend at $P < 0.001$.

Nagel and Börner revealed a clear variability among different coloured beans, which, apparently, played a part in contributing to that variation. Furthermore, collard, barley, lupin, poppy, wheat and maize seeds reached an SD above 10% in the same period and showed a higher inter-varietal difference than the other crops. Due to various harvest seasons of the different crops, genotypic differences cannot be associated with the harvest year. In addition, the experimental design did not allow an analysis of variance concerning the factors harvest year, species or genotype. But experiments on aged seeds belonging to double-haploid barley mapping populations outlined a high intraspecific variation, which is inherited quantitatively. Four quantitative trait loci (QTL) were detected and associated with genes controlling fertility, abiotic and biotic stress response, plant development and morphology (Nagel *et al.*, 2009). Debeaujon *et al.* (2000) also showed an impact of structure and pigmentation on storability of *Arabidopsis* mutants. Therefore mainly the morphology and factors occurring during the vegetative period in a harvest year, as well as the ability of plants to respond to these factors, seem to influence the genotypic behaviour.

Furthermore, by calculating the mean P50 values among the crops, the results presume that the predominant seed storage compound (cereals = carbohydrates; legumes = proteins; oil crops = lipids) may determine seed longevity, with proteins > carbohydrates > lipids (Table 3). Average data by Priestley *et al.* (1985) give the same ranking for longevities (P50) with proteins (13.1 years), carbohydrates (8.4 years) and lipids (7.1 years). However, data from Earle and Jones (1962) showed that legume species had a wide range of protein content from 12 to 55% and an oil content between 1 and 45%. A correlation between absolute longevity, P50 and averaged starch, protein and oil content data by Earle and Jones (1962), Jones and Earle (1966) and Sinclair and DeWit (1975) revealed no significant relationship between the storability of seeds and the starch and protein content (Tables 1 and 5). Only the oil content appeared to have a strong influence ($R^2 = 0.64$; $P < 0.01$) on the absolute longevity but not on P50. At storage conditions of −20°C and various RHs, Pritchard and Dickie (2004) could confirm a more rapid ageing in oily seed, but further proved that at a constant RH the dependence of predicted longevity on oil content is weak. A similar correlation was also found by Walters *et al.* (2005) but the relationship did not hold when families with varying chemical composition, such as legumes, were considered separately in those studies. Based on the findings of Priestley (1986), Vertucci and Leopold (1987), Gidrol *et al.* (1989), Steadman *et al.* (1996), Sun and Leopold (1997), Medeiros *et al.* (1998) and Sinniah *et al.* (1998), Pritchard and Dickie (2004) assumed that a major effect of chemical compounds on seed longevity is related to: (1) the sorption properties of seeds; (2) the potential sites for free radical attack; and (3) the presence of protective compounds and their activity. However, regarding the seed full survival curve, oil content seemed to influence longevity under open storage conditions, so further analysis will be necessary to clarify these mechanisms.

Apart from the relationship between compositional and longevity data, a negative correlation between oil content and thousand kernel weight ($R^2 = 0.49$; $P < 0.05$) appeared. Assuming decreasing seed longevity with increasing oil content, a shorter survival curve could be estimated with lower seed weight. Except for non-existent correlations between these factors, a direct analysis between different seed mass and longevities in dry storage revealed no significant relationship (Pritchard and Dickie, 2004).

Regarding field emergence, significant correlations ($R^2 > 0.54$, $P < 0.01$) with germination results appeared among all species except for lupin and common vetch. Improved germination performance under field conditions, as compared to controlled laboratory conditions, was found for oat, barley, common bean, pea, common vetch and sunflower seeds. It seems probable that the basis of this improvement lies in the activity of soil microflora, which can interact with moisture availability to promote germination (Powell, 1988). This was especially marked in 1-year-old common bean seeds, which showed 90.4 ± 4.3% emergence in the field whereas the seeds in the lab gave only 66.4 ± 17.9% germination. In contrast, chive, carrot and lettuce

Table 5. Correlations between P50, absolute longevities, chemical composition (% of dry mass) and 1000 kernel weight (weight/1000 in g) taken from data of Earle and Jones (1962), Jones and Earle (1966) and Sinclair and DeWit (1975). Correlation coefficients (R^2) are given for each linear regression, with the number of species considered given in parentheses

	P50	Carbohydrates	Proteins	Lipids	Weight/1000
Absolute longevity	0.902 (12)***	0.107 (8)	0.036 (11)	0.640 (11)**	0.567 (6)
P50		0.091 (8)	0.007 (15)	0.090 (15)	0.169 (10)
Carbohydrates			0.683 (8)*	0.714 (8) **	0.740 (3)
Proteins				0.198 (15)	0.271 (10)
Lipids					0.488 (10)*

* Indicates a significant trend at $P < 0.05$; **, a significant trend at $P < 0.01$; ***, a significant trend at $P < 0.001$.

Table 6. Recommendations (in years) for various species stored under ambient conditions (20.3°C and 50.5% RH) and 95% confidence limits (in years) of storage period until germination drops to 80%

Crop	Species	Estimation (80%)	95% Confidence limits (years)		Recommendation
			Lower bound	Upper bound	
Cereals					
	Avena sativa L.	4.0	3.1	4.8	4.0
	Hordeum vulgare L.	3.4	1.3	4.8	4.0
	Secale cereale L.	3.4	2.7	4.0	3.0
	× *Triticosecale* Wittm.	2.9	1.9	3.6	3.0
	Triticum aestivum L.	3.1	1.6	4.3	3.0
	Zea mays L.	7.8	6.6	8.8	8.0
Legumes					
	Lupinus albus L.	6.3	4.1	7.8	6.0
	Phaseolus vulgaris L.	3.5	1.4	5.1	3.0
	Pisum sativum L.	10.0	8.5	11.1	10.0
	Vicia sativa L.	8.2	7.5	8.8	8.0
Oil crops					
	Brassica oleracea L.	4.1	2.7	5.0	4.0
	Helianthus annuus L.	2.2	1.2	2.9	2.0
	Linum usitatissimum L.	7.1	6.3	7.9	7.0
	Papaver somniferum L.	3.7	−10.7	5.2	3.0
Miscellaneous crops					
	Allium schoenoprasum L.	0.6	−0.7	1.2	0.5
	Cucumis sativus L.	10.7	8.4	12.2	8.0
	Daucus carota L.	1.5	−4.5	3.2	1.0
	Lactuca sativa L.	3.9	3.3	4.3	3.0

seeds behaved worse in the field as compared with lab results, which could be an artefact of the design (standardized conditions for all species, timing) of the field experiment. However, the germination test reveals basic information about the seed behaviour in the field and provides important data about viability of seeds after long-term storage.

In general, the data contribute to a broader understanding of seed behaviour in storage. Although most genebanks are working on the basis of FAO/IPGRI Genebank Standards (FAO, 1994) with a recommended temperature of −18°C and 3–7% seed moisture content, seed companies, breeders, farmers and genebanks store seed mostly under unfavourable conditions. Additionally, the *ex situ* genebank for agricultural and horticultural crop plants in Gatersleben includes a room designed for storing remnant seeds which are provided to users during the first years after harvest. Questions about duration of the provision of seeds may be answered by this study. For seed keepers in general, storage temperature and seed moisture content are important in dealing with seed longevity (Ellis and Roberts, 1980). Considering the storage conditions of about 20.3 ± 2.3°C and 50.5 ± 6.3% RH, this study can be a useful guideline for viability estimations under these conditions. Therefore, Table 6 provides recommendations of storage times until germination falls to 80%. As previously assumed, genotypic differences and initial germination can provoke strong differences in germination behaviour and were taken into account for these recommendations. But for decision-making of genebank managers these factors need to be considered when a monitoring frequency is chosen.

Acknowledgements

The authors wish to acknowledge past and current staff of the Gatersleben genebank who collected, regenerated, stored and monitored the seed material. Special thanks are due to Sibylle Pistrick, the head of the IPK germination testing laboratory. Finally, we are grateful to Professor Christina Walters, USDA-ARS, National Center for Genetic Resources Preservation, Fort Collins, USA, Professor Michael Kruse, University of Hohenheim, Germany and Professor Rolf Rauber, University of Göttingen, Germany, for their advice on statistical analysis and constructive comments.

References

Börner, A. (2006) Preservation of plant genetic resources in the biotechnology era. *Biotechnology Journal* **1**, 1393–1404.

Debeaujon, I., Léon-Kloosterziel, K.M. and Koornneef, M. (2000) Influence of the testa on seed dormancy, germination, and longevity in Arabidopsis. *Plant Physiology* **122**, 403–413.

Earle, F.R. and Jones, Q. (1962) Analyses of seed samples from 113 plant families. *Economic Botany* **15**, 221–250.

Ellis, R.H. and Roberts, E.H. (1980) Improved equations for the prediction of seed longevity. *Annals of Botany* **45**, 13–30.

FAO (1994) *Genebank standards*. Rome, Food and Agriculture Organization of the United Nations; Rome, International Plant Genetic Resources Institute.

Gidrol, X., Serghini, H., Noubhani, A., Mocquot, B. and Mazliak, P. (1989) Biochemical changes induced be accelerated aging in sunflower seeds. I. Lipid peroxidation and membrane damage. *Physiologia Plantarum* **76**, 591–597.

Haferkamp, M.E., Smith, L. and Nilan, R.A. (1953) Studies on aged seeds I – relation of age of seed to germination and longevity. *Agronomy Journal* **45**, 434–437.

ISTA (2005) *International rules for seed testing*. Bassersdorf, International Seed Testing Association.

Jones, Q. and Earle, F.R. (1966) Chemical analyses of seeds II: oil and protein content of 759 species. *Economic Botany* **20**, 127–155.

Linington, S.H. and Pritchard, H.W. (2001) Gene Banks. pp. 165–181 *in* Levin, S.A. (Ed.) *Encyclopedia of biodiversity*. Vol. 3. San Diego, Academic Press.

Liu K., Eastwood R.J., Flynn S., Turner R.M. and **Stuppy W.H.** (2008) Seed information database (release 7.1, May 2008). Available at http://www.kew.org/data/sid (accessed 27 October 2009).

Maxted, N., Ford-Lloyd, V.V. and Hawkes, J.G. (1997) *Plant genetic conservation. The in situ approach*. London, Chapman & Hall.

Medeiros, A.C.S., Probert, R.J., Sader, R. and Smith, R.D. (1998) The moisture relations of seed longevity in *Astronium urundeuva* (Fr. All.) Engl. *Seed Science and Technology* **2**, 289–298.

Moore, F.D. and Roos, E.E. (1982) Determining differences in viability loss rates during seed storage. *Seed Science and Technology* **10**, 283–300.

Nagel, M., Vogel, H., Landjeva, S., Buck-Sorlin, G., Lohwasser, U., Scholz, U. and Börner, A. (2009) Seed conservation in ex situ genebanks – genetic studies on longevity in barley. *Euphytica* **170**, 5–114.

Powell, A.A. (1988) Seed vigour and field establishment. *Advances in Research and Technology of Seeds* **11**, 29–61.

Priestley, D.A. (1986) *Seed ageing: implications for seed storage and persistence in the soil*. New York, Comstock.

Priestley, D.A., Cullinan, V.I. and Wolfe, J. (1985) Differences in seed longevity at the species level. *Plant, Cell and Environment* **8**, 557–562.

Pritchard, H.W. and Dickie, J.B. (2004) Predicting seed longevity: the use and abuse of seed viability equations. pp. 653–722 *in* Smith, R.D.; Dickie, J.B.; Linington, S.H.; Pritchard, H.W.; Probert, R.J. (Eds) *Seed conservation: turning science into practice*. Kew, London, Royal Botanic Gardens Kew.

Roberts, E.H. (1972) *Viability of seeds*. London, Chapman & Hall.

Ruckenbauer, P. (1971) Keimfähiger Winterweizen aus dem Jahre 1877 [Germinable winterwheat by the 1877]. *Die Bodenkultur* **22**, 372–386.

Sallon, S., Solowey, E., Cohen, Y., Korchinsky, R., Egli, M., Woodhatch, I., Somchoni, O. and Kislev, M. (2008) Germination, genetics, and growth of an ancient date seed. *Science* **320**, 1464.

Sinclair, T.R. and DeWit, C.T. (1975) Photosynthate and nitrogen requirements for seed production by various crops. *Science* **189**, 565–567.

Sinniah, U.R., Ellis, R.H. and John, P. (1998) Irrigation and seed quality development in rapid-cycling brassica: soluble carbohydrates and heat-stable proteins. *Annals of Botany* **82**, 647–655.

Steadman, K.J., Pritchard, H.W. and Dey, P.M. (1996) Tissue-specific soluble sugars in seeds as indicators of storage category. *Annals of Botany* **77**, 667–674.

Steiner, A.S. and Ruckenbauer, P. (1995) Germination of 110-year-old cereal and weed seeds, the Vienna sample of 1877. Verification of effective ultra-dry storage at ambient temperature. *Seed Science Research* **5**, 195–199.

Sun, W.Q. and Leopold, A.C. (1997) Cytoplasmic vitrification and survival of anhydrobiotic organisms. *Comparative Biochemistry and Physiology* **117**, 327–333.

Vertucci, C.W. and Leopold, A.C. (1987) Oxidative processes in soybean and pea seeds. *Plant Physiology* **84**, 1038–1043.

Walters, C., Wheeler, L.M. and Grotenhuis, J.M. (2005) Longevity of seeds stored in a genebank: species characteristics. *Seed Science Research* **15**, 1–20.

3.2 PAPER 2:

Machine learning links seed composition, glucos-inolates and viability of oilseed rape after 31 years of long-term storage

by

Manuela Nagel, Katharina Holstein, Evelin Willner and Andreas Börner

Published in

Seed Science Research (2018), 28, 340 -348

https://doi.org/10.1017/S0960258518000259

To view supplementary material for this article, please visit

https://doi.org/10.1017/S0960258518000259

Seed Science Research

cambridge.org/ssr

Research Paper

Cite this article: Nagel M, Holstein K, Willner E, Börner A (2018). Machine learning links seed composition, glucosinolates and viability of oilseed rape after 31 years of long-term storage. *Seed Science Research* **28**, 340–348. https://doi.org/10.1017/S0960258518000259

Received: 12 November 2017
Accepted: 1 June 2018
First published online: 12 July 2018

Keywords:
artificial neural networks; *Brassica napus*; fatty acids; gene bank; germination; multivariate regression; seed conservation; seed longevity; seed viability

Author for correspondence:
Manuela Nagel, Email: Nagel@ipk-gatersleben.de

Machine learning links seed composition, glucosinolates and viability of oilseed rape after 31 years of long-term storage

Manuela Nagel[1], Katharina Holstein[2], Evelin Willner[1] and Andreas Börner[1]

[1]Genebank Department, Leibniz-Institute of Plant Genetics and Crop Plant Research (IPK Gatersleben), Seeland, Germany and [2]Fraunhofer Institute for Factory Operation and Automation (IFF), Magdeburg, Germany

Abstract

Seed longevity is influenced by many factors, a widely discussed one of which is the seed lipid content and fatty acid composition. Here, linear and non-linear regressions based on machine learning were applied to analyse germinability and seed composition of a set of 42 oilseed rape (*Brassica napus* L.) accessions grown under the same single environment and at the same time following a period of up to 31 years storage at 7°C. Mean viability was halved after 27.0 years of storage, but this figure concealed a major influence of genotype. There was also wide variation with respect to fatty acid composition, particularly with respect to oleic, α-linolenic, eicosenoic and erucic acid. Linear regression (r_L) revealed significant correlation coefficients between normal seedling appearance and the content of α-linolenic acid (+0.52) and total oil (+0.59). Multivariate regression using artificial neural networks including a radial basis function (RBF), a multilayer perceptron (MLP) and a partial least square (PLS) recognized underlying structures and revealed high significant correlation coefficients (r_M) for oil content (+0.87), eicosenoic acid (+0.75), stearic acid (+0.73) and lignoceric acid (+0.97). Oil content or a combination of oleic, α-linolenic, arachidic, eicosenoic and eicosadienoic acids and glucosinolates resulted in highest model fitting parameters R^2 of 0.90 and 0.88, respectively. In addition, the glucosinolate content, predominantly in the Brassicaceae family and ranging from 4.6 to 79.5 μM, was negatively correlated with viability ($r_L = -0.43$). Summarizing, oil content, some fatty acids and glucosinolates contribute to variations in average half-life (15.2 to 50.7 years) of oilseed rape seeds. In contrast to linear regression, multivariate regression using artificial neural networks revealed high associations for combinations of parameters including underestimated minor fatty acids such as arachidic, stearic and eicosadienoic acids. This indicates that genetic and seed composition factors contribute to seed longevity. In addition, multivariate regressions might be a successful approach to predict seed viability based on fatty acids and seed oil content.

Introduction

The concept of machine learning was introduced in the late 1960s as a means to detect patterns in data (Shalev-Shwartz and Ben-David, 2014). This field of computational statistics aims for a deeper exploration of datasets by clustering data into groups or solving a classification or (multivariate) regression tasks.

Depending on the complexity of the dataset, different approaches are used to solve regression tasks. For a simple regression analysis, a linear model might be already sufficient to describe the dataset completely and obtain valid predictions. If the dataset is more complex, e.g. various inputs that are not independent from each other, non-linear approaches have shown to be superior to describe the underlying principles between input and output data. The aim of a multivariate regression analysis is to predict the output behaviour based on the dataset correctly. For such cases, the input data are a set of numerical features, e.g. metabolite or substrate concentrations, chemical compounds or viability data (Andre, 2003; Hall, 2011; Rivas-Ubach *et al.*, 2012) and is referenced as a set of feature vectors. These input vectors can be based on different units and may be continuous or discrete. Therefore, input data are normalized to cope with varying magnitudes and specific properties. The desired output vector can be a discrete quantity or continuous (Worley and Powers, 2013), e.g. a correlation value between zero and one, where larger values indicate higher correlations. Note, using only one input vector in a multivariate regression analysis still differs from a standard linear regression analysis as the correlation between input and output vector is a non-linear function. Next to the multivariate regression analysis, clustering is an unsupervised approach and the principal component (PCA) or linear discriminant analysis (LDA) are well-known examples. In PCA or LDA, data are divided into groups based on a similarity measure in the dataset (Rojas, 1996). In contrast, classification of data is usually a supervised approach, where a desired output to input data is known in advance. This enables mapping of inputs to outputs to assign data of unknown classes to one of the given input classes. Thus, class membership in a classification task is a binary decision, e.g. seeds are viable or not.

Machine learning offers a variety of different algorithms to solve complex regression tasks. The simplest one would be a partial least squares (PLS) algorithm, where the sum of square residuals is minimized (Wold *et al.*, 2001). More complex models using non-linear relations are modelled on biological neurons, e.g. in multilayer perceptrons (MLP) the input data are propagated through multiple layers of activation functions (neurons) (Cybenko, 1989). Other artificial neural networks incorporate radial basis functions (RBF) and group similar feature vectors by approximation before regression is performed (Moody and Darken, 1989). To represent dependencies between input and output of a given dataset correctly, first, a mathematical model is chosen, then adapted during a training phase using an optimization algorithm, such as stochastic gradient descent or back propagation, and finally applied on validation dataset.

To assess the quality of the multivariate regression analysis the correlation coefficient r_M can be used to describe the relationship between the input value or matrix and the desired output. A further quality measure is the distance between the desired output and the predicted output, often referred to as R^2. The correlation coefficient r_M is restricted to values between -1 and 1, where values equal to -1 describe a negative correlation, values equal to 1 describe a positive correlation and values equal to 0 show no correlation. A quality measure $R^2 < 0$ shows that a horizontal line drawn in a correlation plot explains the correlation better than the results of the chosen model, $R^2 = 0$ represents an equal explanation as the line, and $R^2 > 0$ provides a better explanation than the horizontal line.

Multivariate regression analysis can produce false positive correlations. To validate the results from non-linear analysis, a cross validation scheme is employed. The available dataset is split randomly into multiple distinct subsets, where analysis is performed on each subset independently. Choosing a subset in a random way guarantees that the procedure does not produce good results by coincidence. Achieving high correlation coefficients for all cross-runs ensures that there is indeed a non-linear correlation for the overall dataset and not just the chosen random test sample. Another problem in machine learning algorithms is overfitting when models are perfectly adapted to the training data but the generalization of the obtained model is not feasible. Overfitting of artificial neural networks can be indicated by coefficients equal to 1 or highly disparate r and R^2 (Worley and Powers, 2013). To avoid overfitting effects, it is crucial to select model parameters appropriate to the structure and size of the available data.

Seed longevity, i.e. the ability of seeds to remain viable over certain storage periods, is determined by an intricate network of genetic and environmental factors. The genetic factors are associated with seed morphology and composition, whereas the environment affects by a combination of conditions prevailing during seed development, ripening, at harvest and during storage.

The major cause of the deterioration of seed quality over time is the oxidative stress, which results from a build-up of reactive oxygen species (Bailly, 2004; Kranner *et al.*, 2010; Waterworth *et al.*, 2015). Certain antioxidant compounds act to scavenge these molecules and thereby are able to protect lipids and proteins from degradation. A prominent such protectant in seeds is the fat-soluble tocopherol (Hwang *et al.*, 2014; Sattler *et al.*, 2004) which react with lipid peroxy and alkoxy radicals and so terminate the chain reaction of lipid peroxidation (Falk and Munné-Bosch, 2010). The assumption that oil-rich seeds are particularly sensitive to deterioration has been present for many years. Supporting evidence was originally based on the finding that auto-oxidation of polyunsaturated fatty acids produces free radicals, thereby compromising membrane integrity (Priestley and Leopold, 1979). The rate of oxidation is strongly dependent on oxygen concentration and temperature (Crapiste *et al.*, 1999). However, in general, the correlation between seed oil content and longevity has been described as weak (Nagel and Börner, 2010; Priestley *et al.*, 1985; Walters *et al.*, 2005).

Oilseed rape (*Brassica napus* L.) provides a substantial quantity of the world's vegetable oil production; about 44% of seed dry matter is oil (http://faostat.fao.org, 2015). The most abundant fatty acids present are linolenic (C18:3), linoleic (C18:2), oleic (C18:1) and erucic (22:1) acid. Other classes of compounds present are tocopherol (vitamin E), cellulose, phenolic acids, phytate and glucosinolates (Wittkop *et al.*, 2009). Seed longevity at cold storage (-18°C) is relatively low with a half-viability of about 25 years (Walters *et al.*, 2005).

Here, the longevity following long-term storage (7°C, 5.0% seed moisture content) of 42 oilseed rape gene bank accessions have been investigated. Accessions were grown in the same field, harvested in 1983 and stored at comparable storage conditions until 2014. The aims were: (1) to obtain an estimate of how long oilseed rape takes for seed viability to fall to 50% (P50, half-viability period) during storage at 7°C; (2) to compare key seed components and fatty acid composition of the stored material; and (3) to link seed viability with the content of key fatty acids and/or seed compounds by linear and non-linear correlation analyses using machine learning.

Materials and methods

Four replicate batches of 50 seeds of a set of 42 *B. napus* ssp. *napus* var. *napus* f. *biennis* accessions were tested for their ability to germinate in 2014. The accessions had last been multiplied together in a single field and experienced the same maternal environment in 1983. Fully mature seeds were harvested, cleaned and have been maintained at 7 ± 3°C and 5.0 ± 0.3% seed moisture content at the IPK gene bank Satellite Collection North (Malchow, Germany). The viability tests were conducted by laying seeds on moist filter paper, then keeping them under a 12 h photoperiod at 22°C. The proportion of normal seedlings (% NS) which emerged was counted, following the protocol recommended by ISTA (2014), while the overall proportion of germinated seed (%TG, total germination) was assessed after 7 days had elapsed. The %TG were compared with historic viability data collected from the same accessions in 1983, 1990, 1993 and 2009. For %NS, a comparison was only possible for historic data from 2009. Based on available %TG data, a probit analysis was conducted to estimate the half-viability periods (P50) for each and overall accessions. The probit germination percentage at storage time p_s was given by the expression $K_i - (1 \times \sigma^{-1})\ p_s$, where K_i represented the initial probit germination percentage and σ the standard deviation of the distribution of dead seeds in time (Ellis and Roberts, 1980). Percentage values of 100% response (three accessions) were corrected using 99.997% corresponding to 4.01 probit.

Seed fatty acid composition (% of total fatty acids present) was investigated in 2014, based on three replicate 200 mg samples of

37 out of the 42 accessions, each representing the progeny of three to five plants. Following Rücker and Röbbelen (1996), seed oil was extracted using petroleum benzine and triglycerides transmethylated by isooctane to fatty acid methyl esters, which were subjected to gas liquid chromatography (GC) using a polyethyleneglycol-2-nitroterephtalsacidester column. The concentrations of oil, protein, moisture and the content of glucosinolate in µmol g^{-1} dry weight (DW) were measured using near infrared reflectance spectroscopy (NIRS, Foss-NIRSystem 5000, Foss GmbH, Germany). NIRS calibration was adjusted by the Thuringian State Institute for Agriculture (TLL, Jena, Germany) and both NIRS and GC results are frequently evaluated by the Canadian Grain Commission.

Statistical analysis was carried out using routines implemented in GenStat software (VSN International, 2013). In particular, the data were tested for normal distribution and subjected to analysis of variance, least significant differences at $P < 0.05$ (LSD5%) and linear correlation analysis were calculated between accessions. Correlation coefficients (r_L) are only given for significant correlations at $P < 0.05$.

Non-linear statistical analysis was carried out using the neural network toolbox in MATLAB. To compute principal components, the dataset was minimized to a complete set. Here, not all data points could be provided for all accessions over a storage period of 31 years, e.g. %TG for 1990 was not available for CR 743, CR 818 and CR 822. These accessions were left out of PCA and further non-linear analyses. To determine associations between seed viability and seed composition a linear regression analysis and a multivariate regression using artificial neural networks were applied. Both correlation analyses were performed between %NS and %TG from 2014 and a single compound as input vector x. Note, in mathematical definitions lower case characters (x) are used for single values or vectors and upper case characters (X) are used for matrices. In a second step, various combinations of fatty acids and/or major components and/or historic viability results (%TG for 1983, 1990, 1993 and %NS for 2009) were used as input matrix X for a multivariate regression based on artificial neural networks (Krzanowski, 2000). Traits were chosen in a biological meaningful manner: a combination of unsaturated fatty acids, oil content and glucosinolates were expected to have highest predictability for %NS 2014 and %TG 2014. Further combinations were partly based on results of the linear regression analysis. To account for variance in the dataset due to sampling, sample size, and calibrations, each input vector x was normalized using vector L2 norm (Euclidian norm):

$$\|x\|_2 = \left(\sum_i |x_i|^2\right)^{\frac{1}{2}}$$

Furthermore, to avoid biased analysis during correlation analysis, e.g. due to different concentrations or units, all inputs were standardized (S) using z scoring as follows:

$$S = \frac{x - E[x]}{\sigma(x)},$$

with $E[x]$ the expected value and $\sigma(x)$ the standard deviation of the input vector (Worley and Powers, 2013). The significance level P was set to values ≤ 0.01.

An RBF using 3, 5, 7 and 10 centres, a MLP using 5 or 10 in a single hidden layer, as well as 3 and 5 neurons in the first and second layer, and a PLS using 5 components were tested for

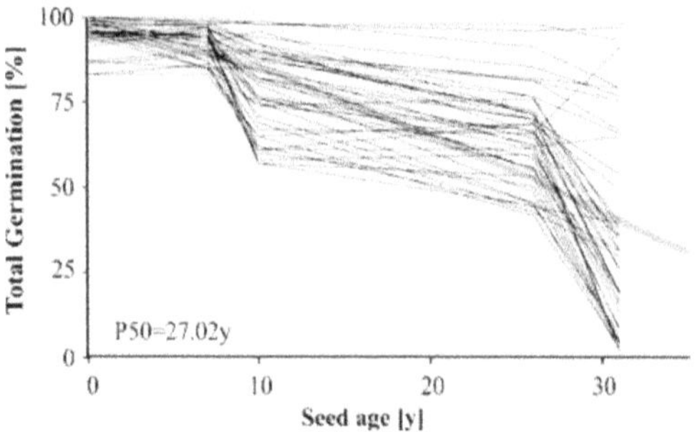

Figure 1. The decline in seed viability of oilseed rape after 31 years in long-term storage. The data are shown as the arithmetic mean derived from four replicates of 42 accessions. The curve derived from probit analysis of the data appears as a bold yellow line with the equation $y = -0.06x + 1.67$, where total germination (%) is on a probit scale. The half-viability period (P50) calculated from the probit curve is 27.02 years.

best performance. All artificial neural networks were run on the same unreduced data for reasons of comparability, where RBF models failed for some input vectors (where %TG data was missing). Each mathematical model (RBF, MLP, PLS) was run using a 10-fold cross validation. For each mathematical model and their different layouts, datasets were split into a training and generalization set (ratio 0.7). Models were trained on the training dataset. Afterwards, models were validated on the generalization dataset. Comparing results for all test runs on the training, generalization and cross-validation datasets ensures a minimization of overfitting. To evaluate performance results for the multivariate regression, the correlation coefficient r_M and the fitting parameter R^2 were chosen. Here, R^2 does not correspond to the square value of r_M but describes the correlation between predicted output and input vector, i.e. a quality measure. Still, values of r_M and R^2 should be similar. Furthermore, r_M and R^2 over all cross-validations should be comparable, with a desired R^2 between 0.4 and 1. With Y_a, the desired output vector, e.g. %NS 2014, Y_p, the predicted output vector by the neural network and mean values indicated by bars, r_M is given by:

$$r = \frac{\overline{(Y_p - \overline{Y_p})(Y_a - \overline{Y_a})}}{\sqrt{\mathrm{Var}(Y_p)\mathrm{Var}(Y_a)}},$$

and R^2 is given by:

$$R^2 = 1 - \frac{\sum_i (y_{p,i} - y_{a,i})^2}{\sum_i (y_{a,i} - \overline{Y_a})^2}.$$

Next to r_M and R^2, mean values and standard deviations of these two factors were computed and checked for overall behaviour of the non-linear correlation. In addition, all cross-runs and correlation plots for all input configurations were examined. If these were found to lie in a certain range ($r_M > 0.6$, $R^2 > 0.4$), results were considered to be mathematical meaningful (Johnson and Wichern, 2007) and overfitting could be excluded.

Results

The viability of freshly harvested seed in 1983 was 80–100% for the 42 accessions (mean 97.4 ± 5.4%) and between 2 and 97% by 2014 after 31 years of storage (Fig. 1, Table S1). The probit

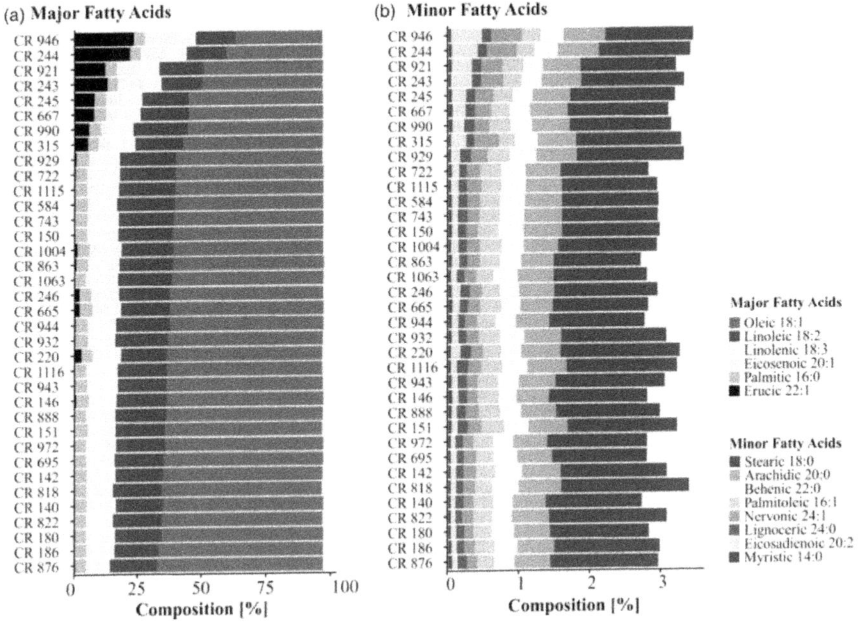

Figure 2. Fatty acid composition of 37 oilseed rape accessions measured after 31 years of storage. Fatty acid compositions are given as percentage (%) of the total fatty acid content. (a) Major fatty acids were declared when content was >4%. (b) Minor fatty acids.

analysis based on %TG of the five germination tests produced estimated half-viability periods between 15.2 and 50.7 years for an overall mean of 27.0 years. Highest correlation ($r_L = 0.71$, $P < 0.001$) was found between %TG performed in 2014 and 2009 and lowest correlation ($r_L = 0.30$, $P < 0.01$) between %TG performed in 2014 and initial germination tested in 1983 (Fig. S1). Although the accessions were submitted to the same maternal and storage environment, the low correlation coefficient between the initial and final germinability implied that factors other than initial germination were also involved.

Seed composition was significantly different ($P < 0.05$) between accessions (Table S1) and is visualised in Figs 2a,b and S2. A predominant group (29 of the 37 accessions analysed) contained high amounts of palmitic (4.7 ± 0.3%), oleic (60.4 ± 1.9%), linoleic (19.9 ± 1.4%) and α-linolenic (9.7 ± 0.9%) acids. The remaining eight accessions contained less oleic (47.1 ± 7.8%), linoleic (17.8 ± 1.8%) and α-linolenic (8.1 ± 0.7%) acids, along with more eicosenoic (7.6 ± 3.0%), eicosadienoic (0.3 ± 0.1%) and erucic (12.0 ± 6.9%) acids. The content of oil (44.6 ± 1.9%), protein (24.2 ± 1.9%) and moisture (5.0 ± 0.3%) hardly varied between accessions, while the glucosinolate content ranged from 9.1 to 85.1 µmol g^{-1} DW.

Linear regression revealed moderate associations between individual compounds and seed viability after storage (Fig. 3). Highest significant correlations were found between %NS and the contents of α-linolenic acid ($r_L = +0.52$, $P < 0.01$), oil content ($r_L = +0.59$, $P < 0.01$) and glucosinolate ($r_L = -0.61$, $P < 0.001$). Correlation coefficients were slightly lower when %TG and P50 were compared. In addition, there was an unexpected significant negative correlation between %NS and thousand seed weight ($r_L = -0.49$, $P < 0.05$).

In contrast to the linear regression, multivariate regression analysis revealed higher correlations between input vectors and seed viability. Standard approaches of machine learning were used including RBF, MLP and PLS to analyse underlying non-linear functions between the input vector or matrix and desired output vector. Networks were set up for MLP with 5, 10, or 3 and 5 neurons, PLS with 5 components, RBF with 5 centres. Due to incomplete datasets for %TG 2014, no correlation could be found for RBF networks for some compounds (Table S2). In addition, a tendency to lower R^2 for larger network set-ups could be seen for MLP. Therefore, five components were chosen for all networks and values were in the desired range of $r_M > 0.6$ and $R^2 > 0.4$ after cross-validations over all datasets (Table S3). In Tables 1, S4 and S5, superscripts indicate which network set-up produced the given results. Thereby, a high correlation coefficient r_M shows a strong relation between the input vectors and %TG 2014 or %NS 2014 whereas R^2 is a fitting parameter for the model and describes the correlation between the input vector and the predicted output vector. A negative R^2 indicates a bias in fitting of the model. The highest correlations with %TG 2014 were found with the following input vectors: lignoceric acid ($r_M = 0.97$, $R^2 = -0.03$, Fig. 4), nervonic acid ($r_M = 0.96$, $R^2 = 0.33$), glucosinolates ($r_M = 0.96$, $R^2 = 0.58$) and oil content ($r_M = 0.87$, $R^2 = 0.39$) (Table 1). Likewise, with %NS 2014, the highest correlations were found with oil content ($r_M = 0.99$, $R^2 = 0.90$), eicosenoic ($r_M = 0.99$, $R^2 = 0.62$) and stearic acid ($r_M = 0.97$, $R^2 = 0.42$) as single input vectors. Overall, correlation varied for %TG 2014 and %NS 2014 from $r_M = 0.24$ to 0.97 and from $r_M = 0.04$ to 0.99, respectively (Table 1). Correlation and R^2 increased strongly when %NS 2014 (Table S4) and %TG 2014 (Table S5) were combined with thousand seed weight (TSW) and historic viability data, respectively. In these cases, correlation ranged for %TG 2014 and %NS 2014 between $r_M = 0.74$ and 1.00 and between $r_M = 0.56$ and 1.00, respectively (Tables S4 and S5).

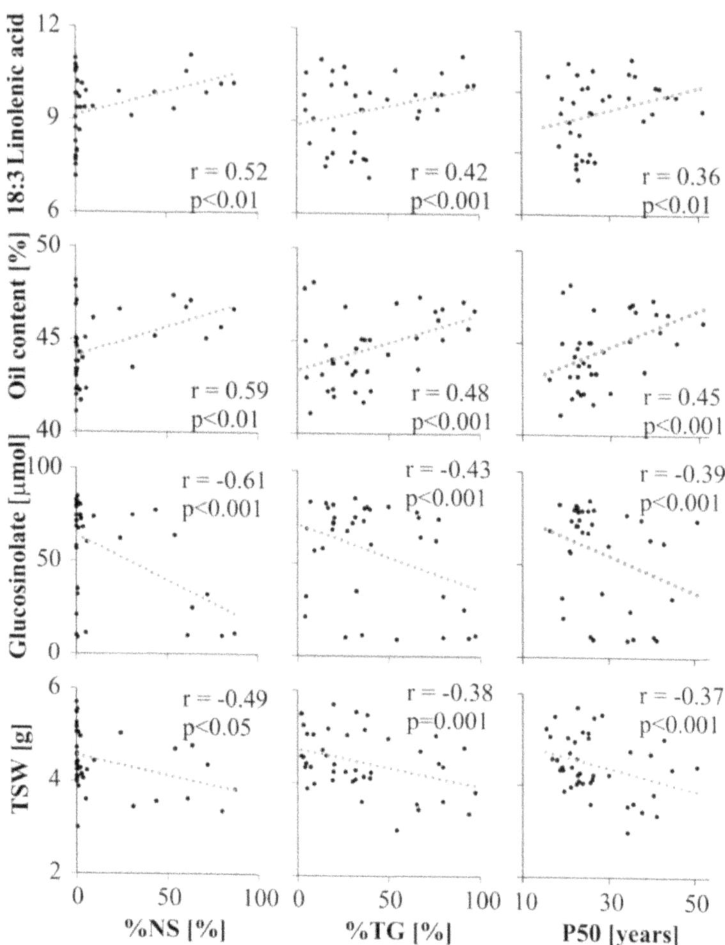

Figure 3. Significant correlations between seed viability and linolenic acid, seed components and thousand seed weight (TSW) among a set of oilseed rape accessions stored for 31 years. Seed viability is represented by normal seedlings, total germination and the half-viability period (P50).

A PCA on the complete dataset, i.e. the dataset without incomplete %TG 1990 data, provided a measure for the data structure (Fig. S3). In general, no cluster for certain accessions or the oil contents can be found. In addition to the composition performed by PCA, datasets are highlighted according to highest non-linear correlation coefficients r_M for %TG 2014 and %NS 2014. Individual P50 values, calculated for each access, were not used as an input variable because r_L between P50 and viability was not higher than the measured input parameters. Furthermore, the P50 is only an estimate of the half-viability period and standard deviation is high, and thus might lead to inaccuracies in the predictions.

Using a combination of parameters as an input matrix further improved results of the multivariate regression. The best combinations to predict seed viability after 31 years of storage (i.e % TG 2014) were: (1) myristic, stearic, oleic, α-linolenic, arachidic, eicosenoic, eicosadienoic, lignoceric and nervonic fatty acids, oil content and glucosinolates ($r_M = 0.90$, $R^2 = 0.76$); (2) α-linolenic acid, oil and glucosinolates content ($r_M = 0.84$, $R^2 = 0.56$); and (3) stearic, linoleic, arachidic, eicosadienoic, erucic fatty acids and glucosinolates ($r_M = 0.84$, $R^2 = 0.64$). The best predictive combinations for %NS 2014 were: (1) a combination of all fatty acids and all compounds ($r_M = 0.97$, $R^2 = 0.56$); (2) oleic acid, α-linolenic acid, arachidic acid, eicosenoic acid and glucosinolates ($r_M = 0.95$, $R^2 = 0.88$); and (3) myristic acid, stearic acid, oleic acid, α-linolenic acid, arachidic acid, eicosenoic acid, eicosadienoic acid, lignoceric acid and nervonic acid ($r_M = 0.96$, $R^2 = 0.85$) (Table 1). Also input matrix combinations including historic viability data (namely %TG for 1983, 1990, 1993 and %NS for 2009) and/or TSW showed a better predictability of longevity then only using the 2014 values (Fig. 4, Tables S2 and S3). By including historic viability data and TSW the fitting R^2 of the model for lignoceric acid content increased from initially −0.03 to 0.91 (Fig. 4a,b,d). In general, an input vector using only two values, e.g. TSW and myristic acid, showed high correlation values together with high R^2.

Discussion

The estimated average P50 of the current *B. napus accessions* was 27 years, which matches well with estimates obtained both from a small sample of *B. napus* (25 years, 12 accessions) and a large one of *B. oleracea* (23 years, 370 accessions) seed stored for about 40 years at −18°C (Walters *et al.*, 2005). When seeds were stored

Table 1. Multivariate regression reveals correlation coefficients for %NS 2014 and %TG 2014 for a combination of fatty acids and compounds

Method	NS% 2014		%TG 2014	
	r_M	R^2	r_M	R^2
Fatty acids	$0.71^{\times}$	0.46	0.63^{*}	0.37
Compounds	$0.91^{\times}$	0.40	$0.77^{\times}$	0.51
Fatty acids and compounds	$0.97^{\times}$	0.56	0.64^{+}	−0.14
14:0 Myristic acid in %	0.57^{*}	0.29	0.86^{+}	0.32
16:0 Palmitic acid in %	0.51^{+}	0.02	0.66^{+}	−0.04
16:1 Palmitoleic acid in %	0.04^{*}	−0.01	0.63^{+}	0.34
18:0 Stearic acid in %	0.99^{+}	0.42	0.73^{+}	0.29
18:1 Oleic acid in %	0.88^{+}	0.25	0.60^{*}	0.05
18:2 Linoleic acid in %	0.39^{+}	0.10	0.87^{+}	0.02
18:3 α-Linolenic acid in %	0.81^{*}	0.32	0.76^{+}	0.16
20:0 Arachidic acid in %	0.61^{+}	0.31	0.23^{+}	0.05
20:1 Eicosenoic acid in %	0.99^{+}	0.62	0.75^{+}	0.53
20:2 Eicosadienoic acid in %	0.89^{+}	0.41	0.75^{+}	0.43
22:0 Behenic acid in %	0.22^{*}	−0.12	0.39^{+}	0.03
22:1 Erucic acid in %	0.59^{+}	0.28	0.80^{+}	−0.02
24:0 Lignoceric acid in %	0.56^{+}	0.02	0.97^{*}	−0.03
24:1 Nervonic acid in %	0.60^{*}	0.02	0.91^{*}	0.33
Oil in %	0.99^{+}	0.90	0.87^{*}	0.39
Glucosinolates in μmol g^{-1} DW	0.73^{*}	0.31	0.96^{+}	0.58
Protein in %	0.84^{*}	0.24	0.80^{*}	0.42
H_2O in %	0.63^{*}	0.20	0.40^{+}	0.10
Stearic, linoleic, arachidic, eicosadienoic and erucic acid in %, glucosinolates in μmol	$0.92^{\times}$	0.52	0.84^{+}	0.64
Oleic, α-linolenic, arachidic, eicosenoic and eicosadienoic acid in %, glucosinolates in μmol	0.95^{+}	0.88	0.77^{+}	0.31
α-Linolenic acid in %, oil in % and glucosinolates in μmol	$0.75^{\times}$	0.49	0.84^{*}	0.56
Myristic, stearic, oleic, α-linolenic, arachidic, eicosenoic, eicosadienoic, erucic, lignoceric and nervonic acid in %	0.96^{+}	0.85	$0.67^{\times}$	0.32
Myristic, stearic, oleic, α-linolenic, arachidic, eicosenoic, eicosadienoic, erucic, lignoceric and nervonic acid in %, oil in %, glucosinolates in μmol	$0.83^{\times}$	0.67	$0.90^{\times}$	0.76
Myristic, stearic, oleic, α-linolenic, arachidic, eicosenoic, eicosadienoic, erucic, lignoceric and nervonic acid in %, oil, proteins and H_2O in %, glucosinolates in μmol	0.78^{+}	0.44	0.81^{*}	0.64

Values for r_M represent best fit for all networks at $P < 0.01$ using either $^{+}$MLP with five neurons in a single layer, $^{\times}$RBF with five centres or *PLS with five components. Further input matrix combinations, i.e. historic viability data and thousand seed weight (TSW), can be found in Tables S2 and S3. Compounds include oil, glucosinolates, protein and H_2O content. Fatty acids include all individual fatty acids, compare Table S1.

under non-controlled ambient conditions (~20°C and 50% relative humidity), P50 of *B. napus* was only 13.9 (Priestley *et al.*, 1985) and that of *B. oleracea* only 7.3 years (Nagel and Börner, 2010). Considering that seed maternal and storage environment were comparable, a range in P50 values between 15.2 and 50.7 years suggests that genetic factors may contribute to variation in seed longevity.

Lipid composition determines oil quality and membrane structure and has profound effects on seed viability in the dry state (Hoekstra, 2005). Here, seed components varied between accessions. Contents in α-linolenic fatty acid and total oil and glucosinolates in particular were shown to correlate with seed viability after 31 years of dry storage. However, as coefficients of correlation of the linear regression were only between ±0.49 and ±0.61, a multivariate regression using several input vectors was applied to convey complex relations between seed composition and viability. In doing so, saturated myristic, stearic, arachidic, lignoceric and unsaturated eicosenoic, eicosadienoic, oleic, α-linolenic and nervonic fatty acid were identified as predominant fatty acids influencing seed viability. Until now, very long fatty acids, e.g. lignoceric and nervonic acid, have never been discussed in relation to seed viability except in the context of human diseases and biomarkers (Lemaitre *et al.*, 2015; Clark *et al.*, 2016). Hence, multivariate regression detects relationships in complex molecular networks and enables the study of novel complex pathways.

Lipids are heterogeneously distributed across oilseed rape seed, which may facilitate different degradation processes during

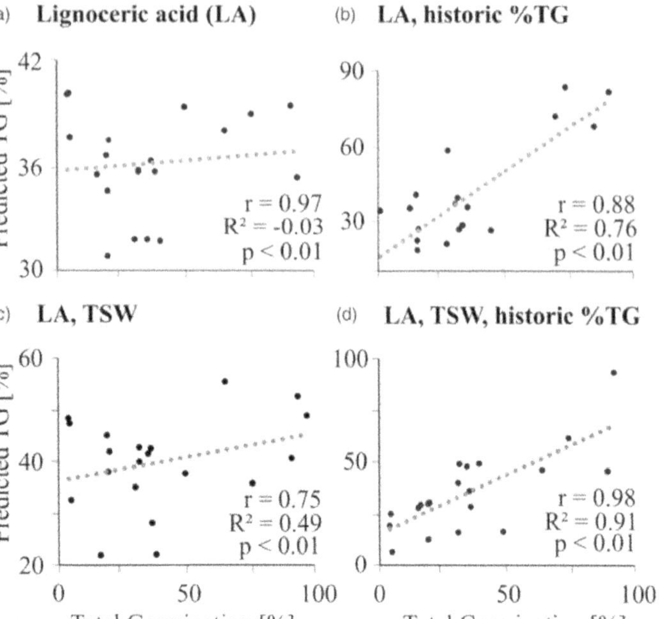

Figure 4. Prediction curve to estimate total germination (%TG) in 2014. Total germination and lignoceric acid were evaluated for the best fitted model based on multivariate regression using a partial least square (PLS) with five components. Non-linear regression was applied to (a) lignoceric acid (LA) content (%); (b) to the input matrix of lignoceric acid content (%) and historic viabilities (%TG); (c) of lignoceric acid content (%) and thousand seed weight (TSW, g); and (d) of lignoceric acid content (%), thousand seed weight (g) and historic viabilities (%TG). The non-linear regression based on artificial neural networks shows a high correlation between predicted total germination and measured total germination. Here, the importance of not only a high *r* but also R^2 can be seen. A higher fitting parameter ensures a better correlation between target values and actual total germination (compare improving correlation from a to d).

storage. Highest levels of palmitic acid (16:0) were found in the embryonic axis, while high levels of linoleic (18:2) and α-linolenic (18:3) acid were shown (Woodfield *et al.*, 2017) in the seed coat/aleurone layer as well as at the outer cotyledon. It may well be that the outer layers are first targets of non-enzymatic oxidation for α-linolenic acid. Indeed, there is a correlation between longevity and the average number of double bonds per polar lipid molecule (Hoekstra, 2005) which is supported by Ponquett *et al.* (1992). Although the conclusions are biased by the number of non-oily seed species, Ponquett *et al.* (1992) indicated that α-linolenic fatty acid per unit tocopherol determines the rate of oxidation. Tocopherol is a lipophilic antioxidant that is particularly abundant in oily seeds (Sattler *et al.*, 2004). Therefore, the localization as well as concentration of certain fatty acids in conjunction with antioxidants may be important in determining the rate of oxidative processes during storage.

In contrast, polyunsaturated fatty acids, such as linoleic and α-linolenic acid, are found extensively in all membranes (Harwood, 1997). Although observations were not possible here over time, Riewe *et al.* (2017) and Oenel *et al.* (2017) demonstrated in wheat and Arabidopsis that hydrolytic processes occurred during long-term storage of dry seeds. In rice, the down-regulation of different lipoxygenases reduced the production of malondialdehyde and lipid peroxides and contributed to increased tolerance against accelerated ageing (Ma *et al.*, 2015). In the present study, stearic, oleic, α-linolenic, eicosenoic, eicosadienoic, lignoceric and nervonic fatty acids, of which most are unsaturated, showed positive correlations with both %TG and %NS. It is assumed that all seed compartments are affected by oxidative processes and higher amounts of unsaturated fatty acids in particular may contribute to higher viability.

The glucosinolates found in many Brassicaceae are thought to function as a defence against herbivores. However, correlations between glucosinolate content and seed survival in the soil have never been reported (de Jong *et al.*, 2013). Linear and multivariate regressions showed that accessions accumulating a high level of glucosinolate tended to perform less well with respect to %NS. The speculation is that since glucosinolate hydrolysis products compromise bacterial membrane integrity (Borges *et al.*, 2015) it may equally damage plant cell membranes. However, antimicrobial activity facilitating glucosinolate hydrolysis is unlikely at water activity below 0.75 (Bewley *et al.*, 2013) and other mechanisms might be responsible.

Using multivariate regressions with a single input vector uncovered new significant correlations between seed composition and storability. In contrast to single input vectors, a small number of input vectors, e.g. combining one fatty acid with historic viability data or TSW (compare Fig. 4) improved predictability enormously. It seems logical to expect that historic viability data would improve r_M. However, linear regression between seed components and P50, based on historic viability data, did not increase correlation coefficients. In the seed industry, it is common to test the initial seed germination. These data may be used to estimate storability. In addition, whether TSW has an effect on seed vigour is not clear (Nagel *et al.*, 2013). In kale higher vigour was related to either bigger or smaller seeds depending on the seed lot properties (Komba *et al.*, 2007). However, oilseed rape seedlings from large seeds tend to be more vigorous and tolerant to insect damage due to a higher initial shoot biomass and higher growth rate (Bettey *et al.*, 2000; Elliott *et al.*, 2008). Therefore, there tends to be valuable information on historic viability and TSW results that supports final predictability.

High correlations for %TG 2014 and %NS 2014 can be established for quite a few network configurations. Parameters larger than five or various layers of neurons resulted in overfitting and forced the neural network to represent approximately three to four data points per centre (Johnson and Wichern, 2007). Therefore, a small network size seemed to be sufficient to convey correlations between input values and viability. However, using all available data did not always yield the best results (compare Tables S2 and S3). This would only be the case if all compounds and fatty acids were significant for the correlation. Thus, focus should be laid on set-ups where, next to high correlation r_M, high fitting parameters R^2 are also achieved. However, further studies would be interesting to follow compositional changes during storage and to apply multivariate regression for seed longevity predictions by including a higher number of accessions, thus providing a larger dataset after splitting and for cross-validation.

In conclusion, the complexity of seed longevity was shown with a unique viability dataset acquired during long-term storage for 31 years. In addition to environmental factors, seed composition was assumed to affect seed viability after long-term storage. Linear regression was shown to support this assumption but coefficients were too low to make confident interpretations. In contrast, the multivariate approaches based on machine learning were able to simultaneously analyse the impact of several parameters to reveal some important seed components, predominantly myristic, stearic, oleic, α-linolenic, arachidic, eicosenoic, eicosadienoic, lignoceric and nervonic fatty acids, oil content and glucosinolates that influence seed viability. These may be considered good candidates for developing viability prediction tools.

Supplementary material. To view supplementary material for this article, please visit https://doi.org/10.1017/S0960258518000259

Acknowledgements. The authors thank Veronika Miehe, Helga Schmalfeldt and Norddeutsche Pflanzenzucht Hans-Georg Lembke KG for conducting germination tests and seed composition analysis. Andrea Matros, Friedrich Melchert, Udo Seiffert and the anonymous reviewers are gratefully acknowledged for their support and helpful and critical comments on the manuscript.

Financial support. Financial support was provided by the EU [FP7 grant (311840) EcoSeed].

References

Andre M (2003) Multivariate analysis and classification of the chemical quality of 7-aminocephalosporanic acid using near-infrared reflectance spectroscopy *Analytical Chemistry* **75**, 3460–3467.

Bailly C (2004) Active oxygen species and antioxidants in seed biology. *Seed Science Research* **14**, 93–107.

Bettey M, Finch-Savage WE, King GJ and Lynn JR (2000) Quantitative genetic analysis of seed vigour and pre-emergence seedling growth traits in *Brassica oleracea*. *New Phytologist* **148**, 277–286.

Bewley JD, Bradford KJ, Hilhorst HWM and Nonogaki H (2013) *Seeds: Physiology of Development, Germination and Dormancy*, 3rd edition. New York: Springer.

Borges A, Abreu AC, Ferreira C, Saavedra MJ, Simoes LC and Simoes M (2015) Antibacterial activity and mode of action of selected glucosinolate hydrolysis products against bacterial pathogens. *Journal of Food Science and Technology-Mysore* **52**, 4737–4748.

Clark SR, Baune BT, Schubert KO, Lavoie S, Smesny S, Rice SM, Schäfer MR, Benninger F, Feucht M, Klier CM, McGorry PD and Amminger GP (2016) Prediction of transition from ultra-high risk to first-episode psychosis using a probabilistic model combining history, clinical assessment and fatty-acid biomarkers. *Translational Psychiatry* **6**, e897.

Crapiste GH, Brevedan MIV and Carelli AA (1999) Oxidation of sunflower oil during storage. *Journal of the American Oil Chemists Society* **76**, 1437–1443.

Cybenko G (1989) Approximation by superpositions of a sigmoidal function. *Mathematics of Control, Signals, and Systems (MCSS)* **2**, 303–314.

de Jong TJ, Isanta MT and Hesse E (2013) Comparison of the crop species *Brassica napus* and wild *B. rapa*: characteristics relevant for building up a persistent seed bank in the soil. *Seed Science Research* **23**, 169–179.

Elliott RH, Franke C and Rakow GFW (2008) Effects of seed size and seed weight on seedling establishment, vigour and tolerance of Argentine canola (*Brassica napus*) to flea beetles, *Phyllotreta* spp. *Canadian Journal of Plant Science* **88**, 207–217.

Ellis RH and Roberts EH (1980) Improved equations for the prediction of seed longevity. *Annals of Botany* **45**, 13–30.

Falk J and Munné-Bosch S (2010) Tocochromanol functions in plants: antioxidation and beyond. *Journal of Experimental Botany* **61**, 1549–1566.

Hall RD (2011) Biology of plant metabolomics. *Annual Plant Reviews* **43**, 407–420.

Harwood JL (1997) Plant lipid metabolism, in Dey PM and Harborne JB (eds), *Plant Biochemistry*. San Diego, CA: Academic Press.

Hoekstra FA (2005) Differential longevities in desiccated anhydrobiotic plant systems. *Integrative and Comparative Biology* **45**, 725–733.

Hwang JE, Ahn JW, Kwon SJ, Kim JB, Kim SH, Kang SY and Kim DS (2014) Selection and molecular characterization of a high tocopherol accumulation rice mutant line induced by gamma irradiation. *Molecular Biology Reports* **41**, 7671–7681.

ISTA (2014) International Rules for Seed Testing. Bassersdorf, Switzerland: International Seed Testing Association.

Johnson RA and Wichern DW (2007) *Applied Multivariate Statistical Analysis* (6th edition). New York, Pearson Book.

Komba CG, Brunton BJ and Hampton JG (2007) Effect of seed size within seed lots on seed quality in kale. *Seed Science and Technology* **35**, 244–248.

Kranner I, Minibayeva FV, Beckett RP and Seal CE (2010) What is stress? Concepts, definitions and applications in seed science. *New Phytologist* **188**, 655–673.

Krzanowski WJ (2000) *Principles of Multivariate Analysis: A User's Perspective*. New York, Oxford University Press.

Lemaitre RN, Fretts AM, Sitlani CM, Biggs ML, Mukamal K, King IB, Song X, Djoussé L, Siscovick DS, McKnight B, Sotoodehnia N, Kizer JR and Mozaffarian D (2015) Plasma phospholipid very-long-chain saturated fatty acids and incident diabetes in older adults: the Cardiovascular Health Study. *The American Journal of Clinical Nutrition* **101**, 1047–1054.

Ma L, Zhu FG, Li ZW, Zhang JF, Li X, Dong JL and Wang T (2015) TALEN-based mutagenesis of lipoxygenase LOX3 enhances the storage tolerance of rice (*Oryza sativa*) seeds. *PLoS One* **10**), e0143877. https://doi.org/10.1371/journal.pone.0143877

Moody J and Darken CJ (1989) Fast learning in networks of locally-tuned processing units. *Neural Computation* **1**, 281–294.

Nagel M, Behrens A-K and Börner A (2013) Effects of Rht dwarfing alleles on wheat seed vigour after controlled deterioration. *Crop and Pasture Science* **64**, 857–864.

Nagel M and Börner A (2010) The longevity of crop seeds stored under ambient conditions. *Seed Science Research* **20**, 1–12.

Oenel A, Fekete A, Krischke M, Faul SC, Gresser G, Havaux M, Mueller MJ and Berger S (2017) Enzymatic and non-enzymatic mechanisms contribute to lipid oxidation during seed aging. *Plant and Cell Physiology* **58**, 925–933.

Ponquett RT, Smith MT and Ross G (1992) Lipid autoxidation and seed ageing: putative relationships between seed longevity and lipid stability. *Seed Science Research* **2**, 51–54.

Priestley DA, Cullinan VI and Wolfe J (1985) Differences in seed longevity at the species level. *Plant Cell and Environment* **8**, 557–562.

Priestley DA and Leopold AC (1979) Absence of lipid oxidation during accelerated aging of soybean seeds. *Plant Physiology* **63**, 726–729.

Riewe D, Wiebach J and Altmann T (2017) Structure annotation and quantification of wheat seed oxidized lipids by high resolution LC-MS/MS. *Plant Physiology* **175**, 600–618.

Rivas-Ubach A, Sardans J, Pérez-Trujillo M, Estiarte M and Peñuelas J (2012) Strong relationship between elemental stoichiometry and metabolome in plants. *Proceedings of the National Academy of Sciences of the USA* **109**, 4181–4186.

Rojas R (1996) *Neural Networks: A Systematic Introduction*. Berlin: Springer-Verlag.

Rücker B and Röbbelen G (1996) Impact of low linolenic acid content on seed yield of winter oilseed rape (*Brassica napus* L.). *Plant Breeding* **115**, 226–230.

Sattler SE, Gilliland LU, Magallanes-Lundback M, Pollard M and DellaPenna D (2004) Vitamin E is essential for seed longevity, and for preventing lipid peroxidation during germination. *Plant Cell* **16**, 1419–1432.

Shalev-Shwartz S and Ben-David S (2014) *Understanding Machine Learning: From Theory to Algorithms.* New York: Cambridge University Press.

VSN International (2013) *GenStat for Windows* (17th edition). Hemel Hempstead, UK.

Walters C, Wheeler LM and Grotenhuis JM (2005) Longevity of seeds stored in a genebank: species characteristics. *Seed Science Research* **15**, 1–20.

Waterworth WM, Bray CM and West CE (2015) The importance of safe-guarding genome integrity in germination and seed longevity. *Journal of Experimental Botany* **66**, 3549–3558.

Wittkop B, Snowdon RJ and Friedt W (2009) Status and perspectives of breeding for enhanced yield and quality of oilseed crops for Europe. *Euphytica* **170**, 131–140.

Wold S, Sjöström M and Eriksson L (2001) PLS-regression: a basic tool of chemometrics. *Chemometrics and Intelligent Laboratory Systems* **58**, 109–130.

Woodfield HK, Sturtevant D, Borisjuk L, Munz E, Guschina IA, Chapman K and Harwood JL (2017) Spatial and temporal mapping of key lipid species in *Brassica napus* seeds. *Plant Physiology* **173**, 1998–2009.

Worley B and Powers R (2013) Multivariate analysis in metabolomics. *Current Metabolomics* **1**, 92–107.

3.3 PAPER 3:

Changes in tocochromanols and glutathione reveal differences in the mechanisms of seed ageing under seedbank conditions and controlled deterioration in barley

by

Thomas Roach[1], **Manuela Nagel[1]**, Andreas Börner, Caterina Eberle and Ilse Kranner

[1]These authors contributed equally to this work.

Published in

Environmental and Experimental Botany (2018), 156, 8-15

https://doi.org/10.1016/j.envexpbot.2018.08.027

To view supplementary material for this article, please visit

https://doi.org/10.1016/j.envexpbot.2018.08.027

Environmental and Experimental Botany 156 (2018) 8–15

Contents lists available at ScienceDirect

Environmental and Experimental Botany

journal homepage: www.elsevier.com/locate/envexpbot

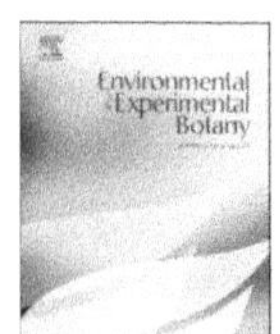

Changes in tocochromanols and glutathione reveal differences in the mechanisms of seed ageing under seedbank conditions and controlled deterioration in barley

Thomas Roach[a,1], Manuela Nagel[b,1,*], Andreas Börner[b], Caterina Eberle[a], Ilse Kranner[a]

[a] *Department of Botany and Centre for Molecular Biosciences Innsbruck (CMBI), University of Innsbruck, Sternwartestraße 15, A-6020, Innsbruck, Austria*
[b] *Genebank Department, Leibniz Institute of Plant Genetics and Crop Plant Research (IPK), Corrensstraße 3, D-06466, Seeland, Germany*

ARTICLE INFO

Keywords:
antioxidant
seedbank
lipid peroxidation
oxygen
ROS
vitamin E

ABSTRACT

"Orthodox" seeds are desiccation tolerant, enabling dry storage for agricultural and conservational purposes. Protocols of "controlled deterioration" (CD) using elevated temperatures and seed water contents (WCs) are frequently used to study mechanisms of seed ageing. However, the relevance of using CD to understand how seeds age at lower water contents and temperatures has been disputed. Here, comparisons in antioxidant changes are drawn between barley seed stored for up to 15 years in a seedbank with seeds aged using CD (45 °C, 75% RH; WC = 13% FW) under normoxia (21%) or elevated (75%) O_2. Regardless of ageing method, glutathione (GSH) levels decreased concomitantly with an oxidative shift in the glutathione redox state ($E_{GSSG/2GSH}$) as viability was lost. An earlier oxidative shift in $E_{GSSG/2GSH}$ and more rapid loss of seed vigour was found in seeds aged by CD under 75% O_2, compared to 21% O_2, but longevity was not shortened by elevated O_2. Seedbank-aged seeds contained more glutathione disulphide (GSSG) levels, leaked more electrolytes upon imbibition and contained less tocochromanols than non-stored seeds. In contrast, in response to CD, levels of GSSG fell, electrolyte leakage did not increase and tocochromanol levels increased (21% O_2), or remained stable (75% O_2). In a further set of 24 barley genotypes stored for up to 15 years at 20 °C (WC = 11% FW) and 0 °C (WC = 8% FW), tocochromanol levels positively (r = 0.42) correlated with total germination, which ranged between 1-78% (20 °C) and 65-97% (0 °C). Of all tocochromanol species, α-tocopherol correlated most with total germination (r = 0.49) and γ-tocotrienol least (r = 0.08). It is concluded that CD is not a perfect model for understanding mechanisms of ageing that occur under seedbank conditions.

1. Introduction

Seed ageing comprises various molecular processes that result in damage to seed macromolecules, affecting germinability and seed vigour. Seed deterioration is governed by storage temperature, seed water content (WC) and can also be influenced by O_2 contents in the storage environment (Groot et al., 2014; Harrington, 1963; Ibrahim et al., 1983; Roberts, 1973). At room temperature, the aqueous domain of the seed becomes 'glassy' when seed WC decreases below a certain threshold, typically 0.1 g H_2O g^{-1} dry mass, restricting metabolism (Buitink et al., 1998; Fernandez-Marin et al., 2013) and extending seed longevity (Vertucci, 1989; Walters et al., 2005). Glass formation and weakening are also temperature dependent. Protocols of "artificial ageing" are frequently applied by seed scientists and industry to assess seed longevity upon storage. Two commonly used methods of artificial ageing are "controlled deterioration" (CD) and "accelerated ageing". CD involves seed exposure to elevated temperatures (e.g. 30 to 45 °C) and relative humidities (RH), such as 60–75%, that achieve seed WCs that

Abbreviations: CD, controlled deterioration; DW, dry weight; EPPO, elevated partial pressure of oxygen; $E_{GSSG/2GSH}$, glutathione half-cell reduction potential; FW, fresh weight; GSH, glutathione; GSSG, glutathione disulphide; HGGT, homogentisate geranylgeranyl transferase; HPT1, homogentisate phytyltransferase; LMW, low-molecular-weight; mBBr, monobromobimane; PSSG, protein-bound glutathione; PUFA, polyunsaturated fatty acid; r, Pearson's correlation coefficient; R^2, coefficient of determination; RH, relative humidity; ROS, reactive oxygen species; TG, total germination; VTE1, tocopherol cyclase; VTE2, vitamin E2; VT3, vitamin E3 or methyl trasferase; WC, water content

* Corresponding author.
E-mail addresses: thomas.roach@uibk.ac.at (T. Roach), Nagel@ipk-gatersleben.de (M. Nagel), Boerner@ipk-gatersleben.de (A. Börner), caterina@edith-eberle.at (C. Eberle), Ilse.Kranner@uibk.ac.at (I. Kranner).

[1] These authors contributed equally to this work.

https://doi.org/10.1016/j.envexpbot.2018.08.027
Received 12 July 2018; Received in revised form 22 August 2018; Accepted 22 August 2018
Available online 23 August 2018
0098-8472/

Environmental and Experimental Botany 156 (2018) 8–15

do not change during CD, and is often used as a proxy for seedbank ageing. Accelerated ageing involves treatment of dry seeds with high RH (close to 100%) and temperatures (such as 45 °C), and seed WCs change dramatically during the treatment, which is often used to quickly (days to weeks) produce seed lots with low viability. Although it has been established that seed WC and temperature affect rates of seed ageing, the precise underlying biochemical processes of seed deterioration under various conditions are yet to be resolved. In addition, it is believed that reactive oxygen species (ROS) are involved in seed deterioration upon ageing (Bailly et al., 2008; Kranner et al., 2010), but how storage conditions influence the susceptibility of the hydrophilic and hydrophobic (i.e. membranes / storage lipids) domains of seeds to ROS is unclear.

In 15 year-stored wheat seeds, several hundred different oxidised lipid-associated species were identified using high resolution LC-MS/MS (Riewe et al., 2017). The accumulation of oxygenated triacylglycerols upon ageing in Arabidopsis seeds was shown to be mainly based on non-enzymatic oxidation of seed storage lipids (Oenel et al., 2017). Non-enzymatic oxidation can be initiated by free radical attack of a polyunsaturated fatty acid (PUFA), which readily reacts with O_2, forming a peroxyl radical that can attack another PUFA, leading to a self-propagating process known as lipid peroxidation (McDonald, 1999). Peroxyl radicals can be scavenged by tocochromanols, the collective name for lipid-soluble tocopherols, tocotrienols and tocomonoenols, which are structurally distinguished by the presence of three trans-double bonds in the hydrocarbon tail, forming four isomers, α, β, γ and δ (Munné-Bosch and Alegre, 2002). Tocochromanol synthesis is shared by plastoglobules and the chloroplast envelope, with leaf plastids predominantly storing α-tocopherol (DellaPenna and Pogson, 2006; Esteban et al., 2009; Vidi et al., 2006). γ-Tocopherol has been reported to be the dominant tocopherol in the seeds of many eudicot species (Falk and Munné-Bosch, 2010; Fernández-Marín et al., 2017) alongside a small fraction (typically < 2.5%) of tocomonoenols (Pellaud et al., 2018). Tocotrienols have been almost exclusively found in seeds, particularly in the endosperm of caryopses (Falk et al., 2004; Horvath et al., 2006). Tocochromanol synthesis involves homogentisate phytyltransferase (HPT1 or VTE2), tocopherol cyclase (VTE1), methyl transferase (VTE3) and homogentisate geranylgeranyl transferase (HGGT) (Dormann, 2007). In most eudicot seeds and monocot embryos α- or γ-tocopherol predominate (Falk and Munné-Bosch, 2010; Matthaus et al., 2016), whereas tocotrienols accumulate in the seed endosperm of monocots (Horvath et al., 2006; Moreau et al., 2007).

Seeds of *vte1* Arabidopsis and *hpt* or *hggt*-knock-out tomato mutants had reduced longevity and suffered more lipid peroxidation than wild-type seeds (Chen et al., 2016; Mène-Saffrané et al., 2010; Sattler et al., 2004). A rice mutant over-expressing the *OsVTE2* gene also had improved seed longevity (Hwang et al., 2014), supporting a role for tocochromanols in seed longevity. However, a decrease in tocochromanols does not always occur during seed ageing (Lee et al., 2017; Morscher et al., 2015; Nagel et al., 2015; Priestley et al., 1980; Seal et al., 2010b). Electrolyte leakage from imbibing seeds, and the consequential increase in leachate conductivity, is used as an indicator of seed quality (ISTA, 2010), whereby aged seeds leak more electrolytes due to damaged membranes. Since tocochromanols are involved in protecting membranes from lipid peroxidation, electrolyte leakage could be associated to ageing-induced losses of tocochromanols.

Tocochromanol levels can influence levels of glutathione (GSH), a water soluble low-molecular-weight (LMW) thiol-based antioxidant. The tocopherol-deficient *vte1* Arabidopsis mutant produces more GSH in its leaves, whereas GSH levels are decreased in *vte1*-overexpressing plants (Kanwischer et al., 2005). GSH is highly abundant in all seeds and can scavenge ROS directly or via electron donation to ROS-detoxifying enzymes, such as glutathione-*S*-transferases. Thiols are prone to oxidation, ultimately leading to post-translational modifications, such as disulphide bond formation (i.e. glutathione disulphide; GSSG) or glutathionylation of protein thiol groups (PSSG). Glutathionylated proteins are protected from irreversible oxidation, and both GSSG and PSSG contents increase during seed maturation drying, and may further accumulate during seed ageing (Kranner and Grill, 1996). Glutathione is a major cellular redox buffer, in plants and animal cells alike, and the GSH/GSSG redox state can be quantified via the glutathione half-cell reduction potential ($E_{GSSG/2GSH}$) (Kranner et al., 2006; Schafer and Buettner, 2001), which has been used as a marker of seed viability (Birtic et al., 2011; Kranner et al., 2006; Morscher et al., 2015; Nagel et al., 2015; Roach et al., 2010; Seal et al., 2010a).

Barley (*Hordeum vulgare* L.) is an important crop used for food, feed, beer and whisky production, and has a fully sequenced genome (Mascher et al., 2017). Much attention is given to the prevention of unwanted lipid oxidations in barley and other grain products during processing, which lead to rancidity and undesired flavours. Moreover, α-tocopherol, referred to as vitamin E in the diet, is a commonly added food preservative (E307) to prevent rancidity. To illuminate aspects of different seed deterioration mechanisms imposed by the storage environment, we investigated the relationship between individual tocochromanol species, GSH levels and $E_{GSSG/2GSH}$ with viability of barley seed aged at different rates and methods.

2. Material and Methods

2.1. Seed material, ageing and germination conditions, and conductivity measurements

For comparing ambient storage with CD, seeds of the six-row winter barley HOR 11311 were harvested in 1) 1999 and 2) 2010 and stored under ambient conditions (50.5 ± 6.3% RH, 20.3 ± 2.3 °C) until 2014 and 3) harvested in 2016 and stored for 2 months under ambient conditions. For CD, seeds harvested in 2010 were aged by CD in 2014, identically to Morscher et al. (2015) in a hermetically sealed box (28 × 28 x 12 cm) at 45 °C and at 75% RH, as checked with a RH meter (HC2-AW-USB, Rotronic, Switzerland), after equilibration to a WC of 13.1% ± 0.5% fresh weight (FW), using non-saturated LiCl salt solution. Seeds were aged for 0, 6, 9 and 14 d under ambient O_2 atmosphere (21%) or elevated O_2 (75%). At each ageing interval, all boxes (one for each replicate) were opened to allow seed removal and gas exchange. Before returning to CD conditions, the O_2 of half the boxes were adjusted with pure O_2 gas to 75%, as measured with a fibre-optic optode (Fibox 3, PreSens, Regensburg, Germany) using temperature compensation. For each ageing interval, each replicate (n = 3) consisted of 20 seeds for germination and 20 seeds for biochemical analyses. Total germination (TG) was carried out on Whatman® seed testing paper (Grade 3644) moistened with 40 mL distilled H_2O in a sealed box 14 × 20 x 3 cm, and kept at 20 °C. Electrolyte leakage was measured by imbibing surface-rinsed desiccated seeds in 2 mL of H_2O for 0.5 h (n = 5 replicates of 6 seeds) and measuring the conductivity of the leachate with a Cond 330i (WTW Xylem Analytics, Weilheim, Germany).

Furthermore, seeds from 24 genotypes were selected according to their variable longevity after ambient storage. To avoid effects of row-type, annuity, country of origin and growth environment, two-row [*H. vulgare* L. convar. *distichon* (L.) Alef. var. *nutans* (Rode) Alef., 17 genotypes] and six-row (*H. vulgare* L. convar. *vulgare* var. *hybernum* Viborg, seven genotypes) spring and winter barleys from 16 countries were multiplied at the IPK Gatersleben gene bank, Germany, between 1995 and 2003 (Supplemental Table S1). Freshly harvested seeds were desiccated to a WC of 7.7 ± 0.3% FW and were either stored in cold storage (0 ± 1 °C) or in paper bags at ambient storage (50.5 ± 6.3% RH, 20.3 ± 2.3 °C) and equilibrated to a WC of 10.9 ± 0.5% FW until 2009, when they were tested for TG on moistened filter paper at 25:20 °C (14:10 h, day:night), or transferred to -20 °C for use in biochemical analyses. Genotypes were regenerated again in 2008, stored under ambient conditions for 12 months (to break dormancy), before testing TG or storage at -20 °C. Seeds of each genotype were collected

Environmental and Experimental Botany 156 (2018) 8–15

from multiple plants and pooled for storage in the IPK gene bank. Individual replicates (n = 4-5), consisting of individual seeds (n = 50 for germination and 50 for biochemical analyses), were made from each seed lot.

2.2. Biochemical analyses

For biochemical analyses, seeds were freeze-dried for 5-7 d and ground in liquid nitrogen-cooled Teflon capsules, together with two 10 mm quartz balls, using a Mikro-Dismembrator laboratory ball mill (S B. Braun Biotech International) at 2,000 rpm for 7 min. Prior to analyses, the resulting seed powder was stored at -80 °C together with silica gel and storage boxes were not opened before they had reached room temperature. Seed WC was determined by the difference between seed FW and dry weight (DW), calculated before and after freeze drying, respectively, and calculated as WC = (FW-DW)/FW*100.

Tocochromanols were isolated and separated following a modified procedure of Bagci et al. (2004). Fifty mg of seed powder was extracted in 1 mL of ice-cold heptane by shaking at 30 Hz for 15 s and incubating for 5 min in an ice-cold ultrasonic bath. Initial trials showed that full extraction was achieved and no further tocochromanols could be extracted by a second extraction. Extracts were centrifuged (21,000 g, 45 min at 4 °C) immediately before HPLC analysis using an Agilent 1100 HPLC system (Agilent, Germany). α-, β- and γ- tocopherols and tocotrienols were separated by isocratic HPLC on a 250 x 4.6 mm LC-Diol column with 5 μm particle size (Supelco Analytical, Supelcosil™), using a mobile phase of heptane : tert-butyl-methyl-ether (Acros Organics), 97.5% : 2.5% (v/v) at a flow rate of 1 mL min^{-1} and were detected with a fluorescence detector (Ex: 295 nm, Em: 325 nm).

For seeds of HOR 11311, GSH and GSSG concentrations were determined by HPLC and $E_{GSSG/2GSH}$ was calculated as described earlier (Kranner et al., 2006; Schafer and Buettner, 2001). Briefly, 40 mg of seed powder was extracted in 1 mL of ice-cold 0.1 M HCl. After centrifugation (21,000 g, 45 min at 4 °C) the supernatant was divided for the separate determination of 1) total glutathione (GSH + GSSG) and 2) GSSG only, using dithiothreitol to reduce disulphides, and blockage of thiol groups with *N*-ethylmaleimide. Thiols were labelled with monobromobimane (mBBr) and GSH was separated from other mBBr-labelled thiols by reversed-phase HPLC with a ChromBudget 120-5-C18 column (5.0 μm, BISCHOFF GmbH, Leonberg, Germany), and detected by a fluorescence detector (ex: 380 nm, em: 480 nm). The concentration of GSH was calculated by subtracting the amount of GSSG (in GSH equivalents) from the amount of total glutathione. For seeds of the other 24 genotypes stored in the IPK Gatersleben gene bank, data was recalculated from Nagel et al. (2015), shown in Fig. 5, for ease of comparison with the tocochromanol data set.

2.3. Statistics

Significant differences between ageing intervals were calculated in SPSS Statistics 23 using ANOVA and a Tukey post-hoc test with 95% confidence limit or a Students T-test for calculating *p*-values. Correlations between TG and tocochromanols were calculated in OriginPro 2015 using Pearson's correlation coefficient (*r*) and coefficient of determination (R^2). *P*-values for the regression coefficients between TG and tocochromanol amounts were calculated using a linear regression model.

3. Results

3.1. Tocochromanol contents increased while GSH and GSSG contents decreased in seeds aged by CD

In seeds aged by CD (45 °C and 75% RH), either under 21% O_2 ("normoxia") or 75% O_2 (termed "elevated oxygen"), TG declined to below 10% after 14 d (Fig. 1A). No significant differences were observed in TG between the two O_2 treatments, but elevated O_2 decreased seed vigour, i.e. germination was delayed (Fig. 1A): either the time to reach 50% germination was significantly longer or radicle lengths were significantly shorter 6 d after imbibition was initiated (Fig. 2). Three tocopherols, α, β and γ and their corresponding tocotrienols were detected in barley seeds, with α-tocotrienol being the dominant form, which is characteristic of barley (Do et al., 2015; Lachman et al., 2018). As viability was lost within 14 d in response to CD under normoxia, total tocochromanol contents increased (p = 0.020) and particularly tocotrienols (p = 0.006), neither of which was apparent during CD under elevated oxygen (Fig. 1B). However, regardless of O_2 concentration in the storage environment, CD reduced both GSH and GSSG contents (Fig. 1C). An oxidative shift in the $E_{GSSG/2GSH}$ occurred after 14 d of CD in seeds aged under normoxic conditions. In comparison, a smaller decline in GSSG, a larger loss of GSH and an earlier oxidative shift in the $E_{GSSG/2GSH}$ occurred after 9 d CD under elevated oxygen (Fig. 1A and C).

3.2. Tocochromanol and GSH contents decreased, while GSSG levels increased in seeds aged during ambient storage

Seeds of genotype HOR 11311 stored for 15 years under ambient conditions (20.3 ± 2.3 °C and 50 ± 6.3% RH) had reduced vigour and viability, requiring 11 days to reach a final TG of 57%, whereas after storage of 2 months or 4 years TG was above 95% within 2 days (Fig. 3A). In contrast to CD (Fig. 1B), total tocochromanol amounts decreased after 4 and 15 years of ambient storage, with significant decreases in all isoforms, except β-tocopherol, the least abundant tocochromanol (Fig. 3B). However, both ambient storage and CD under either oxygen tension led to a decline in GSH, but only ambient storage led to an increase in GSSG (Fig. 3C), in contrast to the significant GSSG decline during CD (Fig. 1C).

3.3. Loss of membrane integrity in ambient-stored seed, but not after CD

No significant increase in the conductivity of leachates during seed imbibition was found in seeds aged by CD to below 10% TG (Fig. 4A), whereas a three-fold increase in electrolyte leakage occurred in seeds that had been stored for 11 years under ambient conditions (Fig. 4B), but which had retained a TG of $>$ 50% (Fig. 4A).

3.4. Variability in seed longevity of 24 barley genetypes under seedbank conditions was reflected in losses of tocochromanols and GSH

Seeds of 24 barley genotypes were stored under "ambient conditions" (20 °C) or under "cold conditions" (0 °C) for 5-13 years (Supplemental Table S1). TG of these seeds ranged between 1-76% (20 °C) and 79-95% (0 °C) (Fig. 5A). Plants were regenerated from seeds stored under cold conditions, and freshly harvested seeds were used as non-stored controls. Comparable to HOR 11311, the levels of tocochromanols decreased with storage time, and for all seedbank-stored and non-stored seeds a positive correlation ($r = 0.46$; $p < 0.001$) was observed between TG and total tocochromanol contents (Fig. 5B). β-tocopherol was at the detection limit (Supplemental Fig. S1) and could only be quantified in 11% of samples. Linear correlation coefficients of the other tocochromanols with TG were $r = 0.49$ for α-tocopherol, $r = 0.40$ for β-tocotrienol, $r = 0.37$ for γ-tocopherol, $r = 0.36$ for α-tocotrienol and $r = 0.08$ for γ-tocotrienol, with *p*-values for all linear models < 0.01 (Supplemental Fig. S2). Glutathione contents were also lower in seedbank-stored seed than in non-stored controls (Fig. 5C), particularly after storage at 20 °C, increasing the $E_{GSSG/2GSH}$ to less negative values, with a sigmoidal relationship between $E_{GSSG/2GSH}$ and TG (Fig. 5D).

Environmental and Experimental Botany 156 (2018) 8–15

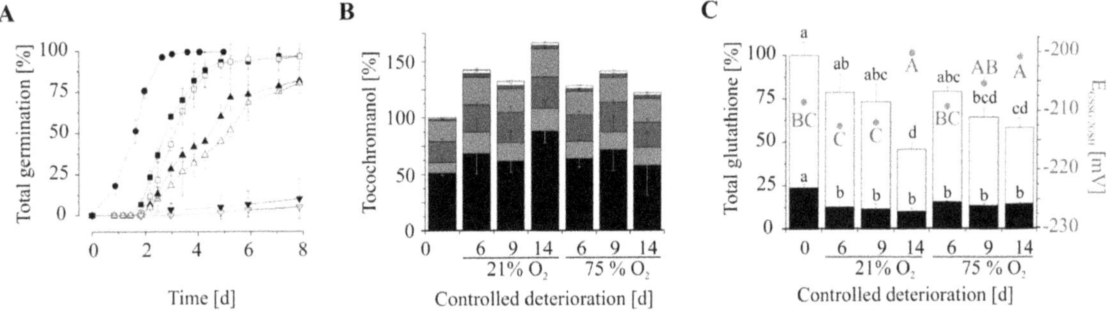

Fig. 1. Effects of CD on germination, tocochromanols and glutathione contents in barley genotype HOR 11311. (A) Germination after CD for 0 d (circles), 6 d (squares), 9 d (upward triangles) or 14 d (downward triangles) under normoxia (21% O_2, filled symbols) or elevated oxygen (75% O_2, open symbols). Amounts of (B) individual tocochromanols, i.e. α-tocotrienol (black), α-tocopherol (red), γ- (blue) and β- (green) tocotrienol, γ- (purple) and β- (white) tocopherol and (C) GSH (white bars), GSSG (black bars), normalized to 0 d CD, and $E_{GSSG/2GSH}$ (grey circles; right Y axis) are shown, with significant differences ($p < 0.05$) indicated by different letters, n = 3 replicates of 50 seeds ± SD.

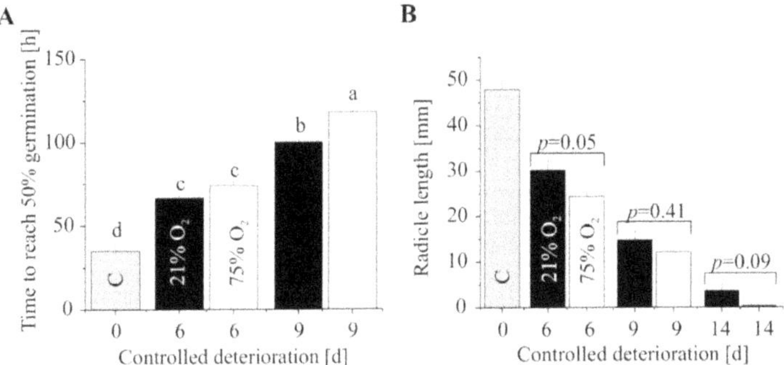

Fig. 2. Effect of elevated O_2 during CD on seedling vigour in barley genotype HOR 11311. (A) Time to reach 50 % germination in seeds aged at CD for 6 d or 9 d under normoxia (21% O_2, black bars) or elevated oxygen (75% O_2, white bars). Different letters indicate significant differences ($p < 0.05$), n = 3 ± SE. (B) Radicle length after 6 d since the onset of germination. The *p* values are given for differences in O_2 concentrations at the same ageing interval, as calculated with the T-test, n = 60 replicates ± SE.

4. Discussion

4.1. Relevance of tocochromanols to seed ageing in barley

In the seeds of the 25 barley genotypes studied here, α-, β- and γ-tocopherols and their corresponding tocotrienols were detected, in agreement with previous reports (Do et al., 2015; Ehrenbergerová et al., 2006; Falk et al., 2004), and there was considerable genotypic influence on total tocochromanol contents, which ranged from 30 to 60 nmol g^{-1} DW in non-aged seeds (Figs. 3B and 5). The most abundant forms were α-tocotrienol and γ-tocotrienol, and 75% of total seed tocochromanols were tocotrienols (Supplemental Fig. S1). Tococtrienols are produced by HGGT activity, which desaturates the isoprenoid side chain in three locations. In barley seeds, HGGT is expressed in the endosperm, leading to tocotrienol formation (Cahoon et al., 2003; Falk et al., 2004). In contrast, the barley germ (embryo and scutellum) contains mainly tocopherols (Falk et al., 2004). Considering that 15% of total tocochromanols in whole seeds were tocopherols and that the germ only comprises 6.7 ± 0.1% of the barley seed DW, the germ is considerably more enriched in tocochromanols, on a DW basis, than the endosperm, and particularly in α-tocopherol (Falk et al., 2004). This finding agrees with many other studies in diverse plant taxa, showing that the embryonic axis or germ are enriched in tocochromanols relative to seed energy reserves (Bellani et al., 2012; Bruni et al., 2002; Gerna et al.,

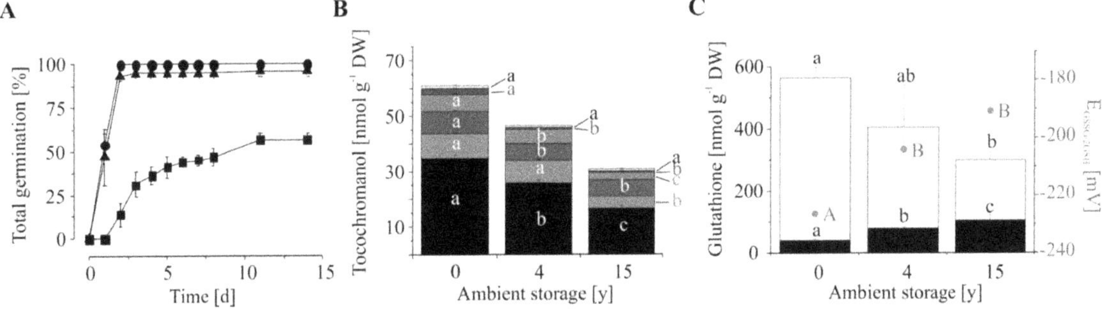

Fig. 3. Effects of long-term ambient seed storage on TG, tocochromanol and glutathione levels in barley genotype HOR 11311. (A) Germination after ambient storage at 20 °C and 50% relative humidity (8.7% WC) for 0 (triangles), 4 (circles) and 15 (squares) years. Amounts of (B) individual tocochromanols, i.e. α-tocotrienol (black), α-tocopherol (red), γ- (blue) and β- (green) tocotrienols, γ- (purple) and β- (white) tocopherols and (C) GSH (white bars), GSSG (black bars) and $E_{GSSG/2GSH}$ (grey circles; right Y axis), with significant differences ($p < 0.05$) shown by different letters, n = 4 replicates of 50 seeds ± SD.

Environmental and Experimental Botany 156 (2018) 8–15

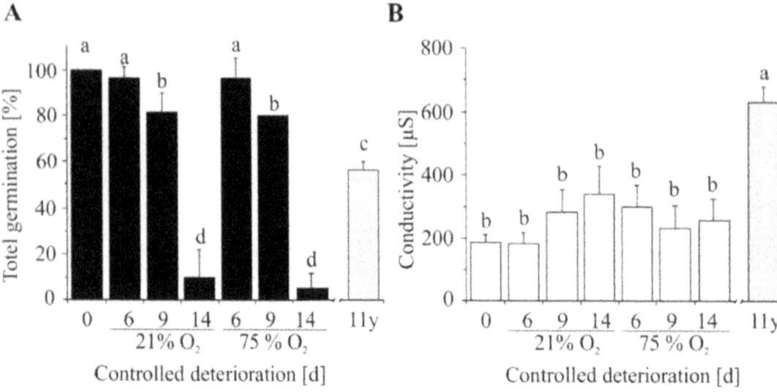

Fig. 4. Changes in TG and membrane permeability, as measured by electrolyte leakage, in genotype HOR 11311. (A) TG and (B) leachate conductivity of seeds aged for 0, 6, 9, and 14 d in CD under 21% or 75% O_2 (black/white bars) or after 11 years ambient storage without CD (grey bars). Significant differences are indicated by different letters ($p < 0.05$).

2017; Pukacka and Ratajczak, 2005; Roach et al., 2010; Torres et al., 1997; Wang et al., 2013). The barley germ contains five times more lipids than the endosperm, which also influences the contents of the lipid-soluble tocochromanols (Anness, 1984; Barnes, 1982). Therefore, the germ - which is essential for germination and seedling development - seems to be better protected against oxidative stress than the endosperm.

4.2. Tocochromanol levels decrease during seedbank storage

Under ambient conditions (20 °C, 50% RH), the time for barley seed viability to fall to 50% is estimated to be 9.2 years (Nagel and Börner, 2010), although there is a strong underlining genotypic influence on seed longevity (Nagel et al., 2015). Interestingly, HOR 11311 seeds stored under ambient conditions for 4 years showed significant losses of tocochromanols and GSH without germination being affected (Fig. 1). Reduction in tocochromanols has also been shown to have occurred in buried seeds of *Vellozia alata* and *Xyris bialata* without viability loss (Garcia et al., 2012; Munne-Bosch et al., 2011). Nonetheless, correlating the contents of individual tocochromanols of the 24 genotypes with TG of either non-aged seeds or seeds that had been aged under cold or ambient storage revealed that α-tocopherol correlated the most with seed viability, and γ-tocotrienol the least (Supplemental Fig. S2).

Of all tocochromanols, α-tocopherol and α-tocotrienol have been suggested to have more antioxidant power than γ-tocotrienol (Munné-Bosch and Alegre, 2002). Regardless, other endosperm-bound (Falk et al., 2004) tocotrienols (α-tocotrienol and β-tocotrienol) positively correlated with TG, indicating that tocochromanols in seed parts other than the embryo may also be associated with viability loss of seedbank-stored seeds. The starchy endosperm of Poaceae seeds consists of mainly dead tissue, and therefore, tocochromanols may have less of a role in maintaining seed viability. However, in the living structures, the protein-rich aleurone and lipid-rich scutellum, which are required for breaking down reserves in the endosperm and transferring energy to the embryo, would require antioxidant protection to support membrane integrity. The increase in electrolyte leakage of seedbank-aged seeds, not found in CD-aged seed (Fig. 4), indicates that more membrane damage occurred during seedbank ageing and that membrane damage in the aleurone layer could have contributed to increased electrolyte leakage in seedbank-aged seeds.

4.3. Molecular mobility and antioxidant metabolism

In the glassy state, metabolic processes cease and only non-enzymatic oxidations are assumed to occur (Fernandez-Marin et al., 2013; Vertucci, 1989), and LMW antioxidants are required to buffer against

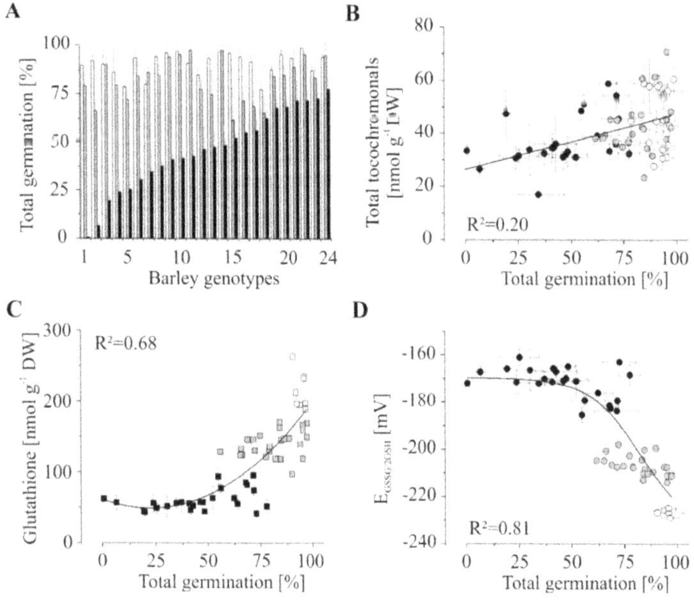

Fig. 5. Relationships between TG and total tocochromanols, GSH levels or $E_{GSSG/2GSH}$ in 24 barley genotypes with varying longevity. (A) Total germination (TG) of seeds after ambient storage (20 °C; black bars), cold storage (0 °C; grey bars) or without storage (white bars). Correlations of TG with (B) total tocochromanol contents, (C) GSH contents, or (D) $E_{GSSG/2GSH}$ in seeds from ambient storage (black symbols), cold storage (grey symbols) or without storage (white symbols), n = 5 reps of 50 seeds ± SD.

non-enzymatic oxidations (Smirnoff, 2010). In comparison to declining tocochromanol levels found in seeds stored for up to 15 years at cold (0 °C; WC = 7.7%) or ambient storage (20 °C; WC = 10.9%), tocochromanol contents increased in seeds rapidly aged with CD (45 °C; WC = 13.1%). Similar increases during CD have been observed in another study on barley (Do et al., 2015), in radish (Bellani et al., 2012), soybean (Giurizatto et al., 2012; Priestley et al., 1980), lettuce and cabbage (Groot et al., 2012), ironweed (Seal et al., 2010b) and maize (Doria et al., 2009). At the temperatures and RH used for CD, the glassy state already weakens towards a "rubbery state", and molecular mobility increases (Ballesteros and Walters, 2011), enabling some enzymatic processes (Fernandez-Marin et al., 2013), including repair mechanisms (Bray and Dasgupta, 1976; Sen and Osborne, 1974, 1977) and transcription (Leubner-Metzger, 2005), as well as activation of programmed cell death (Kranner et al., 2006). According to the increase in tocochromanols during CD of barley, enzymes involved in tocopherol synthesis may also be active in the "rubbery state". Potentially relevant is that leaves of heat-stressed plants tend to increase tocopherol contents (Buchner et al., 2017; Dong et al., 2007), which could also explain why tocochromanols increased during CD (i.e. due to heat-stress). There was higher synthesis of tocotrienols (endosperm / aleurone-bound) and no increases in electrolyte leakage occurred in CD-aged seed, indicating that membranes of the aleurone were protected. At lower WCs (i.e. during ambient storage), levels of tocochromanols decreased (Figs. 3B and 5B), similar to other studies of seepweed (*Suaeda maritima*) and Scots pine (*Pinus sylvestris*) (Seal et al., 2010a; Tammela et al., 2005), where seed viability loss over years was accompanied by loss of α-tocopherol. In summary, artificial ageing of seeds with elevated temperatures and RH (with high seed WCs; longevity of days – weeks) can lead to an increase in tocochromanol levels as seeds lose viability, whereas slower ageing (years – decades under lower seed WCs) appears to be associated with a loss of tocochromanols. Therefore, it appears that at higher WCs seeds may actively respond to the conditions used for CD and accumulate tocochromanols, whereas under the dry seedbank conditions enzymatic activity for this process is excluded.

Regardless of whether barley seeds were aged by CD or in the seedbank, they suffered losses of GSH before TG was affected (Figs. 1, 3 and 5). Seed lots stored at 0 °C all retained more than 50% TG and an $E_{GSSG/2GSH}$ more negative than − 200 mV, whereas half of all genotypes stored at 20 °C had less than 50% TG and $E_{GSSG/2GSH}$ values less negative than − 180 mV, a critical oxidative threshold above which seed and cell viability typically decline (Kranner et al., 2006; Schafer and Buettner, 2001). However, there were differences in glutathione concentrations and redox state between seeds aged under seedbank conditions and CD. GSSG accumulated in seedbank-aged seeds, whereas it was lost in CD-aged seeds (although less so when CD included elevated O_2), likely governed by the physical properties of the seeds. Under high RHs and temperatures, low levels of GSSG could be a consequence of glutathione reductase activity, which recycles GSSG back to GSH. This would further support our hypothesis that in the "rubbery state" induced by CD some enzyme-catalysed defence reactions are active. In the glassy state (i.e. seedbank-stored seeds), tocochromanol levels tended to be depleted in seeds that possessed an oxidative shift in $E_{GSSG/2GSH}$ (Supplemental Fig. S3), which may indicate that the level of damage to lipid domains is connected to the redox state of the aqueous domain in seeds with restricted molecular mobility.

4.4. The influence of oxygen on seed biochemistry during ageing

The relationship between O_2 concentration, seed moisture and temperature is also not well understood. Despite that barley seed longevity was not shortened by additional O_2 during CD, seed vigour was affected, similar as found in sunflower seeds (Morscher et al., 2015). Moreover, similar to results in Fig. 1B, elevated O_2 in the storage environment also prevented a CD-induced increase in tocochromanol contents in sunflower seeds (Morscher et al., 2015). It has been shown that O_2 is more deleterious to seeds at lower WCs (Carta et al., 2018; Ibrahim et al., 1983). In agreement with these findings, seeds desiccated to 40% RH and exposed to elevated partial pressures of O_2 (EPPO) at 18 MPa lost more viability than seeds stored at 20 °C, 40% RH and an atmospheric pressure of about 0.1 MPa (Nagel et al., 2016). Interestingly, α-tocopherol contents of EPPO-aged seeds decreased, indicating that processes comparable to long-term storage in seedbanks took place (Groot et al., 2012). Moreover, GSSG levels accumulated during seedbank storage, whereas during CD they declined, but less so when aged by CD under elevated O_2. Furthermore, since tocochromanol levels increased during CD of barley seeds under normoxia, the lack of increase in tocochromanols under elevated O_2 could be due to greater O_2-dependent tocochromanol breakdown.

5. Summary and Conclusions

Seed ageing under seedbank conditions and CD both led to significant reductions in seed viability and losses of GSH. Lower WCs and temperatures reduced the speed of seed viability loss and prevented a depletion of the tocochromanol pool. A lower level of tocochromanols in the genotypes that lost most TG would agree with a role for tocochromanols in preventing seed deterioration in barley seeds held in seedbanks long-term. However, during CD, an increased tocochromanol pool and lack of increase in GSSG, which occurred in seedbank-stored seeds, could indicate that certain enzymatic processes were operative during CD. Moreover, the discrepancy of tocochromanol contents with viability in response to CD indicates that in barley seeds, damage to the lipid domain unlikely leads to viability loss under these conditions. Therefore, we conclude that CD does not fully reflect the biochemical changes that occur during storage of seeds in a seedbank.

Funding sources

This work was supported by the EU [FP7 grant (311840) EcoSeed].

Declarations of interest

None.

CRediT author statement

Thomas Roach: Conceptualization, Methodology, Formal Analysis, Validation, Writing-Original Draft preparation, Writing-Review & Editing, Visualization, Supervision. **Manuela Nagel**: Conceptualization, Methodology, Investigation, Resources, Validation, Writing-Original Draft preparation, Writing-Review & Editing, Visualization. **Andreas Börner**: Resources, Supervision, Funding Acquisition. **Caterina Eberle**: Investigation. **Ilse Kranner**: Writing-Review & Editing, Supervision, Project Administration, Funding Acquisition.

Acknowledgements

Stefanie Thumm (IPK), Ines Bauer, Bettina Lehr and Birgit Stenzel (University of Innsbruck) are gratefully acknowledged for their excellent technical assistance.

Appendix A. Supplementary data

Supplementary material related to this article can be found, in the online version, at doi:https://doi.org/10.1016/j.envexpbot.2018.08.027.

References

Anness, B.J., 1984. Lipids of barley, malt and adjuncts. J. Inst. Brew. 90, 315–318.

https://doi.org/10.1002/j.2050-0416.1984.tb04282.x.
Bagci, E., Vural, M., Dirmenci, T., Bruehl, L., Aitzetmuller, K., 2004. Fatty acid and tocochromanol patterns of some *Salvia* L. species. Z. Naturforsch. C 59, 305–309.
Bailly, C., El-Maarouf-Bouteau, H., Corbineau, F., 2008. From intracellular signaling networks to cell death: The dual role of reactive oxygen species in seed physiology. C. R. Biol. 331, 806–814. https://doi.org/10.1016/j.crvi.2008.07.022.
Ballesteros, D., Walters, C., 2011. Detailed characterization of mechanical properties and molecular mobility within dry seed glasses: relevance to the physiology of dry biological systems. Plant J. 68, 607–619. https://doi.org/10.1111/j.1365-313X.2011.04711.x.
Barnes, P.J., 1982. Composition of cereal germ preparations. Z Lebensm. Unters. Forsch. A-Food Res. Technol. 174, 467–471.
Bellani, L.M., Salvini, L., Dell'Aquila, A., Scialabba, A., 2012. Reactive oxygen species release, vitamin E, fatty acid and phytosterol contents of artificially aged radish (*Raphanus sativus* L.) seeds during germination. Acta. Physiol. Plant. 34, 1789–1799. https://doi.org/10.1007/s11738-012-0976-0.
Birtic, S., Colville, L., Pritchard, H.W., Pearce, S.R., Kranner, I., 2011. Mathematically combined half-cell reduction potentials of low-molecular-weight thiols as markers of seed ageing. Free Radic. Res. 45, 1093–1102. https://doi.org/10.3109/10715762.2011.595409.
Bray, C.M., Dasgupta, J., 1976. Ribonucleic acid synthesis and loss of viability in pea seed. Planta 132, 103–108. https://doi.org/10.1007/BF00388890.
Bruni, R., Medici, A., Guerrini, A., Scalia, S., Poli, F., Romagnoli, C., Muzzoli, M., Sacchetti, G., 2002. Tocopherol, fatty acids and sterol distributions in wild Ecuadorian *Theobroma subincanum* (Sterculiaceae) seeds. Food Chem. 77, 337–341. https://doi.org/10.1016/s0308-8146(01)00357-0.
Buchner, O., Roach, T., Gertzen, J., Schenk, S., Karadar, M., Stoggl, W., Miller, R., Bertel, C., Neuner, G., Kranner, I., 2017. Drought affects the heat-hardening capacity of alpine plants as indicated by changes in xanthophyll cycle pigments, singlet oxygen scavenging, alpha-tocopherol and plant hormones. Environ. Exp. Bot. 133, 159–175. https://doi.org/10.1016/j.envexpbot.2016.10.010.
Buitink, J., Claessens, M.M.A.E., Hemminga, M.A., Hoekstra, F.A., 1998. Influence of water content and temperature on molecular mobility and intracellular glasses in seeds and pollen. Plant Physiol. 118, 531–541. https://doi.org/10.1104/pp.118.2.531.
Cahoon, E.B., Hall, S.E., Ripp, K.G., Ganzke, T.S., Hitz, W.D., Coughlan, S.J., 2003. Metabolic redesign of vitamin E biosynthesis in plants for tocotrienol production and increased antioxidant content. Nat. Biotechnol. 21, 1082. https://doi.org/10.1038/nbt853.
Carta, A., Bottega, S., Spanò, C., 2018. Aerobic environment ensures viability and antioxidant capacity when seeds are wet with negative effect when moist: implications for persistence in the soil. Seed Sci. Res. 28, 16–23. https://doi.org/10.1017/S0960258517000307.
Chen, D.F., Li, Y.L., Fang, T., Shi, X.L., Chen, X.W., 2016. Specific roles of tocopherols and tocotrienols in seed longevity and germination tolerance to abiotic stress in transgenic rice. Plant Sci. 244, 31–39. https://doi.org/10.1016/j.plantsci.2015.12.005.
DellaPenna, D., Pogson, B.J., 2006. Vitamin synthesis in plants: Tocopherols and carotenoids. Annu. Rev. Plant Biol. 57, 711–738. https://doi.org/10.1146/annurev.arplant.56.032604.144301.
Do, T.D., Cozzolino, D., Muhlhausler, B., Box, A., Able, A.J., 2015. Antioxidant capacity and vitamin E in barley: Effect of genotype and storage. Food Chem. 187, 65–74. https://doi.org/10.1016/j.foodchem.2015.04.028.
Dong, G.J., Liu, X.L., Chen, Z.Y., Pan, W.D., Li, H.J., Liu, G.S., 2007. The dynamics of tocopherol and the effect of high temperature in developing sunflower (*Helianthus annuus* L.) embryo. Food Chem. 102, 138–145. https://doi.org/10.1016/j.foodchem.2006.05.013.
Doria, E., Galleschi, L., Calucci, L., Pinzino, C., Pilu, R., Cassani, E., Nielsen, E., 2009. Phytic acid prevents oxidative stress in seeds: evidence from a maize (*Zea mays* L.) low phytic acid mutant. J. Exp. Bot. 60, 967–978. https://doi.org/10.1093/Jxb/Ern345.
Dormann, P., 2007. Functional diversity of tocochromanols in plants. Planta 225, 269–276. https://doi.org/10.1007/s00425-006-0438-2.
Ehrenbergerova, J., Belcrediova, N., Pryma, J., Vaculova, K., Newman, C.W., 2006. Effect of cultivar, year grown, and cropping system on the content of tocopherols and tocotrienols in grains of hulled and hulless barley. Plant Food. Hum. Nutr. 61, 145–150. https://doi.org/10.1007/s11130-006-0024-6.
Esteban, R., Olano, J.M., Castresana, J., Fernandez-Marin, B., Hernandez, A., Becerril, J.M., Garcia-Plazaola, J.I., 2009. Distribution and evolutionary trends of photoprotective isoprenoids (xanthophylls and tocopherols) within the plant kingdom. Physiol. Plant. 135, 379–389. https://doi.org/10.1111/j.1399-3054.2008.01196.x.
Falk, J., Krahnstover, A., van der Kooij, T.A.W., Schlensog, M., Krupinska, K., 2004. Tocopherol and tocotrienol accumulation during development of caryopses from barley (*Hordeum vulgare* L.). Phytochemistry 65, 2977–2985. https://doi.org/10.1016/j.phytochem.2004.08.047.
Falk, J., Munné-Bosch, S., 2010. Tocochromanol functions in plants: antioxidation and beyond. J. Exp. Bot. 61, 1549–1566. https://doi.org/10.1093/jxb/erq030.
Fernandez-Marin, B., Kranner, I., San Sebastian, M., Artetxe, U., Laza, J.M., Vilas, J.L., Pritchard, H.W., Nadajaran, J., Miguez, F., Becerril, J.M., Garcia-Plazaola, J.I., 2013. Evidence for the absence of enzymatic reactions in the glassy state. A case study of xanthophyll cycle pigments in the desiccation-tolerant moss Syntrichia ruralis. J. Exp. Bot. 64, 3033–3043. https://doi.org/10.1093/jxb/ert145.
Fernández-Marín, B., Míguez, F., Méndez-Fernández, L., Agut, A., Becerril, J.M., García-Plazaola, J.I., Kranner, I., Colville, L., 2017. Seed carotenoid and tocochromanol composition of wild Fabaceae species is shaped by phylogeny and ecological factors. Front. Plant Sci. 8. https://doi.org/10.3389/fpls.2017.01428.
Garcia, Q.S., Giorni, V.T., Muller, M., Munne-Bosch, S., 2012. Common and distinct responses in phytohormone and vitamin E changes during seed burial and dormancy in Xyris bialata and X. peregrina. Plant Biology 14, 347–353. https://doi.org/10.1111/j.1438-8677.2011.00505.x.
Gerna, D., Roach, T., Stoggl, W., Wagner, J., Vaccino, P., Limonta, M., Kranner, I., 2017. Changes in low-molecular-weight thiol-disulphide redox couples are part of bread wheat seed germination and early seedling growth. Free Radic. Res. 51, 568–581. https://doi.org/10.1080/10715762.2017.1338344.
Giurizatto, M.I.K., Ferrarese, O., Ferrarese, M.D.L., Robaina, A.D., Goncalves, M.C., Cardoso, C.A.L., 2012. Alpha-Tocopherol levels in natural and artificial aging of soybean seeds. Acta Sci.-Agron. 34, 339–343. https://doi.org/10.4025/actasciagron.v34i3.12660.
Groot, S.P.C., Surki, A.A., de Vos, R.C.H., Kodde, J., 2012. Seed storage at elevated partial pressure of oxygen, a fast method for analysing seed ageing under dry conditions. Ann. Bot.-London 110, 1149–1159. https://doi.org/10.1093/aob/mcs198.
Groot, S.P.C., de Groot, L., Kodde, J., van Treuren, R., 2014. Prolonging the longevity of ex situ conserved seeds by storage under anoxia. Plant. Genet. Resour.-C 13, 18–26. https://doi.org/10.1017/S1479262114000586.
Harrington, J.F., 1963. Practical instructions and advice on seed storage. In Proceedings of the International Seed Testing Association. pp. 989–994.
Horvath, G., Wessjohann, L., Bigirimana, J., Jansen, M., Guisez, Y., Caubergs, R., Horemans, N., 2006. Differential distribution of tocopherols and tocotrienols in photosynthetic and non-photosynthetic tissues. Phytochemistry 67, 1185–1195. https://doi.org/10.1016/j.phytochem.2006.04.004.
Hwang, J.E., Ahn, J.W., Kwon, S.J., Kim, J.B., Kim, S.H., Kang, S.Y., Kim, D.S., 2014. Selection and molecular characterization of a high tocopherol accumulation rice mutant line induced by gamma irradiation. Mol. Biol. Rep. 41, 7671–7681. https://doi.org/10.1007/s11033-014-3660-1.
Ibrahim, A.E., Roberts, E.H., Murdoch, A.J., 1983. Viability of lettuce seeds. J. Exp. Bot. 34, 631–640. https://doi.org/10.1093/jxb/34.5.631.
ISTA, 2010. International rules for seed testing. International Seed Testing Association, Bassersdorf, Switzerland.
Kanwischer, M., Porfirova, S., Bergmuller, E., Dormann, P., 2005. Alterations in tocopherol cyclase activity in transgenic and mutant plants of Arabidopsis affect tocopherol content, tocopherol composition, and oxidative stress. Plant Physiol. 137, 713–723. https://doi.org/10.1104/pp.104.054908.
Kranner, I., Grill, D., 1996. Significance of thiol-disulfide exchange in resting stages of plant development. Bot. Acta 109, 8–14. https://doi.org/10.1111/j.1438-8677.1996.tb00864.x.
Kranner, I., Birtic, S., Anderson, K.M., Pritchard, H.W., 2006. Glutathione half-cell reduction potential: A universal stress marker and modulator of programmed cell death? Free Radical Bio. Med. 40, 2155–2165. https://doi.org/10.1016/j.freeradbiomed.2006.02.013.
Kranner, I., Minibayeva, F.V., Beckett, R.P., Seal, C.E., 2010. What is stress? Concepts, definitions and applications in seed science. New Phytol. 188, 655–673. https://doi.org/10.1111/j.1469-8137.2010.03461.x.
Lachman, J., Hejtmánková, A., Orsák, M., Popov, M., Martinek, P., 2018. Tocotrienols and tocopherols in colored-grain wheat, tritordeum and barley. Food Chem. 240, 725–735. https://doi.org/10.1016/j.foodchem.2017.07.123.
Lee, J.-S., Kwak, J., Yoon, M.-R., Lee, J.-S., Hay, F.R., 2017. Contrasting tocol ratios associated with seed longevity in rice variety groups. Seed Sci. Res. 27, 273–280. https://doi.org/10.1017/S0960258517000265.
Leubner-Metzger, G., 2005. Beta-1,3-glucanase gene expression in low-hydrated seeds as a mechanism for dormancy release during tobacco after-ripening. Plant J. 41, 133–145. https://doi.org/10.1111/j.1365-313X.2004.02284.x.
Mascher, M., Gundlach, H., Himmelbach, A., Beier, S., Twardziok, S.O., Wicker, T., Radchuk, V., Dockter, C., Hedley, P.E., Russell, J., Bayer, M., Ramsay, L., Liu, H., Haberer, G., Zhang, X.Q., Zhang, Q.S., Barrero, R.A., Li, L., Taudien, S., Groth, M., Felder, M., Hastie, A., Simkova, H., Stankova, H., Vrana, J., Chan, S., Munoz-Amatrian, M., Ounit, R., Wanamaker, S., Bolser, D., Colmsee, C., Schmutzer, T., Aliyeva-Schnorr, L., Grasso, S., Tanskanen, J., Chailyan, A., Sampath, D., Heavens, D., Clissold, L., Cao, S.J., Chapman, B., Dai, F., Han, Y., Li, H., Li, X., Lin, C.Y., McCooke, J.K., Tan, C., Wang, P.H., Wang, S.B., Yin, S.Y., Zhou, G.F., Poland, J.A., Bellgard, M.I., Borisjuk, L., Houben, A., Dolezel, J., Ayling, S., Lonardi, S., Kersey, P., Lagridge, P., Muehlbauer, G.J., Clark, M.D., Caccamo, M., Schulman, A.H., Mayer, K.F.X., Platzer, M., Close, T.J., Scholz, U., Hansson, M., Zhang, G.P., Braumann, I., Spannagl, M., Li, C.D., Waugh, R., Stein, N., 2017. A chromosome conformation capture ordered sequence of the barley genome. Nature 544, 426–433. https://doi.org/10.1038/nature22043.
Matthaus, B., Ozcan, M.M., Al Juhaimi, F., 2016. Some rape/canola seed oils: fatty acid composition and tocopherols. Z. Naturforsch. C 71, 73–77. https://doi.org/10.1515/znc-2016-0003.
McDonald, M.B., 1999. Seed deterioration: physiology, repair and assessment. Seed Sci. Technol. 27, 177–237.
Mène-Saffrané, L., Jones, A.D., DellaPenna, D., 2010. Plastochromanol-8 and tocopherols are essential lipid-soluble antioxidants during seed desiccation and quiescence in Arabidopsis. P. Natl. Acad. Sci. USA 107, 17815–17820. https://doi.org/10.1073/pnas.1006971107.
Moreau, R.A., Wayns, K.E., Flores, R.A., Hicks, K.B., 2007. Tocopherols and tocotrienols in barley oil prepared from germ and other fractions from scarification and sieving of hulless barley. Cereal Chemistry 84, 587–592. https://doi.org/10.1094/Cchem-84-6-0587.
Morscher, F., Kranner, I., Arc, E., Bailly, C., Roach, T., 2015. Glutathione redox state, tocochromanols, fatty acids, antioxidant enzymes and protein carbonylation in sunflower seed embryos associated with after-ripening and ageing. Ann. Bot.-London 116, 669–678. https://doi.org/10.1093/aob/mcv108.
Munne-Bosch, S., Onate, M., Oliveira, P.G., Garcia, Q.S., 2011. Changes in

Environmental and Experimental Botany 156 (2018) 8–15

phytohormones and oxidative stress markers in buried seeds of Vellozia alata. Flora 206, 704–711. https://doi.org/10.1016/j.flora.2010.11.012.
Munné-Bosch, S., Alegre, L., 2002. The function of tocopherols and tocotrienols in plants. Crit. Rev. Plant Sci. 21, 31–57. https://doi.org/10.1080/0735-260291044179.
Nagel, M., Börner, A., 2010. The longevity of crop seeds stored under ambient conditions. Seed Sci. Res. 20, 1–12. https://doi.org/10.1017/S0960258509990213.
Nagel, M., Kranner, I., Neumann, K., Rolletschek, H., Seal, C.E., Colville, L., Fernandez-Marin, B., Börner, A., 2015. Genome-wide association mapping and biochemical markers reveal that seed ageing and longevity are intricately affected by genetic background and developmental and environmental conditions in barley. Plant Cell Environ. 38, 1011–1022. https://doi.org/10.1111/pce.12474.
Nagel, M., Kodde, J., Pistrick, S., Mascher, M., Börner, A., Groot, S.P.C., 2016. Barley seed ageing: genetics behind the dry elevated pressure of oxygen ageing and moist controlled deterioration. Front. Plant Sci. 7, 388. https://doi.org/10.3389/fpls.2016.00388.
Oenel, A., Fekete, A., Krischke, M., Faul, S.C., Gresser, G., Havaux, M., Mueller, M.J., Berger, S., 2017. Enzymatic and non-enzymatic mehanisms contribute to lipid oxidation during seed aging. Plant Cell Physiol. 58, 925–933. https://doi.org/10.1093/pcp/pcx036.
Pellaud, S., Bory, A., Chabert, V., Romanens, J., Chaisse-Leal, L., Doan, A.V., Frey, L., Gust, A., Fromm, K.M., Mene-Saffrane, L., 2018. WRINKLED1 and ACYL-COA:DIACYLGLYCEROL ACYLTRANSFERASE1 regulate tocochromanol metabolism in Arabidopsis. New Phytol. 217, 245–260. https://doi.org/10.1111/nph.14856.
Priestley, D.A., McBride, M.B., Leopold, C., 1980. Tocopherol and organic free radical levels in soybean seeds during natural and accelerated aging. Plant Physiol. 66, 715–719. https://doi.org/10.1104/pp.66.4.715.
Pukacka, S., Ratajczak, E., 2005. Production and scavenging of reactive oxygen species in *Fagus sylvatica* seeds during storage at varied temperature and humidity. J. Plant Physiol. 162, 873–885. https://doi.org/10.1016/j.jplph.2004.10.012.
Riewe, D., Wiebach, J., Altmann, T., 2017. Structure annotation and quantification of wheat seed oxidized lipids by high resolution LC-MS/MS. Plant Physiol. https://doi.org/10.1104/pp.17.00470.
Roach, T., Beckett, R.P., Minibayeva, F.V., Colville, L., Whitaker, C., Chen, H., Bailly, C., Kranner, I., 2010. Extracellular superoxide production, viability and redox poise in response to desiccation in recalcitrant *Castanea sativa* seeds. Plant Cell Environ. 33, 59–75. https://doi.org/10.1111/j.1365-3040.2009.02053.x.
Roberts, E.H., 1973. Predicting the storage life of seeds. Seed Sci. Technol. 1, 499–514.
Sattler, S.E., Gilliland, L.U., Magallanes-Lundback, M., Pollard, M., DellaPenna, D., 2004. Vitamin E is essential for seed longevity, and for preventing lipid peroxidation during germination. Plant Cell 16, 1419–1432. https://doi.org/10.1105/Tpc.021360.
Schafer, F.Q., Buettner, G.R., 2001. Redox environment of the cell as viewed through the redox state of the glutathione disulfide/glutathione couple. Free Radical Bio. Med. 30, 1191–1212. https://doi.org/10.1016/S0891-5849(01)00480-4.
Seal, C.E., Zammit, R., Scott, P., Flowers, T.J., Kranner, I., 2010a. Glutathione half-cell reduction potential and α-tocopherol as viability markers during the prolonged storage of *Suaeda maritima* seeds. Seed Sci. Res. 20, 47–53. https://doi.org/10.1017/S0960258509990250.
Seal, C.E., Zammit, R., Scott, P., Nyamongo, D.O., Daws, M.I., Kranner, I., 2010b. Glutathione half-cell reduction potential as a seed viability marker of the potential oilseed crop *Vernonia galamensis*. Ind, Crop, Prod. 32, 687–691. https://doi.org/10.1016/j.indcrop.2010.06.023.
Sen, S., Osborne, D.J., 1974. Germination of rye embryos following hydration-dehydration treatments - enhancement of protein and RNA-synthesis and earlier induction of DNA-replication. J. Exp. Bot. 25, 1010–1019. https://doi.org/10.1093/jxb/25.6.1010.
Sen, S., Osborne, D.J., 1977. Decline in ribonucleic-acid and protein-synthesis with loss of viability during early hours of imbibition of rye (*Secale cereale* L.) embryos. Biochem. J. 166, 33–38.
Smirnoff, N., 2010. Tocochromanols: Rancid lipids, seed longevity, and beyond. P. Natl. Acad. Sci. USA 107, 17857–17858. https://doi.org/10.1073/pnas.1012749107.
Tammela, P., Salo-Vaananen, P., Laakso, I., Hopia, A., Vuorela, H., Nygren, M., 2005. Tocopherols, tocotrienols and fatty acids as indicators of natural ageing in *Pinus sylvestris* seeds. Scand. J. Forest Res. 20, 378–384. https://doi.org/10.1080/02827580500292063.
Torres, M., De Paula, M., Perez-Otaola, M., Darder, M., Frutos, G., Martinez-Honduvilla, C.J., 1997. Ageing-induced changes in glutathione system of sunflower seeds. Physiol. Plant. 101, 807–814. https://doi.org/10.1111/j.1399-3054.1997.tb01067.x.
Vertucci, C.W., 1989. The effects of low water contents on physiological activities of seeds. Physiol. Plant. 77, 172–176. https://doi.org/10.1111/j.1399-3054.1989.tb05994.x.
Vidi, P.-A., Kanwischer, M., Baginsky, S., Austin, J.R., Csucs, G., Dörmann, P., Kessler, F., Bréhélin, C., 2006. Tocopherol cyclase (VTE1) localization and vitamin E accumulation in chloroplast plastoglobule lipoprotein particles. J. Biol. Chem 281, 11225–11234. https://doi.org/10.1074/jbc.M511939200.
Walters, C., Wheeler, L.M., Grotenhuis, J.M., 2005. Longevity of seeds stored in a genebank: Species characteristics. Seed Sci. Res. 15, 1–20. https://doi.org/10.1079/Ssr2004195.
Wang, X., Song, Y.-e., Li, J.-y., 2013. High expression of tocochromanol biosynthesis genes increases the vitamin E level in a new line of giant embryo rice. J. Agr. Food Chem. 61, 5860–5869. https://doi.org/10.1021/jf401325e.

3.4 PAPER 4:

Wheat seed ageing viewed through the cellular redox environment and changes in pH

by

Manuela Nagel, Charlotte Seal, Louise Colville, Axel Rodenstein, Sun Un, Josefine Richter, Hugh Pritchard, Andreas Börner and Ilse Kranner

Published in

Free Radical Research (2019) 53, 641-654

https://doi.org/10.1080/10715762.2019.1620226

To view supplementary material for this article, please visit

https://doi.org/10.1080/10715762.2019.1620226

FREE RADICAL RESEARCH
https://doi.org/10.1080/10715762.2019.1620226

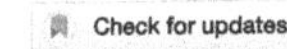

ORIGINAL ARTICLE

Wheat seed ageing viewed through the cellular redox environment and changes in pH

Manuela Nagel[a], Charlotte E. Seal[b], Louise Colville[b], Axel Rodenstein[c], Sun Un[d], Josefine Richter[a], Hugh W. Pritchard[b], Andreas Börner[a] and Ilse Kranner[e]

[a]Genebank Department, IPK, Seeland, Germany; [b]Department of Comparative Plant and Fungal Biology, Kew, UK; [c]Institute of Inorganic Chemistry, University Leipzig, Leipzig, Germany; [d]Department of Biochemistry, Biophysics and Structural Biology, Institute for Integrative Biology of the Cell, I2BC), Université Paris-Saclay, Gif-sur-yvette, France; [e]Department of Botany and Center for Molecular Biosciences (CMBI), University of Innsbruck, Innsbruck, Austria

ABSTRACT
To elucidate biochemical mechanisms leading to seed deterioration, we studied 23 wheat genotypes after exposure to seed bank storage for 6–16 years compared to controlled deterioration (CD) at 45 °C and 14 (CD14) and 18% (CD18) moisture content (MC) for up to 32 days. Under two seed bank storage conditions, seed viability was maintained in cold storage (CS) at 0 °C and 9% seed MC, but significantly decreased in ambient storage (AS) at 20 °C and 9% MC. Under AS and CS, organic free radicals, most likely semiquinones, accumulated, detected by electron paramagnetic resonance, while the antioxidant glutathione (GSH) was partly lost and partly converted to glutathione disulphide (GSSG), detected by HPLC. Under AS the glutathione half-cell reduction potential ($E_{GSSG/2GSH}$) shifted towards more oxidising conditions, from −186 to −141 mV. In seeds exposed to CD14 or CD18, no accumulation of organic free radicals was observed, GSH and seed viability declined within 32 and 7 days, respectively, GSSG hardly changed (CD14) or decreased (CD18) and $E_{GSSG/2GSH}$ shifted to −116 mV. The pH of extracts prepared from seeds subjected to CS, AS and CD14 decreased with viability, and remained high under CD18. Across all treatments, $E_{GSSG/2GSH}$ correlated significantly with seed viability ($r = 0.8$, $p < .001$). Data are discussed with a view that the cytoplasm is in a glassy state in CS and AS, but during the CD treatments, underwent transition to a liquid state. We suggest that enzymes can be active during CD but not under the seed bank conditions tested. However, upon CD, enzyme-based repair processes were apparently outweighed by deteriorative reactions. We conclude that seed ageing by CD and under seed bank conditions are accompanied by different biochemical reactions.

ARTICLE HISTORY
Received 11 October 2018
Revised 9 May 2019
Accepted 10 May 2019

KEYWORDS
Antioxidants; EPR; ESR; gene bank; germination; organic radicals; seed longevity

Introduction

High-quality seeds are essential for agricultural use. Importantly, seeds are also stored long-term in seedbanks to conserve valuable plant genetic resources that would otherwise be irretrievably lost in nature. However, all seeds inevitably lose viability during storage, compromising the use of seeds for industry and seedbanks.

During seed maturation drying, orthodox (desiccation tolerant) seeds dry to about 0.06–0.12 g H_2O g^{-1} dry weight (DW), metabolism ceases and the cytoplasm enters a glassy state [1,2]. Under these conditions, chemical reactions, including those that lead to seed deterioration, are drastically slowed down and seed longevity is extended. The speed of seed deterioration depends on genetic makeup, and growth conditions of the mother plant during seed development, and seed storage conditions. Specifically, temperature and seed moisture content (MC) [3,4] are the main factors that modulate cytoplasmic viscosity [5] and determine the rate of viability loss [6].

Most gene banks store seeds at −18 ± 3 °C after equilibration to 15 ± 3% relative humidity (RH) [7]. Under these conditions, the half-life (P_{50}) for stored seeds was estimated to be between 23 (*Lactuca sativa*) to over 300 years (*Lens culinaris*), by fitting germination time courses to Avrami kinetics [8]. Under ambient storage (AS) at 20.3 °C and 50.5% RH, P_{50} declines to between 4.6 (*L. sativa*) [9] and 10.6 years (*L. culinaris*) [10]. An artificial ageing treatment termed

CONTACT Manuela Nagel Nagel@ipk-gatersleben.de Genebank Department, IPK, Seeland, Germany
Supplemental data for this article can be accessed here.

"controlled deterioration" (CD) is frequently used to estimate seed longevity in storage, whereby seeds are treated with elevated temperatures, typically between 35 and 45 °C, and increased seed MC, such as 11 to over 25% MC [11,12]. Elevating seed MC and/or temperature increases the fluidity of the cytoplasm [1], which influences biochemical processes that determine seed longevity, germinability and dormancy [1,13–15]. Whereas CD is considered an adequate proxy for seed bank conditions by some authors [16], others have disputed its relevance [1,17].

According to the "free radical theory of ageing" [18], cumulative damage incurred by reactive oxygen species (ROS) over the life-span of an organism contributes to viability loss, and this may also apply to seeds [19–21]. In electron transport chains, ROS are continually formed by incomplete reduction of oxygen and are characterised by high chemical reactivity [22,23]. ROS, comprising free radicals such as superoxide ($O_2^{\bullet -}$) and the hydroxyl radical ($HO^{\bullet}$), and non-radical species such as hydrogen peroxide (H_2O_2), can be produced in the cytoplasm and in subcellular compartments [24,25], and in drying biological materials especially, can stimulate the formation of stable organic radicals. Due to their reactivity, ROS have the potential to damage other molecules but are also key elements of signalling pathways [26–29]. Processes leading to increased ROS production can be sensed by compartment-specific systems and regulate gene expression [22]. In hydrated seeds, mitochondria and glyoxysomes [30] have been proposed to be major ROS sources. Furthermore, plasma membrane NADPH oxidases, also termed respiratory burst oxidase homologues, facilitate $O_2^{\bullet -}$ formation by using molecular oxygen and NADPH [31]. In dry seeds, ROS can initiate peroxidation, leading to a free radical chain reaction that may contribute to viability loss [32,33]. ROS and radicals formed downstream can function as pro- and antioxidants, e.g. the oxidation of phenolic compounds can generate semiquinones. In turn, semiquinone radicals can react with ground state oxygen ($^3O_2$) to form $O_2^{\bullet -}$, which can react with the parent compound, e.g. catechol, and generate semiquinone and H_2O_2 [34]. Organic free radicals can accumulate in dry biological materials [35,36]. However, the role of radicals formed during seed maturation drying is still poorly understood, and it is not even established if they are deleterious for seeds, or if they may be required at the early stages of germination when seeds take up water. The latter assumption would agree with the suggestion that germinating seeds need an "oxidative window" to complete germination [20].

To keep ROS at a level required for signalling, without causing irreparable damage, sophisticated defence systems have evolved [22], including ROS-processing enzymes and low-molecular-weight (LMW) antioxidants [19,27]. In dry orthodox seeds, the major LMW, water-soluble antioxidant is glutathione (GSH), acting as ROS scavenger and electron donor for ROS-processing enzymes, such as glutathione-S-transferases (GST). It also binds, via a disulphide bond, to protein thiol groups in a process called protein S-glutathionylation, forming protein-bound glutathione that protects proteins from irreversible oxidation [37]. Upon seed imbibition, glutathionylated proteins can be reduced by glutaredoxins [38]. In the desiccated state, metabolic arrest entails oxidative modifications and consumption of reductants, such as GSH, and accumulation of glutathione disulphide (GSSG) [37,39–41]. Changes in the abundance of GSH and GSSG affect the glutathione half-cell reduction potential ($E_{GSSG/2GSH}$) [42,43] and mark important changes in the viability status during storage [13,15,44,45] and seed germination [46]. The loss of seed viability is accompanied by a depletion of glutathione and a shift towards more oxidising conditions [13,43]. Such redox shifts are pH-dependent [42], but the effect of pH during seed ageing is hardly understood.

The pH influences a plethora of (bio)chemical mechanism in plants cells and their compartments [47]. Regulation of pH is achieved by active or passive membrane transport of H^+ including H^+ ATPases (pumps) and H^+ cotransporters but also of other ions, through transmembrane diffusion of weak acids or bases, by ion exchange or by biochemical reactions [47], also involving light or phytohormones [48]. In higher plants, large differences are found between vacuolar, cytoplasmic and apoplastic pH values, ranging between 5.0 and 7.5 or above [49], required for creating transmembrane gradients that serve to transport ions and molecules across membranes [47]. For example, when leaf cells dehydrate, the pH of the mesophyll apoplast increases, which triggers the release of abscisic acid (ABA) that initiates stomatal closure [50]. However, in dry orthodox seeds, pH regulation is far less comprehensible as their cytoplasm is highly viscous and protons can hardly move. It has been suggested that tissues acidify upon developmental arrest (such as in a mature seed), and when seeds imbibe, changes in pH and redox potential may break seed dormancy and support germination [51].

Globally, wheat (*Triticum aestivum* L. and *Triticum durum* Desf.) is the second most important crop, with 750 million tons of seeds being produced annually (www.fao.org/faostat). Due to the unique combination of starch and storage proteins in the "gluten" fraction, wheat is the most important cereal used for bread, pasta and baked products [52]. Wheat produces caryopses (in which the fruit coat, or "pericarp", and the seed coat, or "testa", are fused), hereafter called "seeds". During maturation drying, approximately 53 ± 5 days after anthesis, wheat seeds develop a capability to survive in the dry state over long periods [53,54]. In hermetic storage at −18 °C, P_{50} for wheat seeds is estimated to be 54 years [8], 7.2 years [9] in AS at 20 °C and 50% RH, but only between days to weeks upon CD.

The aim of this study was to investigate how different storage and ageing conditions in conjunction with genetic background affect free radical production, redox properties, pH and seed deterioration. Seeds of 23 gene bank wheat accessions, hereafter termed "genotypes", were stored under AS or cold storage (CS) at 0 °C for 6–16 years and then multiplied. Freshly harvested (FH) seeds were exposed to CD at 45 °C and 14% (CD14) and 18% MC (CD18) and viability reduction assessed. To examine differences in the underlying biochemical reactions during seed ageing, changes in the GSSG/2GSH redox couple and LMW thiol-disulphide redox couples involved in GSH metabolism and their half-cell reduction potentials were analysed together with changes in pH and the abundance of stable organic radicals, and results are discussed in relation to cytoplasmic viscosity.

Materials and methods

Seed material and ageing treatments

Seeds of 23 winter wheat (*Triticum aestivum* L.) genotypes from 15 countries were multiplied at the Gatersleben gene bank between 1992 and 2002 (Supplementary Table S1). Seeds were harvested, cleaned and dried at 15% RH before either CS (0 ± 1 °C, 8.5 ± 0.1% seed MC) or AS (20.3 ± 2.3 °C, 8.7 ± 0.1% seed MC). In 2008, seeds of all genotypes from CS were multiplied under optimum conditions at the Gatersleben estates, harvested at full maturity and cleaned. FH seeds of six genotypes were chosen based on their high total germination (TG) of 97.1 ± 2.2% after storage in AS (20.3 ± 2.3 °C, 9.2 ± 0.2% seed MC) for 1 year, which was applied to release dormancy, and were then exposed to CD. For each of these six genotypes, 400 dry seeds were equilibrated to 14 or 18% MC [11] and sealed in aluminium bags. At 44 ± 0.5 °C, seeds with 18% MC were aged for up to 7 days (CD18) and seeds with 14% MC for up to 32 days (CD14) to produce TGs between 97 and 0%. Three replicates of 100 seeds of each treatment (FH, CS, AS, CD14 and CD18) were germinated on moist filter paper in a Jacobson Apparatus (Rubarth Apparate GmbH, Laatzen, Germany) using a photoperiod of 14 h:10 h (55 µmol s^{-1} m^{-2}) and a light:dark cycle at 25 ± 2:23 ± 2 °C. TG was assessed as radicle emergence of at least 2 mm after 8 days. The remaining 100 seeds were frozen in liquid nitrogen and freeze-dried for 7 days [55]. Seed MC was determined before and after freeze drying. Seeds were ground to a fine powder with agate balls in a liquid nitrogen-cooled grinding capsule using a Retsch MM400 ball mill (RETSCH GmbH, Haan, Germany).

Measurement of LMW thiols and disulphides

GSH, GSSG and the related LMW thiols and disulphides [cysteine (Cys)/cystine, γ-glutamyl-cysteine (γ-Glu–Cys)/bis-γ-glutamyl-cystine (bis-γ-Glu–Cys) and cystinyl-glycine (Cys–Gly)/cysteinyl-bis-glycine (Cys-bis-Gly)] were extracted in 0.1 M HCl from 50 mg ($n = 3$ replicates) of freeze-dried and finely ground powder produced from 100 seeds and analysed by reversed-phase HPLC (Supplementary Figure S1) as described earlier [55,56]. The half-cell reduction potential of each individual LMW thiol-disulphide couple (E_i), $E_{GSSG/2GSH}$, $E_{Cystine/2Cys}$, $E_{bis-\gamma-Glu-Cys/2-\gamma-Glu-Cys}$ and $E_{Cys\text{-}bis\text{-}Gly/2Cys\text{-}Gly}$, respectively, were calculated using the Nernst equation (Equation (1)). Seed MC, expressed as g H_2O g^{-1} DW, was used to estimate molar concentrations of GSH and GSSG (considering g water g^{-1} seed DW), and intracellular pH was initially assumed to be 7.3 [42,43].

$$E_i = E^{0\prime} - \frac{RT}{nF} \ln \frac{[\text{LMW thiol}]^2}{[\text{LMW disulphide}]} \quad (1)$$

where R is the gas constant (8.314 J K^{-1} mol^{-1}); T, temperature in K; n, number of transferred electrons (2GSH → GSSG + $2H^+ + 2e^-$); F, Faraday constant (9.6485 × 10^4 C mol^{-1}); $E^{o\prime}$, standard half-cell reduction potential of a thiol-disulphide redox couple at an assumed cellular pH of 7.3 ($E^{o\prime}_{GSSG/2GSH} = -258$ mV, $E^{o\prime}_{Cystine/2Cys} = -244$ mV, $E^{o\prime}_{Cys-bis-Gly/2Cys-Gly} = -244$ mV, $E^{o\prime}_{bis-\gamma-Glu-Cys/2\gamma-Glu-Cys} = -252$ mV) [43,57]. The individual E_i values were then mathematically combined into the LMW thiol-disulphide redox environment ($E_{thiol\text{-}disulphide}$, Equation (2)), as described by Birtic et al. [57], where E_i is the half-cell reduction potential of an individual thiol-disulphide redox couple i, and [reduced species]$_i$ is the concentration of the thiols in that redox pair.

$$E_{thiol-disulphide} = \sum_{i=1}^{n(couple)} E_i \cdot x[\text{reduced species}]_i \quad (2)$$

Determination of pH

Using a modified method of Pfeifhofer [58], 75 mg wheat powder ($n = 4$ replicates of freeze-dried and finely ground powder produced from 100 seeds) was extracted in 1.5 mL deionised water at 100 °C for 10 min, then cooled to ambient temperature and centrifuged at 11 750 g for 30 min. The pH of the supernatant was measured using a digital multi metre (ProLab2000, SI Analytics GmbH, Mainz, Germany).

EPR measurements of total free radicals

For the unspecific estimation of total radical concentration, freeze-dried and finely ground powder of the seeds of all genotypes were used after the FH, AS and CS treatments, and in addition, genotypes TRI 4442 and TRI 7371 were also analysed after CD14 and CD18.

Free radicals were measured at room temperature using an X-band (9.6 GHz) electron paramagnetic resonance (EPR) spectrometer (Bruker Corporation, Billarica, USA). Freeze-dried and finely ground wheat powder (15–20 mg) was placed into thin quartz glass tubes of 3 × 160 mm. The EPR spectra were recorded at the microwave power of 1.0 mW and analysed by WinEPR and SimFonia software (Bruker Corporation). Amplitudes (A) and line width (ΔBpp) were determined. Integral intensities (I) as the areas under the absorption curves were calculated by double integration of the first derivative EPR spectra (Equation (3)).

$$I = A \cdot \Delta Bpp^2 \quad (3)$$

Total free radical concentration (N) was determined as the number of free radicals in 1 g of sample (Equation (4)). N is proportional to the integral intensity (I). Ultramarine blue (1:100; 1.76 11:100; 1.76 10^{18} spins g^{-1}) was used as a reference for N and measured simultaneously in a double resonance cavity.

$$N = N_u \left[\frac{I}{I_u \cdot m}\right] \quad (4)$$

where N_u is the number of paramagnetic centres in the ultramarine reference (1.76 11:100; 1.76 10^{18} spins g^{-1}), I and I_u are the integral intensities of the sample and ultramarine blue and m is the sample mass.

Radical identification by high-field EPR measurements

Organic radical spectra were obtained at 285 GHz and 4 K using a locally-built high-field EPR (HFEPR) spectrometer [59]. Unlike in the X-band measurements described above, the spectrometer was not designed to measure the absolute amplitudes of spectra. Nonetheless, the intensities of the spectra were reproducible to a factor of two. Except where indicated, all of the samples measured had the same amplitude to within this factor. The magnetic field values were reliable to ± 0.0001 T and the corresponding *g*-values were calculated by Equation (5). All spectra were taken under nonsaturating conditions.

$$g\text{-values} = \frac{f_{obs}}{B \cdot \beta} \quad (5)$$

where the observation frequency (*fobs*) was 285.090 GHz and *B* is the magnetic field and *b* is 13.996246 GHz T^{-1}.

For technical reasons, conventional X-band EPR measurements cannot be made on large samples such as whole wheat seeds, so we used finely ground seed powder instead. By contrast, it was possible to use whole wheat seeds for the HFEPR measurements. Therefore, it could be assessed whether the grinding procedure influenced the concentration of radicals in the seed. Figure 1 shows 4K HFEPR spectra of 400 mg of intact or ground seed. As expected, the absolute intensity was significantly lower for the intact seeds due to the poorer sample packing in the intact seeds that had poorer microwave properties due to their irregular shapes. By comparison, the uniformly packed ground seed samples were more ideal in both

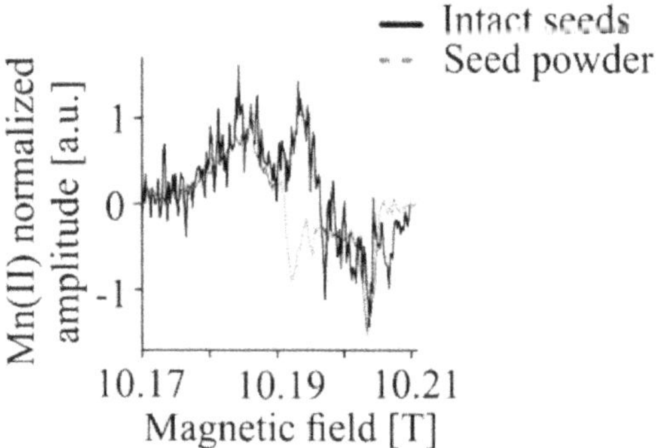

Figure 1. Effect of grinding on organic radical appearance. Intact seeds (solid line) and freeze-dried, finely ground seeds (dashed line) of genotype TRI 18326 aged at 14% MC and 44°C for 32 days measured by HFEPR. The spectra were normalized to the Mn(II) resonances that are also visible.

respects. To compensate for these differences, the two spectra were normalised to the intensity of the EPR resonances arising from the Mn(II) that become very intense above 23 K. Since Mn(II) is chemically and mechanically stable, its EPR signals were well-suited for this purpose. Once the intact and ground seed spectra were normalised in this way, it became apparent that the seed radical concentrations were the same in the intact and ground seed samples. Hence, we concluded that the radical that we report here (between 10.16 and 10.18 T) was not the result of grinding and that grinding did not affect the quantity of this radical. However, there appeared to be a second radical (between 10.19 and 10.21 T) in intact seeds that was much smaller in the powder (Figure 1). To characterise the radical between 10.16 and 10.18 T, freeze-dried and finely ground powder of the seeds of genotypes TRI 4442 and TRI 7970 were used after the following treatments: FH, AS, CS, CD14 after 32 days and CD18 after 7 days.

Statistical treatments

Data were tested for statistical significance using GenStat 18 (VSN International Ltd. Hemel Hempstead, UK). Significant differences were estimated using ANOVA followed by least significant differences at 5% (LSD5%). Pearson or Spearman correlation analyses were conducted to correlate TG with pH and/or redox data and radical concentrations for the treatments FH, AS, CS, CD14 and CD18. Probit analysis [60] was used to determine values of $E_{GSSG/2GSH}$ at which 50% of the seed population died (referred to as LD_{50}).

Results

Variations in seed viability and redox poise in response to different seed storage treatments

The time to seed viability loss was dependent on seed MC and temperature during storage or CD, and was strongly affected by genotype (Figure 2(a)). For example, genotype TRI 73 maintained 100 ± 0% TG after 13 years in CS, whereas TRI 7970 stored for 11 years had the lowest TG of 66.0 ± 5.0%. In AS, TRI 7986 had the highest TG of 89.0 ± 3.0% after 7 years and TRI 1003, TRI 1688, TRI 6928 and TRI 7970 were close to 0% TG after 16, 13, 10 and 11 years, respectively (Supplementary Table S1). To visualise the relationship between seed viability and biochemical changes, seeds of the 23 genotypes were sorted according to their TG at the end of AS. We chose three categories of high, median and low TGs assessed by median values. Genotypes with high TG in AS (82.7%) had TGs of 98.7% in both FH and CS, respectively; genotypes with medium TG in AS (28.2%), had TGs of 96.0 and 95.0% in FH and CS, respectively, and genotypes with low TG in AS (1.7%) had TGs of 91.0 and 79.0% in FH and CS, respectively (Figure 2(a)). Furthermore, FH seeds of six selected genotypes (with TGs between 94 and 99%) were subjected to CD14 and CD18 and lost viability within 32 and 7 days, respectively. In summary, within the three TG groups, seed viability was high in FH, maintained or fell by merely a quarter in CS and dropped to between 88 and 0% under AS during 6–16 years in storage, but was lost within days in CD14 and CD18 (Figure 2(a)).

Changes in GSH, GSSG and $E_{GSSG/2GSH}$ varied with storage or CD conditions (Figure 2(b–e)). Across all 23 genotypes, FH seeds contained a median concentration of 256 nmol g^{-1} DW total glutathione (188 nmol g^{-1} DW GSH and 34 nmol g^{-1} DW GSSG; Figure 2(b,c,d)), resulting in the most negative $E_{GSSG/2GSH}$ values of −186 mV (Figure 2(e)). While TG dropped by a quarter under CS compared to FH seeds (Figure 2(a)), GSH was 74 nmol g^{-1} DW lower (Figure 2(c)) and GSSG 15 nmol g^{-1} DW higher in CS seeds (differences between median values of all three TG groups), resulting in a median $E_{GSSG/2GSH}$ of −175 mV for CS (Figure 2(e)), i.e. slightly more oxidising than in FH seeds. Total glutathione (GSH + GSSG; note that 2 mol GSH are converted into 1 mol GSSG, Figure 2(b)) changed significantly to 210 nmol g^{-1} DW, suggesting that GSH was partly converted to GSSG and partly lost.

In the genotypes with the lowest TG under AS (Figure 2(a), TG group I), GSH was 154 nmol g^{-1} DW lower than in FH seeds (Figure 2(c)), but GSSG levels were only 31 nmol g^{-1} DW higher compared to FH seeds (Figure 2(d)). This resulted in an increase in $E_{GSSG/2GSH}$ to −137 ± mV, suggesting that GSH was partly converted to GSSG and partly lost (total glutathione significantly decreased by 35.0%). In the seeds of the six genotypes subjected to CD14, which lost viability within 32 days (Figure 2(a)), GSH fell steadily to 18 nmol g^{-1} DW (Figure 2(c)), whereas GSSG hardly changed (Figure 2(d)), and shifted towards highly oxidising conditions to $E_{GSSG/2GSH}$ of −117 mV (Figure 2(e)). In the seeds of the six genotypes subjected to CD18, which lost viability within 7 days, GSH also fell steadily to 15 nmol g^{-1} DW, very similar to the decline observed under CD14. However, in contrast to CD14, CD18 was associated with a decrease in GSSG by 16 nmol g^{-1} DW compared to FH seeds, also leading to a shift in $E_{GSSG/2GSH}$ towards highly oxidising conditions (−115 mV). Across all genotypes,

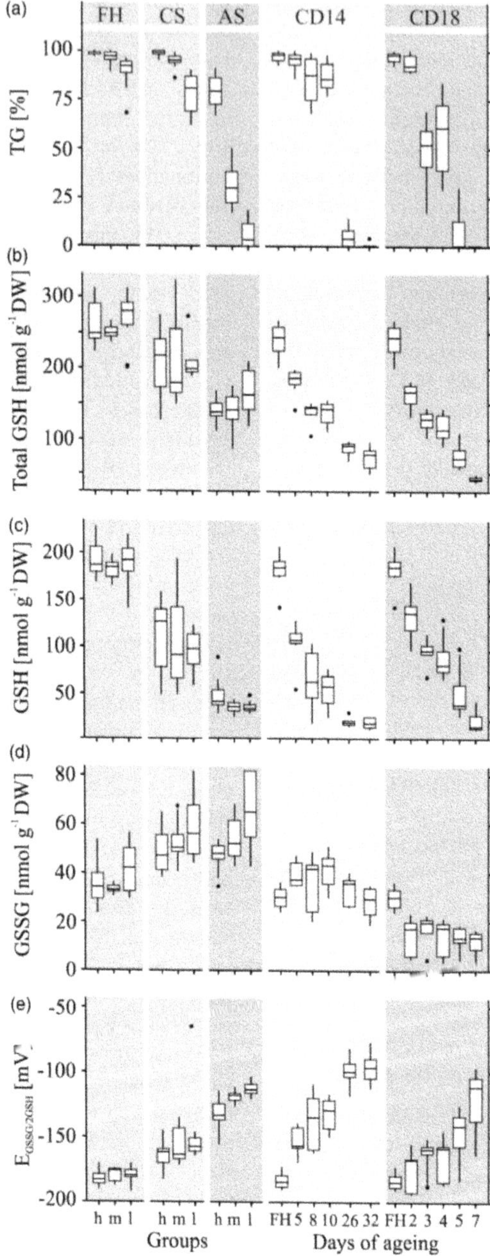

Figure 2. Effects of different seed storage conditions on seed viability and glutathione redox state in 23 wheat genotypes. Panels show (a) seed viability, expressed as total germination (TG); (b) concentrations of total glutathione (GSH); (c) concentrations of GSH; (d) concentrations of glutathione disulphide (GSSG); and (e) the half-cell reduction potential of the GSSG/2GSH redox couple ($E_{GSSG/2GSH}$) at pH 7.3. Shadings indicate freshly harvested seeds (FH); seeds that were kept in cold storage at 0°C for 6 to 16 years (CS); seeds kept in ambient storage conditions at 20°C (AS); seeds that were subjected to controlled deterioration at 14% seed moisture content (MC) (CD14) or 18% seed MC (CD18). For FH, CS and AS data, 23 genotypes were grouped according to their TG at the end of AS into three groups [high (h), medium (m) and low TG (l) on the x-axis, see text for details]; box plots represent 5–10 h, m or l TG genotypes (n = 3 replicates per genotype). For CD14 and CD18, box plots represent six selected genotypes (n = 3 replicates per genotype).

TG of FH seeds and seeds subjected to CS and AS correlated strongly negatively with GSSG ($r = -0.6$, $p < .001$) and $E_{GSSG/2GSH}$ ($r = -0.8$, $p < .001$), and positively with GSH ($r = 0.7$, $p < .001$; Supplementary Table S2).

The $E_{Cystine/2Cys}$, $E_{bis\text{-}\gamma\text{-}Glu\text{-}Cys/2\text{-}\gamma\text{-}Glu\text{-}Cys}$ and $E_{Cys\text{-}bis\text{-}Gly/2Cys\text{-}Gly}$ (Supplementary Figure S2) mostly paralleled $E_{GSSG/2GSH}$, shifting towards more positive values, compared to FH seeds, upon AS (within the three sorted groups of 23 genotypes, as in Figure 2), mostly upon CS and during CD14 and CD18. These trends were most pronounced for $E_{bis\text{-}\gamma\text{-}Glu\text{-}Cys/2\text{-}\gamma\text{-}Glu\text{-}Cys}$ (Supplementary Figure S2(b)). The mathematically combined $E_{thiol\text{-}disulphide}$ was largely dominated by $E_{GSSG/2GSH}$ (shown as an example for genotype TRI 4960 in Figure 3). The $E_{thiol\text{-}disulphide}$ shifted towards oxidising conditions upon CS, and even more so upon AS, compared to FH seeds, as well as during CD14 and CD18. However, during CD14 this oxidative shift in $E_{thiol\text{-}disulphide}$ preceded viability loss, whereas during CD18 it occurred almost simultaneously with viability loss, i.e. the negative correlation between $E_{thiol\text{-}disulphide}$ and TG was stronger for CD18 (Supplementary Table S2).

Variations in seed extract pH in response to different ageing conditions, and implications for the calculation of half-cell reduction potentials

Extracts prepared from FH seeds had a median pH between 6.5 and 6.7, whereas the median pH of extracts of the seed lots with the lowest viability after CS, AS and CD14 were 6.5, 6.3 and 6.4, respectively (Figure 4). For AS and CD14, this was significantly lower than in FH seeds. The pH of extracts prepared from seeds in CS, AS, CD14 decreased with TG, suggesting that seed deterioration under CS, AS, CD14 was accompanied by changes in seed proton concentration. The correlation between TG and extract pH was weakly significant in CS ($r = 0.19$; $p < .05$), the treatment in which the lowest TG was 78%. A highly significant correlation between TG and pH was found for AS ($r = 0.72$; $p < .001$) and CD14 ($r = 0.79$; $p < .001$; Supplementary Table S2), where the lowest values of TG were 2 and 0%, respectively. However, the pH in extracts prepared from seeds subjected to CD18 remained high between 6.6 and 6.7 and did not correlate with viability (Figure 4).

Figure 5(a) shows the relationship between TG and $E_{GSSG/2GSH}$, assuming a pH of 7.3. Across all treatments, seeds with TG between 100 and 85% had mostly $E_{GSSG/2GSH}$ values between −200 and −160 mV (NB: 85% TG represents the threshold TG defined by the FAO gene bank standards [7], below which seed accessions are reproduced). In contrast, seeds with very low viability or dead seeds (TGs between 15 and 0%) had $E_{GSSG/2GSH}$ values between −160 and

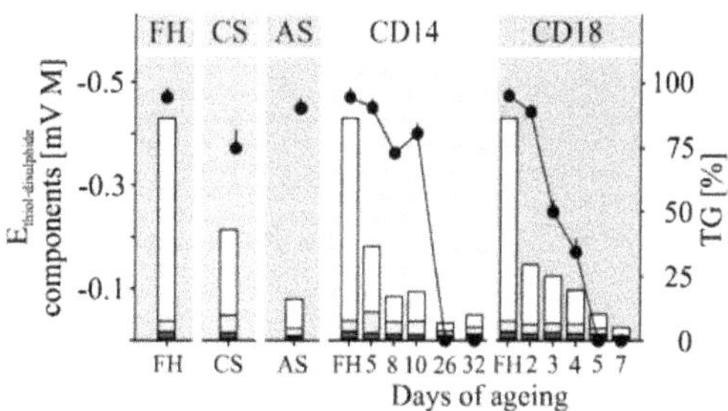

Figure 3. Changes in $E_{thiol\text{-}disulphide}$ during storage. Each bar shows the mathematically combined $E_{thiol\text{-}disulphide}$ with stacked individual E_is, namely $E_{GSSG/2GSH}$ (white), $E_{Cystine/2Cys}$ (dark grey), $E_{bis\text{-}\gamma\text{-}Glu\text{-}Cys/2\text{-}\gamma\text{-}Glu\text{-}Cys}$ (light grey) and $E_{Cys\text{-}bis\text{-}Gly/2Cys\text{-}Gly}$ (black) for genotype TRI 4960. Shadings indicate freshly harvested seeds (FH); seeds that were kept in cold storage at 0°C for 8 years (CS); seeds kept in ambient storage conditions at 20°C (AS); seeds that were subjected to controlled deterioration at 14% seed moisture content (MC) (CD14) or 18% seed MC (CD18); $n = 3$ replicates. For CD14 and CD18, day 0 corresponds to FH seeds. Black symbols show total germination (TG).

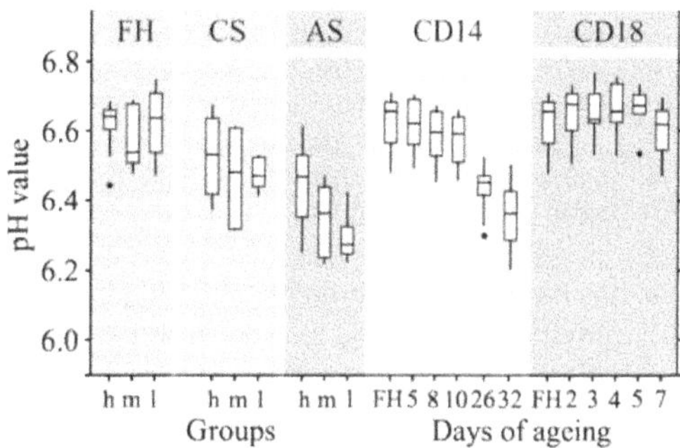

Figure 4. Effects of storage conditions on the pH of seed extracts. pH values were measured in extracts of finely ground seed material. Shadings indicate freshly harvested seeds (FH); seeds that were kept in cold storage at 0°C for 6 to 16 years (CS); seeds kept in ambient storage conditions at 20°C (AS); and seeds that were subjected to controlled deterioration at 14% seed moisture content (MC) (CD14) or 18% seed MC (CD18). For FH, CS and AS data, 23 genotypes were grouped according to their TG in AS into three groups [high (h), medium (m) and low TG (l) on the x-axis, see text for details]; box plots represent 5 to 10 h, m or l TG genotypes ($n = 3$ replicates per genotype). For CD14 and CD18, box plots represent six selected genotypes ($n = 3$ replicates per genotype).

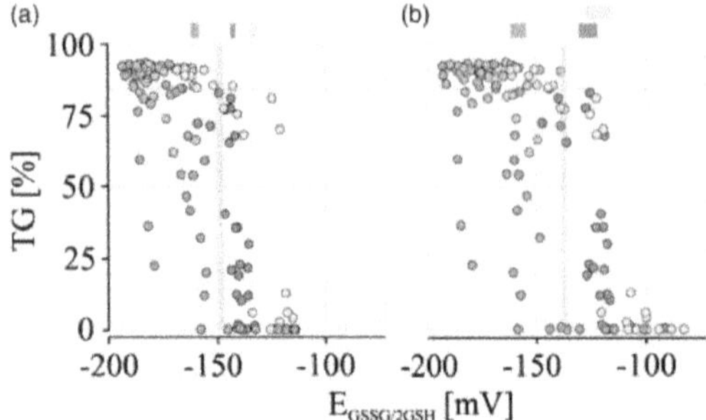

Figure 5. Relationship between total germination and glutathione half-cell reduction potential in dependence on pH. Total germination (TG) showed a sigmoidal relationship with glutathione half-cell reduction potential ($E_{GSSG/2GSH}$), i.e. seeds with high viability had more negative $E_{GSSG/2GSH}$ values than seeds with low or no viability. Data points show single measurements for 23 genotypes of freshly harvested seeds (green; FH), of seeds that were kept in cold storage at 0 °C (blue; CS) or in ambient storage conditions at 20 °C (purple; AS) for 6–16 years; and of seeds of 6 genotypes (as in Figure 2) that were subjected to controlled deterioration at 14% (yellow) or 18% seed MC (red). In panel a) the pH used in the Nernst equation to calculate $E_{GSSG/2GSH}$ was assumed to be in the physiological range (pH 7.3) for all samples and in b) were adjusted to the values shown in Figure 4. For this, an offset correction of + 0.6 (pH 7.3–6.7) was made based on the highest pH values in the extracts of the FH seeds of all genotypes. Red, purple and yellow shadings represent LD_{50} values ± SD for $E_{GSSG/2GSH}$ values taken for the treatments CD18, AS and CD14, respectively, and grey shading represents data across all treatments.

−110 mV. Probit analysis to estimate the P_{50} for $E_{GSSG/2GSH}$, (corresponding to LD_{50}, i.e. the value where 50% of a population is dead and 50% is alive) revealed a value of −164 ± 2 mV for CD18, 134 ± 2 mV for CD14, 143 ± 1 mV for AS. P_{50} values for CS and FH seeds were not calculated because of the high TG of these seed lots. In other words, P_{50} values shifted towards more positive $E_{GSSG/2GSH}$ values from CD18 to AS to CD14.

However, redox potentials depend on pH, and values in Figure 5(a) were calculated assuming physiological pH values of 7.3 in Equation (1). In Figure 5(b), $E_{GSSG/2GSH}$ values were adjusted using the pH values in Figure 4, but with an offset correction of + 0.6 for all values (pH 7.3–6.7; the latter value was the highest pH measured in extracts of FH seeds). This offset was necessary because the pH in the extract of ground seed powder is only a rough estimation of cytoplasmic pH, with likely confounding effects due to coextracted acidic compounds present in organelles, cell wall and seed coat [49]. According to the adjusted pH values, the cluster for seeds subjected to AS and CD14 shifted towards more oxidising conditions. This was also observed for CD18, although less pronounced. The P_{50} values shifted accordingly to 122 ± 2 mV for CD14, 125 ± 1 mV for AS and 161 ± 2 mV for CD18 (Figure 5(b)).

Appearance and identification of free radicals upon seed ageing

Two methods were used to study free radicals during the various seed ageing treatments. First, X-band EPR was used to unspecifically estimate the abundance of total organic radicals. Compared to FH seeds of 23 genotypes, radicals accumulated during AS, with a significant correlation between organic radical accumulation and TG ($r = -0.82$; $p < .001$), GSSG ($r = 0.69$; $p < .01$), GSH ($r = -0.88$; $p < .001$), $E_{GSSG/2GSH}$ ($r = 0.90$; $p < .001$) and $E_{thiol\text{-}disulphide}$ ($r = 0.87$; $p < .001$). However, no change in the amounts of organic radicals was found in seeds subjected to CS, CD14 and CD18 (Figure 6(a,b)). We then used HFEPR (285 GHz) to work towards the identification of the radical species (Figure 6(c)). The HFEPR spectra exhibited six sharp lines arising from Mn(II). At 4 K the amplitude of the radical and Mn(II) resonance was comparable, which impeded the measurement of the former. The amplitudes of the Mn(II) lines are extremely temperature sensitive, reaching a maximum at 23 K, whereas the spectrum of the seed radical diminished significantly with increasing temperature. Using this phenomenon, it was possible to subtract the interfering contribution of the Mn(II) from the 4K data, which subsequently provided a useful means of normalising the spectra without recourse to simulations (Figure 6(c)). For seeds subjected to CD14 and CD18, signals measured by Mn(II) were too poor to draw any conclusion. By contrast, FH seeds and seeds subjected to AS and CS exhibited radical spectra with similar amplitudes. The measured g principal values of the radical were $g_x = 2.0059$, $g_y = 2.0045$ and $g_z = 2.0022$ indicating a semiquinone type radical [61].

Discussion

Storage treatments affect cytoplasmic viscosity

When seed MC decreases below 0.06–0.12 g $H_2O \cdot g^{-1}$ DW at 25 °C, the cytoplasm vitrifies, molecular mobility slows down [1] and this explains, at least partly, that the speed of seed deterioration (Figure 2(a)) is strongly affected by ageing conditions. The seed MC of FH seeds (9.2 ± 0.2% MC) and seeds stored under AS (8.7 ± 0.1% MC) or CS (8.5 ± 0.1% MC) corresponds to about 0.1 g $H_2O \cdot g^{-1}$ DW, comparable with maize seeds that passed the glass transition temperature (T_g)

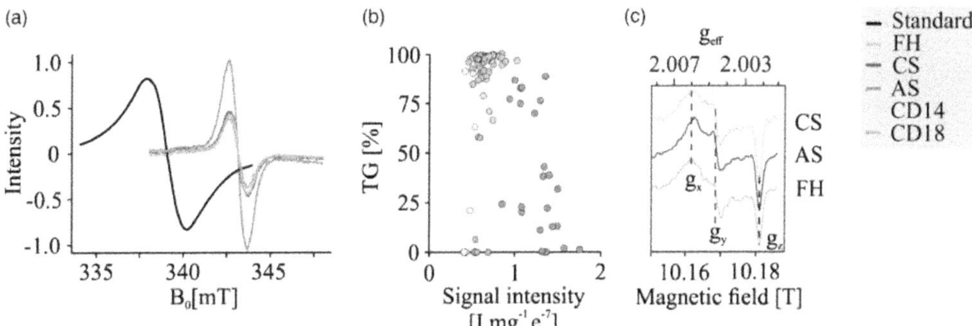

Figure 6. Effects of storage conditions on free radical accumulation, and radical identification. (a) Intensities measured by an X-band EPR spectroscope at 9.3 GHz. Curves show single measurements for genotypes TRI 7970 of freshly harvested seeds (green; FH), of seeds that were kept in cold storage at 0 °C (blue; CS) or in ambient storage conditions at 20 °C (purple; AS) for 6–16 years; and of seeds of genotypes TRI 4442 and TRI 7371 that were subjected to controlled deterioration at 14% seed moisture content (yellow) or 18% seed MC (red). (b) Accumulation of free radicals in dry seeds measured by an X-band EPR spectroscope at 9.3 GHz. Data points show single measurements for 23 genotypes of freshly harvested seeds (green; FH), of seeds that were kept in cold storage at 0 °C (blue; CS) or in ambient storage conditions at 20 °C (purple; AS) for 6–16 years; and of seeds of genotypes TRI 4442 and TRI 7371 that were subjected to controlled deterioration at 14% (yellow) or 18% seed MC (red). (c) Identification of an organic free radical by HFEPR in FH seeds of genotype TRI 7970 and in seeds stored under CS and AS.

at 20 °C and 0.1 g $H_2O \cdot g^{-1}$ DW at which the cytoplasm is highly viscous (Figure 7). Storage at 0 °C under CS will have reduced water activity, thereby further decreasing fluidity [1,62]. Under these conditions, most genotypes roughly follow the "Harrington's rule of thumb", stating that seed longevity doubles with each 5 °C decrease in storage temperature [63] or 1% decrease in seed MC. For example, after 13 years of CS seeds of genotype TRI 73 still showed 100% TG, but only 22% TG after 13 years of AS (Supplementary Table S1). The increased seed MC and temperature during CD [11] led to higher water activity and molecular mobility, together with an enlargement of the molecular space [64]. At storage conditions approaching T_g, the cytoplasm becomes more liquid (Figure 7). In agreement with Harrington [63], seed viability was lost in CD18 after 7 days and in CD14 after 32 days (Figure 2(a)).

Semiquinone-type radicals are present in the visco-elastic cytoplasm during ambient storage

Under AS, seed viability was clearly negatively correlated with accumulated organic free radicals (Figure 6(b)). The g principal values ($g_x = 2.0059$, $g_y = 2.0045$ and $g_z = 2.0022$) obtained by HFEPR (Figure 6(c)) were within the range expected for semiquinone, catechol radicals and radicals derived from trihydroxy-benzene moieties, such as those found in lignin [61] and humic acids [65]. These g principal values can range from $g_x = 2.0042$, $g_y = 2.0035$, $g_z = 2.0024$ corresponding to Klason pine lignin to $g_x = 2.0056$, $g_y = 2.0048$, $g_z = 2.0024$ for dioxane beech lignin [61,65]. The g-tensors of all such radicals are highly sensitive to the local electronic environment. Electrostatically positive interactions, such as cations and hydrogen bonds, will

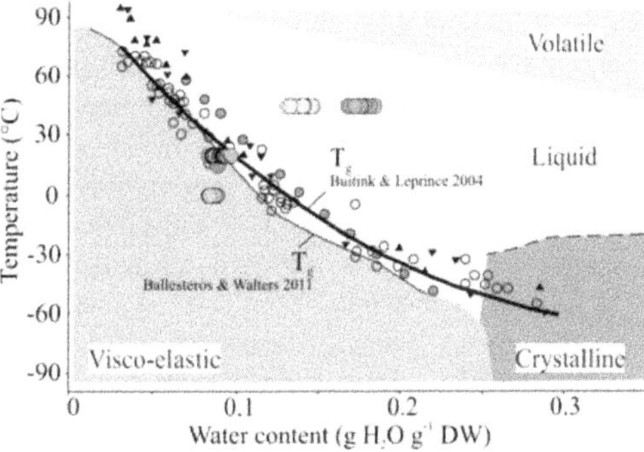

Figure 7. Phase diagram of cytoplasmic viscosity. Glass transition temperature (T_g), as described by the models of Buitink and Leprince [62] and Ballesteros and Walters [1], is a function of temperature and seed water content. The thin black line is redrawn from by Ballesteros and Walters [1] who used data for the seeds of various plant species. The bold black line shows data by Buitink and Leprince [42] for the seeds of pea (downward triangles), soybean (upwards triangles), maize (closed circles) and garden bean (open circles). According to these models, the cytoplasm of the wheat seeds used in the present study would have been in a liquid state when subjected to controlled deterioration at 14 and 18% seed MC (CD14 and CD18; yellow and red data points) at 44 °C, whereas that of freshly harvested seeds (FH, green data points), seeds stored in cold storage (CS, blue data points) at 0 °C and ambient storage (AS, purple data points) at 20 °C was in a visco-elastic state. Figure adapted from Ballesteros and Walters [1] and Buitink and Leprince [62].

tend to shift g_x and g_y downwards. With a value of 2.0022, the g_z values are much less affected. The pH at which lignin is extracted changes the polar/charged environment and significantly affects the g-values. The radical found in humic acid at pH 12 had g-values that ranged from $g_x = 2.0058$, $g_y = 2.0056$, $g_z = 2.0023$ to $g_x = 2.0055$, $g_y = 2.0052$, $g_z = 2.0023$. These values closely match with those of a stable radical found in gallic acid solution above pH 12 (3,4,5-trihydroxybenzoic acid) [65]. These studies strongly suggest that the seed radical that we detected has a similar structure and probably arises from lignin. If it is assumed that the seed radicals exist in a much less polar/charged environment, then the seed radical g-values would be entirely consistent with this assignment.

Semiquinone radicals are formed by oxidation of phenols during lignin degradation and tend to be neither reactive nor toxic. They can even function as antioxidants by receiving electrons, i.e. from $O_2^{\bullet -}$ [66]. However, they can also become problematic by donating electrons to O_2 and forming $O_2^{\bullet -}$ [67]. In seeds, most organic free radicals are located at the seed surface [68,69]. Pericarp and embryo of wheat seeds are more exposed to light and are, therefore, more prone to ROS formation, than the endosperm. In addition, photosynthetic activity in the pericarp ceases only at late stages of seed maturation [70] and quinones are secondary acceptors in photosystem I and II [59]. At this stage, mitochondria are partly impaired by maturation drying, leading to imbalances in the mitochondrial electron transport chain, and lignin to be hydrated at the molecular level, which consequently forms stable organic radicals [71]. Under CS, at 0.09 g $H_2O \cdot g^{-1}$ DW and 0 °C, high seed viability was reflected by a lower level of organic free radicals (Figure 6(b)), indicating that the accumulation of stable radicals is temperature dependent. In conclusion, upon transition of the seed cytoplasm to the glassy state, stable organic radicals are formed. Whether these radicals or other simultaneously occurring oxidative processes contribute to viability loss has still to be elucidated.

Nonenzymatic degradation processes result in a temperature-dependent loss of GSH

When the cytoplasm of seeds is highly viscous, at 0.1 g $H_2O \cdot g^{-1}$ DW in AS, CS, FH, enzymatic reactivity ceases [72]. Therefore, nonenzymatic rather than enzymatic oxidations, also stimulated by semiquinones [73], are assumed to occur [32]. As ascorbate is absent in dry seeds, GSH is the major redox buffer and water-soluble antioxidant available to scavenge any ROS or free radicals [19]. In this process, GSH becomes oxidised to the glutathiyl radical itself [74], which may undergo radical-radical coupling with coexisting radicals, such as other glutathiyl radicals, producing GSSG, or lipid peroxyl radicals [75]. Accordingly, GSSG increased after AS and CS to comparable amounts previously described for barley seeds [13,76]. In CS, contents of GSH were higher, suggesting that other biochemical reactions with the potential to cause a decline in GSSG or GSH in AS, were slowed down (Figure 2(b–d)), consistent with lowered molecular mobility as the cytoplasmic glass is strengthened at lower temperatures [1,5]. Also, other radical scavengers, such as tocochromanols and enzymatic antioxidants such as catalase (CAT) and glutathione reductase (GR), are more abundant under cold storage conditions [76,77]. In barley and maize, tocopherols [78] and GSH [79] accumulate during seed development in the embryo, which requires specific protection [46,80]. Lipids, in total about 2.2%, are mostly located in the embryo. During AS and CS, glycerides, fatty acids, phospholipids, lysophospholipids and galactolipids undergo oxidation and hydrolysis [32], most likely in hydrated/lipid zones or during imbibition [25], leading to membrane damage and leakage [76]. In conclusion, in AS, viability loss is accompanied by an accumulation of oxidised products together with a depleted pool of antioxidants, resulting in strongly oxidising conditions, as seen by an $E_{GSSG/2GSH}$ of −110 mV (Figure 2(e)). By contrast, the low temperatures during CS restrict molecular mobility in the cytoplasm, decelerating biochemical reactions, which helps in maintaining seed viability (Figure 2).

Enzymatic processes in weakening glasses and liquids

At 44 °C and 14 or 18% seed MC, when the cytoplasm becomes more fluid, molecular mobility and the free volume between molecules increases [1] and seed deterioration accelerates. Under CD14 and CD18, most seeds completely lost viability within 32 and 7 days, respectively (Figure 2(a)). The absence of organic free radicals and a weak HFEPR signal (Figure 6) suggests that compounds other than stable free radicals are responsible for deteriorative processes. Semiquinones may quench the reactive molecules. Higher temperatures and seed MCs during CD stimulate respiration [81], and impaired electron transport chains may promote ROS production [71]. In aqueous solutions, H_2O_2 can be rapidly transformed via the Fenton reaction to

HO• [73]. The extremely short-lived HO• radical is prone to react with adjacent molecules, whereas H_2O_2 can move over longer distances [25], but requires transport systems, e.g. aquaporins [22]. This agrees with increased H_2O_2 levels during CD at high seed MC [82], which may have contributed to GSH oxidation. In contrast to AS and CS, seeds under CD14, and even more so under CD18, contained less GSSG, pointing towards enzymatic conversion of GSSG to GSH or consumption of GSSG by other enzymatic processes such as S-glutathionylation. Viability loss and the shift of $E_{GSSG/2GSH}$ to highly oxidising conditions (Figure 2(e): CD14: −117 mV; CD18: −116 mV) within days also occurred in barley seeds [13], which could be explained by a massive burst of ROS counteracting the recycling of antioxidants. In pea seed, prior to any viability loss, degradation of stored mRNA was observed together with a lower expression of antioxidant genes and an altered expression of genes associated with programmed cell death (PCD) [83]. Monitoring different MCs during ageing at 35 °C in sunflower showed that PCD was stimulated in embryonic axis especially, and at higher MCs [84]. In addition, the antioxidant enzymes CAT, superoxide dismutase (SOD) and GR were depleted during accelerated seed ageing at 45 °C and 100% RH, corresponding roughly to 28% MC [82,85]. However, at 30 °C and 75% RH (~14.4% MC), these correlations were absent [82]. Taken together, we conclude that the increased cytoplasmic fluidity supported accelerated ROS production at CD14 and CD18 in the mitochondrial electron transport chains, also stimulating PCD and thereby, rapid seed viability loss.

Seed viability loss is accompanied by acidification

The concept of pH in "dry" systems is an intriguing one. As outlined above, even a glassy cytoplasm is a highly viscous liquid, and it contains biomolecules and also protons that determine its pH, albeit with strongly decelerated molecular mobility. The pH in extracts of seed stored in CS, AS and CD14 decreased in parallel with viability (Figure 4), and the questions arises as to which mechanisms could lead to increases in seed proton concentration, and thereby, acidification of seed extracts. In vegetative tissues, acidification of the cytoplasm was observed in response to abiotic stress factors such as temperature variations and anoxia [47,49]. In leaves, tissue dehydration can result in an increase of the apoplastic pH, triggering ABA release from mesophyll cells [50]. An ABA increase in guard cells activates the gluconeogenic conversion of malate into starch, and alkalinises the cytosol, which is an important signal for stomatal closure [47,86]. Similarly, embryo acidification during seed dormancy release was viewed as an indicator of the termination of developmental arrest [51]. Dry seed ageing is also accompanied by membrane damage [76], which may alter proton concentrations. Plant cell walls act like an ion exchanger, containing high concentrations of uronic acid and cations in the cell wall space. If cation activity is increased or if membranes are damaged, protons are forced from their binding sites [47,87]. Due to the inability of dry seeds (AS, CS) to repair membranes by replacing damaged membrane proteins and lipid constituents, cell wall damage is accompanied by H^+ leakage and severe cytoplasmic acidosis [47]. This is in agreement with a drop in pH shown for *Brassica campestris* seeds in parallel with the decrease of seed viability and respiration rate during storage under AS for 4 years [88]. For barley seeds stored at 6% seed MC and CD at 13 and 18%, indirect evidence for cellular acidification was presented by modifying the pH (between pH 7.3 and 6.8) used for the Nernst equation [13]. However, ageing at CD18 did not change seed extract pH, indicating that protons were neither produced nor translocated by plasma membrane enzymes, i.e. H^+-ATPase. Sealed aluminium bags used for CD [11] may produce hypoxic conditions, which may lead to a ROS burst upon reoxygenation. However, pH regulation is highly complex and natural buffers, i.e. phosphate compounds and several organic acids, might also be involved. In addition, the pH of extracts represents a pooled parameter of mostly cytoplasmic and other cellular compartments that range from pH 4–8 [49]. Unfortunately, measuring the pH in dry seeds is challenging as pH electrodes cannot be used and other methods such as nuclear magnetic resonance (NMR) or EPR require the seed to be imbibed, which greatly changes seed metabolism. Considering the available literature discussed above, damage to membranes and cell walls appear to be the most likely cause for the acidification of the extracts.

Summary and conclusions

Seed MC and storage temperature influence cytoplasmic viscosity and determine biochemical reactions during seed storage. Under CS and AS, the cytoplasm is highly viscous and seed viability loss is accompanied by the accumulation of organic free radicals. During CD14 and CD18, the cytoplasm becomes more fluid and enzymatic activities are more likely, stable free radicals disappear and viability loss cannot be attributed to them. However, ROS may contribute to viability loss. At CD18, the absence of acidification of the cytoplasm points towards a complex mechanism that buffers pH and controls viability loss. Furthermore, the mechanisms that cause viability loss under CD appear to be, at least in part, different from those that lead to seed viability loss under seed bank conditions.

Acknowledgements

We gratefully acknowledge support by Sibylle Pistrick, Anita Winger, Annett Marlow, Stefanie Thumm, Peter Schreiber and Michael Grau for excellent technical support and Gerald Kastberger and Anja Krieger-Liszkay for critical comments on the figures and the manuscript.

Disclosure statement

No potential conflict of interest was reported by the authors.

Funding

This work was supported by the EU (FP7 grant 311840 "EcoSeed" (Impacts of Environmental Conditions on Seed Quality)) and the French Infrastructure for Integrated Structural Biology (FRISBI) ANR-10-INBS-05.

ORCID

Manuela Nagel http://orcid.org/0000-0003-0396-0333
Charlotte E. Seal http://orcid.org/0000-0002-9329-9325
Louise Colville http://orcid.org/0000-0003-4626-4522
Sun Un http://orcid.org/0000-0001-7032-2433
Hugh W. Pritchard http://orcid.org/0000-0002-2487-6475
Andreas Börner http://orcid.org/0000-0003-3301-9026
Ilse Kranner http://orcid.org/0000-0003-4959-9109

References

[1] Ballesteros D, Walters C. Detailed characterization of mechanical properties and molecular mobility within dry seed glasses: relevance to the physiology of dry biological systems. Plant J. 2011;68(4):607–619.
[2] Leprince O, Pellizzaro A, Berriri S, et al. Late seed maturation: drying without dying. J Exp Bot. 2017;68(4): 827–841.
[3] Roberts EH. Viability of seeds. London: Chapman & Hall; 1972.
[4] Sano N, Rajjou L, North HM, et al. Staying alive: molecular aspects of seed longevity. Plant Cell Physiol. 2016;57(4):660–674.
[5] Walters C, Ballesteros D, Vertucci VA. Structural mechanics of seed deterioration: standing the test of time. Plant Sci. 2010;179(6):565–573.
[6] Roberts EH. The viability of cereal seed in relation to temperature and moisture. Ann Bot Lond. 1960; 24(1):12–31.
[7] FAO GenBank. standards for plant genetic resources for food and agriculture. Rome: Food and Agriculture Organization of the United Nations; 2014.
[8] Walters C, Wheeler LM, Grotenhuis JM. Longevity of seeds stored in a GeneBank: species characteristics. Seed Sci Res. 2005;15(1):1–20.
[9] Nagel M, Börner A. The longevity of crop seeds stored under ambient conditions. Seed Sci Res. 2010;20(1):1–12.
[10] Priestley DA, Cullinan VI, Wolfe J. Differences in seed longevity at the species level. Plant Cell Environ. 1985; 8(8):557–562.
[11] Hampton JG, TeKrony DM. Handbook of vigour test methods. Zürich: International Seed Testing Association; 1995.
[12] Hay FR, Adams J, Manger K, et al. The use of non-saturated lithium chloride solutions for experimental control of seed water content. Seed Sci Technol. 2008; 36(3):737–746.
[13] Nagel M, Kranner I, Neumann K, et al. Genome-wide association mapping and biochemical markers reveal that seed ageing and longevity are intricately affected by genetic background and developmental and environmental conditions in barley. Plant Cell Environ. 2015; 38(6):1011–1022.
[14] Kong Q, Mao PS, Yu XD, et al. Physiological changes in oat seeds aged at different moisture contents. Seed Sci Technol. 2014;42(2):190–201.
[15] Morscher F, Kranner I, Arc E, et al. Glutathione redox state, tocochromanols, fatty acids, antioxidant enzymes and protein carbonylation in sunflower seed embryos associated with after-ripening and ageing. Ann Bot. 2015;116(4):669–678.
[16] Rajjou L, Lovigny Y, Groot SPC, et al. Proteome-wide characterization of seed aging in Arabidopsis: a comparison between artificial and natural aging protocols. Plant Physiol. 2008;148(1):620–641.
[17] Schwember AR, Bradford KJ. Quantitative trait loci associated with longevity of lettuce seeds under conventional and controlled deterioration storage conditions. J Exp Bot. 2010;61(15):4423–4436.
[18] Harman D. Aging: a theory based on free radical and radiation chemistry. J Gerontol. 1956;11(3):298–300.
[19] Kranner I, Minibayeva FV, Beckett RP, et al. What is stress? Concepts, definitions and applications in seed science. New Phytol. 2010;188(3):655–673.
[20] Bailly C, El-Maarouf-Bouteau H, Corbineau F. From intracellular signaling networks to cell death: the dual role of reactive oxygen species in seed physiology. C R Biol. 2008;331(10):806–814.
[21] Hendry GAF. Oxygen, free radical processes and seed longevity. Seed Sci Res. 1993;3(3):141–153.
[22] Waszczak C, Carmody M, Kangasjärvi J. Reactive oxygen species in plant signaling. Annu Rev Plant Biol. 2018;69:209–236.
[23] Moller IM. Plant mitochondria and oxidative stress: electron transport, NADPH turnover, and metabolism of reactive oxygen species. Annu Rev Plant Physiol Plant Mol Biol. 2001;52:561–591.

[24] Apel K, Hirt H. Reactive oxygen species: metabolism, oxidative stress, and signal transduction. Annu Rev Plant Biol. 2004;55:373–399.

[25] Jeevan Kumar SP, Rajendra Prasad S, Banerjee R, et al. Seed birth to death: dual functions of reactive oxygen species in seed physiology. Ann Bot. 2015;116(4): 663–668.

[26] Foyer CH, Noctor G. Stress-triggered redox signalling: what's in pROSpect? Plant Cell Environ. 2016;39(5): 951–964.

[27] Foyer CH, Noctor G. Redox homeostasis and antioxidant signaling: a metabolic interface between stress perception and physiological responses. Plant Cell. 2005;17(7):1866–1875.

[28] Viña J, Borras C, Abdelaziz KM, et al. The free radical theory of aging revisited: the cell signaling disruption theory of aging. Antioxid Redox Signal. 2013;19(8): 779–787.

[29] Møller IM, Jensen PE, Hansson A. Oxidative modifications to cellular components in plants. Annu Rev Plant Biol. 2007;58:459–481.

[30] Palma K, Kermode AR. Metabolism of hydrogen peroxide during reserve mobilization and programmed cell death of barley (*Hordeum vulgare* L.) aleurone layer cells. Free Radic Biol Med. 2003;35(10):1261–1270.

[31] Sarath G, Hou G, Baird LM, et al. Reactive oxygen species, ABA and nitric oxide interactions on the germination of warm-season C4-grasses. Planta. 2007;226(3): 697–708.

[32] Riewe D, Wiebach J, Altmann T. Structure annotation and quantification of wheat seed oxidized lipids by high-resolution LC-MS/MS. Plant Physiol. 2017;175(2): 600–618.

[33] Mira S, González-Benito ME, Hill LM, et al. Characterization of volatile production during storage of lettuce (*Lactuca sativa*) seed. J Exp Bot. 2010;61(14): 3915–3924.

[34] Sakihama Y, Cohen MF, Grace SC, et al. Plant phenolic antioxidant and prooxidant activities: phenolics-induced oxidative damage mediated by metals in plants. Toxicology. 2002;177(1):67–80.

[35] Leprince O, Deltour R, Thorpe PC, et al. The role of free radicals and radical processing systems in loss of desiccation tolerance in germinating maize (*Zea mays* L.). New Phytol. 1990;116(4):573–580.

[36] Priestley DA, McBride MB, Leopold C. Tocopherol and organic free radical levels in soybean seeds during natural and accelerated aging. Plant Physiol. 1980; 66(4):715–719.

[37] Kranner I, Grill D. Significance of thiol-disulfide exchange in resting stages of plant development. Bot Acta. 1996;109(1):8–14.

[38] Dixon DP, Skipsey M, Grundy NM, et al. Stress-induced protein S-glutathionylation in Arabidopsis. Plant Physiol. 2005;138(4):2233–2244.

[39] Kranner I, Birtic S. A modulating role for antioxidants in desiccation tolerance. Integr Comp Biol. 2005;45(5): 734–740.

[40] Colville L, Kranner I. Desiccation tolerant plants as model systems to study redox regulation of protein thiols. Plant Growth Regul. 2010;62(3):241–255.

[41] Zagorchev L, Seal CE, Kranner I, et al. Redox state of low-molecular-weight thiols and disulphides during somatic embryogenesis of salt-treated suspension cultures of *Dactylis glomerata* L. Free Radic Res. 2012; 46(5):656–664.

[42] Schafer FQ, Buettner GR. Redox environment of the cell as viewed through the redox state of the glutathione disulfide/glutathione couple. Free Radic Biol Med. 2001;30(11):1191–1212.

[43] Kranner I, Birtić S, Anderson KM, et al. Glutathione half-cell reduction potential: a universal stress marker and modulator of programmed cell death? Free Radic Biol Med. 2006;40(12):2155–2165.

[44] Seal CE, Zammit R, Scott P, et al. Glutathione half-cell reduction potential and α-tocopherol as viability markers during the prolonged storage of *Suaeda maritima* seeds. Seed Sci Res. 2010;20(1):47–53.

[45] Roach T, Beckett RP, Minibayeva FV, et al. Extracellular superoxide production, viability and redox poise in response to desiccation in recalcitrant *Castanea sativa* seeds. Plant Cell Environ. 2010;33(1):59–75.

[46] Gerna D, Roach T, Stöggl W, et al. Changes in low-molecular-weight thiol-disulphide redox couples are part of bread wheat seed germination and early seedling growth. Free Radic Res. 2017;51(6):568–581.

[47] Felle HH. pH regulation in anoxic plants. Ann Bot. 2005;96(4):519–532.

[48] Felle HH. Short-term pH regulation in plants. Physiol Plant. 1988;74(3):583–591.

[49] Kurkdjian A, Guern J. Intracellular pH: measurement and importance in cell activity. Annu Rev Plant Physiol Plant Mol Biol. 1989;40(1):271–303.

[50] Hartung W, Radin JW, Hendrix DL. Abscisic acid movement into the apoplastic solution of water-stressed cotton leaves: role of apoplastic pH. Plant Physiol. 1988;86(3):908–913.

[51] Footitt S, Cohn MA. Seed dormancy in red rice. VIII. Embryo acidification during dormancy-breaking and subsequent germination. Plant Physiol. 1992;100(3): 1196–1202.

[52] Shewry PR. Wheat. J Exp Bot. 2009;60(6):1537–1553.

[53] Nasehzadeh M, Ellis RH. Wheat seed weight and quality differ temporally in sensitivity to warm or cool conditions during seed development and maturation. Ann Bot. 2017;120(3):479–493.

[54] Lehner A, Bailly C, Flechel B, et al. Changes in wheat seed germination ability, soluble carbohydrate and antioxidant enzyme activities in the embryo during the desiccation phase of maturation. J Cereal Sci. 2006;43(2):175–182.

[55] Bailly C, Kranner I. Analyses of reactive oxygen species and antioxidants in relation to seed longevity and germination. In: Kermode AR, editor. Seed Dormancy: Methods and Protocols, Methods in Molecular Biology. 2011/09/08 ed. Volume 773. New York, New York, United States: Humana Press; 2011:343–367.

[56] Kranner I, Grill D. Determination of glutathione and glutathione disulphide in lichens: a comparison of frequently used methods. Phytochem Anal. 1996;7(1): 24–28.

[57] Birtić S, Colville L, Pritchard HW, et al. Mathematically combined half-cell reduction potentials of low-

molecular-weight thiols as markers of seed ageing. Free Radic Res. 2011;45(9):1093–1102.
[58] Pfeifhofer HW. The response of Härtel's turbidity test is controlled by the acidity of the conifer needle extracts rather than by the availability of Ca^{2+} ions. Can J For Res. 1999;29(1):33–39.
[59] Un S, Dorlet P, Rutherford AW. A high-field EPR tour of radicals in photosystems I and II. Appl Magn Reson. 2001;21(3–4):341–361.
[60] Ellis RH, Roberts EH. Improved equations for the prediction of seed longevity. Ann Bot Lond. 1980; 45(1):13–30.
[61] Bährle C, Nick TU, Bennati M, et al. High-field electron paramagnetic resonance and density functional theory study of stable organic radicals in lignin: influence of the extraction process, botanical origin, and protonation reactions on the radical g tensor. J Phys Chem A. 2015;119(24):6475–6482.
[62] Buitink J, Leprince O. Glass formation in plant anhydrobiotes: survival in the dry state. Cryobiology. 2004; 48(3):215–228.
[63] Harrington JF. Practical instructions and advice on seed storage. Proceedings of the International Seed Testing Association. 1963. p. 989–994.
[64] Walters C. Orthodoxy, recalcitrance and in-between: describing variation in seed storage characteristics using threshold responses to water loss. Planta. 2015; 242(2):397–406.
[65] Christoforidis KC, Un S, Deligiannakis Y. High-field 285 GHz electron paramagnetic resonance study of indigenous radicals of humic acids. J Phys Chem A. 2007;111(46):11860–11866.
[66] Valgimigli L, Amorati R, Fumo MG, et al. The unusual reaction of semiquinone radicals with molecular oxygen. J Org Chem. 2008;73(5):1830–1841.
[67] Siraki AG, Klotz LO, Kehrer JP. Free radicals and reactive oxygen species. In: McQueen CA, ed. Comprehensive toxicology. 3rd ed. Oxford: Elsevier; 2018. p. 262–294.
[68] Nakagawa K, Hara H. Investigation of radical locations in various sesame seeds by CW EPR and 9-GHz EPR imaging. Free Radic Res. 2015;49(1):1–6.
[69] Nakagawa K, Epel B. Locations of radical species in black pepper seeds investigated by CW EPR and 9GHz EPR imaging. Spectrochim Acta A Mol Biomol Spectrosc. 2014;131:342–346.
[70] Kong LA, Xie Y, Sun MZ, et al. Comparison of the photosynthetic characteristics in the pericarp and flag leaves during wheat (*Triticum aestivum* L.) caryopsis development. Photosynthetica. 2016;54(1): 40–46.
[71] Leprince O, Atherton NM, Deltour R, et al. The involvement of respiration in free radical processes during loss of desiccation tolerance in germinating *Zea mays* L. (an electron paramagnetic resonance study). Plant Physiol. 1994;104(4):1333–1339.
[72] Fernández-Marín B, Kranner I, San Sebastián M, et al. Evidence for the absence of enzymatic reactions in the glassy state. A case study of xanthophyll cycle pigments in the desiccation-tolerant moss *Syntrichia ruralis*. J Exp Bot. 2013;64(10):3033–3043.
[73] Hu Y-Z, Jiang L-J. Generation of semiquinone radical anion and reactive oxygen ($^1O_2$, $O_2^{\bullet-}$ and $^\bullet OH$) during the photosensitization of a water-soluble perylenequinone derivative. J Photochem Photobiol B Biol. 1996;33(1):51–59.
[74] Smirnoff N. Tocochromanols: rancid lipids, seed longevity, and beyond. P Natl Acad Sci USA. 2010; 107(42):17857–17858.
[75] Munné-Bosch S, Alegre L. The function of tocopherols and tocotrienols in plants. Crit Rev Plant Sci. 2002; 21(1):31–57.
[76] Roach T, Nagel M, Börner A, et al. Changes in tocochromanols and glutathione reveal differences in the mechanisms of seed ageing under seedbank conditions and controlled deterioration in barley. Environ Exp Bot. 2018;156:8–15.
[77] Spanò C, Bottega S, Lorenzi R, et al. Ageing in embryos from wheat grains stored at different temperatures: oxidative stress and antioxidant response. Functional Plant Biol. 2011;38(7):624–631.
[78] Falk J, Krahnstöver A, van der Kooij TAW, et al. Tocopherol and tocotrienol accumulation during development of caryopses from barley (*Hordeum vulgare* L.). Phytochemistry. 2004;65(22):2977–2985.
[79] Rauser WE, Schupp R, Rennenberg H. Cysteine, γ-glutamylcysteine, and glutathione levels in maize seedlings: distribution and translocation in normal and cadmium-exposed plants. Plant Physiol. 1991;97(1): 128–138.
[80] Rhazi L, Cazalis R, Lemelin E, et al. Changes in the glutathione thiol-disulfide status during wheat grain development. Plant Physiol Biochem. 2003;41(10): 895–902.
[81] Bewley JD. Seed germination and dormancy. Plant Cell. 1997;9(7):1055–1066.
[82] Lehner A, Mamadou N, Poels P, et al. Changes in soluble carbohydrates, lipid peroxidation and antioxidant enzyme activities in the embryo during ageing in wheat grains. J Cereal Sci. 2008;47(3):555–565.
[83] Chen HY, Osuna D, Colville L, et al. Transcriptome-wide mapping of pea seed ageing reveals a pivotal role for genes related to oxidative stress and programmed cell death. PLOS ONE. 2013;8(10):e78471.
[84] El-Maarouf Bouteau H, Mazuy C, Corbineau F, et al. DNA alteration and programmed cell death during ageing of sunflower seed. J Exp Bot. 2011;62(14): 5003–5011.
[85] Kibinza S, Bazin J, Bailly C, et al. Catalase is a key enzyme in seed recovery from ageing during priming. Plant Sci. 2011;181(3):309–315.
[86] Planes MD, Niñoles R, Rubio L, et al. A mechanism of growth inhibition by abscisic acid in germinating seeds of *Arabidopsis thaliana* based on inhibition of plasma membrane H^+-ATPase and decreased cytosolic pH, K^+, and anions. J Exp Bot. 2015;66(3):813–825.
[87] Grignon C, Sentenac H. pH and ionic conditions in the apoplast. Annu Rev Plant Physiol Plant Mol Biol. 1991; 42(1):103–128.
[88] Verma SS, Verma U, Tomer RPS. Studies on seed quality parameters in deteriorating seeds in Brassica (*Brassica campestris*). Seed Sci Technol. 2003;31(2): 389–396.

3.5 PAPER 5:

Comparative physiology and proteomics of two wheat genotypes differing in seed storage tolerance

by

Xiuling Chen, Guangkun Yin, Andreas Börner, Xia Xin, Juanjuan He,

Manuela Nagel, Xu Liu and Xinxiong Lu

Published in

Plant Physiology and Biochemistry (2018) 130, 455-463

https://doi.org/10.1016/j.plaphy.2018.07.022

To view supplementary material for this article, please visit

https://doi.org/10.1016/j.plaphy.2018.07.022

Plant Physiology and Biochemistry 130 (2018) 455–463

Contents lists available at ScienceDirect

Plant Physiology and Biochemistry

journal homepage: www.elsevier.com/locate/plaphy

Research article

Comparative physiology and proteomics of two wheat genotypes differing in seed storage tolerance

Xiuling Chen[a,1], Guangkun Yin[a,1], Andreas Börner[b], Xia Xin[a], Juanjuan He[a], Manuela Nagel[b], Xu Liu[a,**], Xinxiong Lu[a,*]

[a] National Crop Genebank, Institute of Crop Sciences, Chinese Academy of Agricultural Sciences, Beijing 100081, China

[b] Genebank Department, Leibniz Institute of Plant Genetics and Crop Plant Research, Corrensstr. 3, 06466, Stadt Seeland, OT Gatersleben, Germany

ARTICLE INFO

Keywords:
Genebank
Longevity
Proteomics
Storability
Wheat

ABSTRACT

The longevity of seeds stored in Genebank is based on their storability. However, the mechanism of seed storability is largely unknown. In previous studies, accelerated ageing treatments were always applied for rapidly acquiring different seed viabilities, which could not reflect the actual situation during seed storage, especially for the seed stored in Genebank. In this study, two wheat genotypes (accession TRI_23248 and TRI_10230) were supplied by IPK-Gatersleben Genebank, Germany, where they were stored for 10 years in the long-term storage (−18 °C) and at ambient conditions (20 °C) The comparison of viability of those seed after this storage period, identified TRI_23248 as storage tolerant (ST) and TRI_10230 as storage sensitive (SS). The abundance patterns of proteins in these seeds identified 93 protein spots in the ST and 105 spots in the SS seeds that were markedly changed; their functions were mainly associated with disease or defense, protein destination and storage, energy, and other. The ST seeds possessed a stronger ability in activating the defense system against oxidative damage, utilizing storage proteins for germination, and maintaining energy metabolism for ATP supply. These results provided novel insights into the mechanism of seed storability, which can facilitate the comprehensive understanding of seed longevity.

1. Introduction

Wheat (*Triticum aestivum* L.) is one of the most important cereals cultivated worldwide and its collection and conservation are the vital components of crop genetic diversity preservation. According to the Food and Agriculture Organization (FAO) statistics in 2010, more than 850,000 accessions of wheat are preserved in global Genebanks (FAO, 2010). As of 2017, a total of 49,000 accessions are stored at −18 °C and 48,000 accessions are stored at 4 °C at the National Genebank of China. The average germination rate of 427 wheat seed accessions that had been stored for 43 years in Genebank of the United States decreased from 93% to 73% (Walters et al., 2005). We monitored 2500 accessions that had been stored for over 20 years in the National Genebank of China and found that the average germination rate decreased from 97% to 94% (Xin et al., 2011). The average germination rate was frequently used to better evaluate viability of seeds in Genebank as viability of individual seed accessions varies greatly. For example, the germination rate of wheat seed (accession ZM005212) reduced from 100% to 27% after 20 years of storage, that of cotton seed (accession 00001085) decreased from 96% to 24% (Xin et al., 2011). Maintaining seed viability is the ultimate aim of Genebank management. Seed longevity is controlled by multiple factors including seed storability, storage conditions, etc. Genebanks follow storage standards and management procedures stated in the Genebank Standard (FAO, 2014). Therefore, under the same conditions of harvest, storage, and conservation in Genebank, the difference in individual longevity is due to the

Abbreviation: AcCoA, Acetyl-CoA; ADH1D, Alcohol dehydrogenase; ALDH, Aldehyde dehydrogenase; APX, Ascorbate peroxidase; AsA, Ascorbate reduced form; CAT, Catalase; DDAPs, down-regulated differential abundance proteins; DHA, Dehydroascorbate; DHAR, Dehydroascorbate reductase; EMP, Glycolytic pathway; GR, Glutathione reductase; GSH, Glutathione reduced form; GSSG, Glutathione oxidized form; GST, Glutathione-S-transferase; MDA, Malondialdehyde; MDH, Malate dehydrogenase; MDHA, Monodehydroascorbate; MDHAR, Monodehydroascorbate reductase; RCS, Reactive carbonyl species; ROS, Reactive oxygen species; SOD, Superoxide dismutase; SS, Storage-sensitive; ST, Storage-tolerant; TCA, Tricarboxylic acid; UDAPs, up-regulated differential abundance proteins

* Corresponding author.
** Corresponding author.
E-mail address: luxinxiong@caas.cn (X. Lu).

[1] These authors contributed equally to this work.

https://doi.org/10.1016/j.plaphy.2018.07.022
Received 16 April 2018; Received in revised form 13 July 2018; Accepted 18 July 2018
Available online 20 July 2018
0981-9428/

differences in seed storability. Thus, the storability is the most important factor for the longevity of seeds stored in Genebank. Furthermore, illustrating the mechanism of seed storability might provide the monitoring and early warning indicators for the seed-safe conservation in Genebank.

Seed storability is a complex trait. Several studies have investigated the genetics and molecular basis of seed storability and identified quantitative trait loci (QTLs) responsible for controlling seed storability in wheat (Landjeva et al., 2010) and rice (Jiang et al., 2011). For example, five QTLs that were identified to control seed longevity in wheat were mapped to chromosomes 1D and 5D (Landjeva et al., 2010), and seven QTLs in Milyang 23/Tong 88-7 and Dasanbyeo/TR22183 recombinant inbred lines of rice were identified to affect seed storability (Jiang et al., 2011). These QTLs could be utilized to further elucidate the mechanism of seed storability and longevity.

Many studies have illustrated the reasons for loss of longevity during storage. During seed storage, oxidative and mitochondrial damage occurs. The ascorbate and glutathione (AsA-GSH) cycle plays extremely important roles in scavenging reactive oxygen species (ROS). During seed storage, the activity of the AsA-GSH cycle is reduced resulting in ROS accumulation (Xin et al., 2011, 2014) and carbonylation of proteins in seeds (Nguyen et al., 2015; Yin et al., 2017). Besides, the activity of mitochondria, the organelles that supply energy for seed germination, is significantly decreased during storage, thereby inhibiting seed germination (Xin et al., 2014; Yin et al., 2016). Thus, the level of antioxidants and the level of energy supply are key factors in the regulation of seed longevity. Proteomics has been utilized to investigate protein expression patterns during seed storage (Wang et al., 2015; Yin et al., 2017). The expression of proteins related to disease defense and energy is largely altered during seed storage. However, the effect of such changes on storage tolerance and sensitivity of seeds is largely unknown. Studies have utilized accelerated aging treatments, such as high temperature and high humidity, to prepare seeds with different viability. Nevertheless, those treatments could not replicate the actual seed status during storage, especially for the seed stored in Genebank under reduced temperature and moisture conditions. Thus, further studies are necessary to investigate the mechanisms of seed storability under such conditions.

In the present study, we assessed seed storage tolerance in two wheat genotypes that were stored under cold (−18 °C) and ambient (20 °C) conditions for 10 years. Furthermore, proteomics was utilized to compare protein expression patterns between storage-tolerant and storage-sensitive seeds. The results might provide novel insights into the mechanism of seed storability and longevity.

2. Materials and methods

2.1. *Plant material and treatment*

Two wheat genotypes were supplied by IPK Gatersleben Genebank, Germany, with accession numbers TRI_23248 and TRI_10230. Seeds were harvested in 2004 and stored in sealed glass jars in the cold rooms of the Genebank at −18 ± 2 °C. Surplus seeds of the same harvest were stored under ambient conditions (20.3 ± 2.4 °C). Germination of the seed used for the experiments was tested after harvest in 2004 and again in 2014 according to the rules of the International Seed Testing Association.

2.2. *Measurement of malondialdehyde (MDA) content and relative conductivity*

MDA content was determined in 24 h imbibed wheat embryos to evaluate lipid peroxidation level as described by Heath and Packer (1968). Briefly, the embryo was homogenized at 4 °C in trichloroacetic acid buffer (5% w/v). The homogenate was centrifuged for 5 min at 20,000 × *g* and 0.5 mL of the supernatant was blended with thiobarbituric acid solution (0.5% w/v) and then boiled for 15 min at 100 °C. The reaction solution was centrifuged for 10 min at 3000 × *g* and 25 °C. Finally, the absorbance of the supernatant was immediately detected at 450, 532, and 600 nm.

Ten embryos of wheat seeds were soaked in 15 mL of Milli-Q water (Millipore, Milford, MA, USA) for 24 h at 25 °C. The electric conductivity of the embryo leachate was determined at 0 h and 24 h with an electrical conductivity meter (Delta 326; METTLER-TOLEDO, Greifensee, Switzerland). The absolute conductivity was obtained after soaking the embryos in a water bath at 100 °C for 30 min. Relative conductivity was presented as the percentage (%) of absolute conductivity.

2.3. *Quantitative reverse-transcription PCR assay (RT-qPCR)*

Total RNA was isolated from 24 h imbibed wheat embryos with an RNAprep pure plant kit (Tiangen, Beijing, China) following manufacturer's instructions and converted to first-strand cDNA using a FastQuant RT kit (Tiangen, China). Gene expression was quantified by RT-qPCR using Tiangen SuperReal premix plus (SYBR Green) in a 20 μL reaction system comprised of 10 μL 2 × SuperReal PreMix Plus, 10 μM primer, and about 200 ng cDNA. The reactions were run in a Roche LightCycler 480 (Applied Biosystems, Foster City, CA, USA). The thermal cycling program consisted of one step at 95 °C for 15 min, followed by 40 cycles at 95 °C for 10 s, 60 °C for 30 s, and 72 °C for 20 s. The coding sequence of the target gene was searched in the NCBI database, and gene-specific primers were designed by using the Primer express software (Applied Biosystems, USA) (Supplementary Table 1). The wheat *actin* (AB18199.1) gene was used as an internal control and relative quantification of gene expression was calculated with the $2^{-\Delta\Delta Ct}$ method. All samples were assayed in three biological replicates.

2.4. *Ascorbic acid and glutathione content*

Contents of the ascorbic acid and glutathione were analyzed by ultra-high performance liquid chromatography (WGQ1S) equipped with BEH G8. Briefly, dry powder (about 30 mg) was suspended in extraction buffer (500 μL, 10% (w/v) metaphosphoric acid, 2 mM EDTA) and incubated on ice for 5 min, followed by centrifugation for 5 min at 16,000 × g. The test solution of total ascorbic acid was obtained by adding 0.5 M Tris-HCl (97 μL) and 1 M DTT (3 μL), incubating it at 25 °C for 5 min, and adding 0.2 M H_2SO_4 (100 μL). The test solution of ASA, GSH, and glutathione oxidized form GSSG was acquired by mixing supernatant with equal volume of 0.125 M Tris-HCl. All the mixtures were filtered through a membrane filter (0.22 μm Whatman Germany). The separation was conducted at 0.2 mL/min flow rate of mobile phase (20 mM, potassium phosphate buffer), 2 μL injection volume of sample and, column temperature of 30 °C. Standard samples (ASA, GSH, and GSSG: Sigma- Aldrich, St. Louis, MO, USA) were used to identify ASA and GSH molecules in wheat embryo by comparing retention times of each molecule in standard vs. test solution; their contents were calculated from standard curve (concentration: 5, 15, 25, 50 nM).

2.5. *Proteomic analysis*

Total protein was extracted from 24 h imbibed wheat embryos in 50 mM Tris-HCl buffer (pH 7.5) containing 0.8 M sucrose, 5 mM EDTA, and 65 mM DTT. Subsequently, the obtained suspensions were fractionated with Tris-saturated phenol by incubation on a shaker at 4 °C for 15 min according to the modified protocol. Protein pellets were collected by precipitation in 0.1 M ammonium acetate/methanol at −20 °C for 3 h. After washing with ice-cold ammonium acetate/methanol and acetone separately, the pellets were air-dried and dissolved in rehydration buffer (7 M urea, 2 M thiourea, 2% CHAPS, 1% DTT, and 0.2% Bio-Lyte [pH 5–8]) by incubation on a shaker for 1 h at 20 °C and centrifugation at 40,000 × g.

Total proteins were separated by two-dimensional electrophoresis as described by Yin et al. (2016). Each protein sample was adjusted to 1 mg protein, loaded onto an immobilized linear pH gradient strip (17 cm, pH 5–8; Bio-Rad, Hercules, CA, USA) and fully absorbed for 1.5 h at 25 °C. The first dimension isoelectric focusing was conducted on a flatbed electrophoresis unit (PROTEAN IEF Cell; Bio-Rad, USA) as follows: rehydration for 12 h, 0 V–150 V in 15 min, 150 V to 1000 V in 1 h, 1000 V to 8000 V in 5 h, and 8000 V until 60 kVh. After isoelectric focusing, strips were equilibrated for 15 min, first in buffer A containing 6 M urea, 20% (v/v) glycerol, 2% (w/v) sodium dodecyl sulfate (SDS), 375 mM Tris-HCl (pH 8.8), and 2% (w/v) DTT; and then, in buffer B with 2.5% (w/v) iodacetamide instead of DTT. Subsequently, equilibrated strips were transferred to homogenous 12% (v/v) polyacrylamide gel to complete the second dimension SDS polyacrylamide gel electrophoresis, which ran at 200 V for about 6 h with circulating cooling buffer containing 25 mM Tris (pH 8.3), 195 mM glycine, and 0.1% (w/v) SDS. Ultimately, gels were dyed overnight with Coomassie Brilliant Blue G250 solution (0.125% w/v of G250 in 3:1:2 v/v methanol: acetic acid: ultrapure water; Sigma-Aldrich, USA) and destained with a 3:1:2 (v/v) methanol: acetic acid: ultrapure water solution with several changes, until a colorless background was achieved. All gel images were acquired with a scanner, and images were analyzed with PDQ$_{UEST}$ 8.0.1 (Bio-Rad, USA). Firstly, the spot abundance expressions were compared the difference in the TRI_23248 and TRI_10230 seeds which were stored in − 18 °C and 20 °C conditions for 10 years, and then compared the difference between the differences in each genotype seeds. Only spots that consistently changed more than 1.5-fold in all three independent sets of gels were finally selected and identified by mass spectrometry (MS).

Protein digestion was performed as follows. Excised gel particles were cleaned with acetonitrile (50% v/v) in ammonium bicarbonate (25 mM) to remove Coomassie Brilliant Blue, and then dehydrated for 10 min. After drying under vacuum, the samples were incubated in 10 µL of trypsin solution (MERCK, Darmstadt, Germany) for 30 min. Finally they were overlaid with 10 µL of ammonium bicarbonate (25 mM) to digest at 37 °C for 14 h.

Two hundred and ten protein spots were identified by a matrix-assisted laser desorption/ionization time-of-flight (MALDI–TOF)/TOF mass spectrometer (UltrafleXtreme; Bruker Daltonics, Billerica, MA, USA) at the Institute of Botany, The Chinese Academy of Science. Peptides that were resuspended in acetonitrile were spotted onto a MALDI target plate and covered with 1 µL of matrix solution (1 mg/mL α-cyano-4-hydroxycinnamic acid in 70% acetonitrile containing 0.1% trifluoroacetic acid) to be loaded on the instrument. MS/MS fragmentation spectra were obtained by 1500 laser shots after selecting the 15 strongest peaks of MS spectra that were acquired with 400 laser shots per spectrum. MS data were uploaded to Mascot (Matrix Science, London, UK) public web site (http://www.matrixscience.com) and searched against NCBI nr protein databases (version 20,130,501 comprising 25,010,123 sequences, 8,625,376,125 residues and version 20,150,104 comprising 54,183,042 sequences, 19,531,459,180 residues). Search parameters included proteolytic enzyme (trypsin), max missed cleavages (1), fixed modifications (carbamidomethyl), variable modifications (oxidation), peptide mass tolerance (100 ppm), and fragment mass tolerance (0.5 Da). Proteins were identified by individual ions scores, which indicate identity or extensive homology ($p < 0.05$).

2.6. Statistical analyses

Mean separations were performed with one-way analysis of variance and differences at $p < 0.05$ were considered significant. Results were pooled across repeated experiments. Correlation analysis (bivariate correlation, two-tailed test) was performed using SPSS (IBM, Armonk, NY, USA).

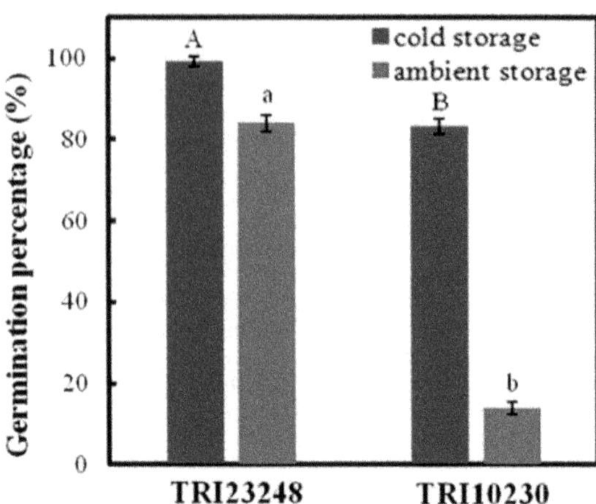

Fig. 1. Germination percentage of two wheat seed genotypes stored in cold (-18 °C) and ambient (20 °C) conditions for 10 years. The results were expressed as a mean ± SD of three independent experiments ($n = 3$). Different letters indicate statistically significant difference between ST and SS genotypes according to Student's t-test ($p < 0.05$).

3. Results

3.1. Comparison of seed storability between two wheat genotypes

The initial seed germination rate of the two wheat genotypes TRI_23248 and TRI_10230 was 100% and 98%, respectively, before their storage at − 18 °C and 20 °C in 2004. Their vigor was tested after being stored for 10 years. The germination rate of TRI_23248 and TRI_10230 seeds stored at − 18 °C was 100% and 83%, respectively, and when stored at 20 °C, the germination rate of TRI_23248 was 84%, but that of TRI10230 was only 14% (Fig. 1). Thus, the vigor loss of TRI_23248 seeds during storage was significantly lower than that of TRI_10230, indicating that TRI_23248 can be regarded as a storage-tolerant (ST) and TRI_10230 a storage-sensitive (SS) genotype. After storage in cold (− 18 °C) and ambient (20 °C) conditions for 10 years, the embryonic MDA and conductivity of leachate in ST seeds were significantly lower than those in SS seeds (Fig. 2), indicating that ST seeds have stronger defense against oxidative injury during storage.

3.2. Analysis of antioxidant activities in ST and SS genotypes

To investigate antioxidant enzyme activities in response to seed storability, we used RT-qPCR to assess the transcript abundance of 17 genes encoding antioxidant enzymes in ST and SS seeds stored under cold and ambient conditions (Fig. 3). Gene expression was expressed relative to the expression under cold storage for both accessions. Three Cu/Zn superoxide dismutase (SOD) isoenzyme genes (*Cu/ZnSOD1*, *Cu/ZnSOD2*, and *Cu/ZnSOD3*) were differentially expressed in the two accessions. In ST seeds, *Cu/ZnSOD1* transcript presented a constant level, whereas *Cu/ZnSOD2* and *Cu/ZnSOD3* transcription was lower in seeds stored at ambient temperature than in those stored at cold temperature. In SS seeds, *Cu/ZnSOD1* and *Cu/ZnSOD3* transcription was significantly lower, whereas that of *Cu/ZnSOD2* was similar. *FeSOD1* transcription was significantly higher under ambient than under cold storage in ST, but lower in SS seeds. The transcription of three *MnSODs* (*MnSOD1*, *MnSOD2*, and *MnSOD3*) exhibited no significant difference among storage conditions in ST seeds. Similarly, in SS, *MnSOD2* and *MnSOD3* transcription were similar between the two storage conditions, but that of *MnSOD1* was higher under ambient storage conditions. The expression of catalase genes *CAT1* was significantly higher under ambient

Plant Physiology and Biochemistry 130 (2018) 455–463

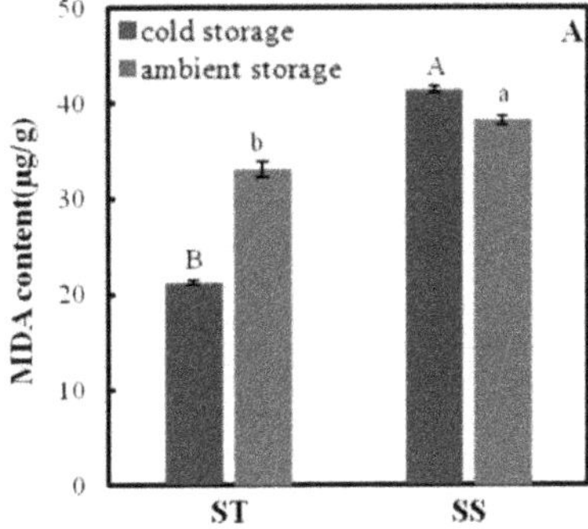

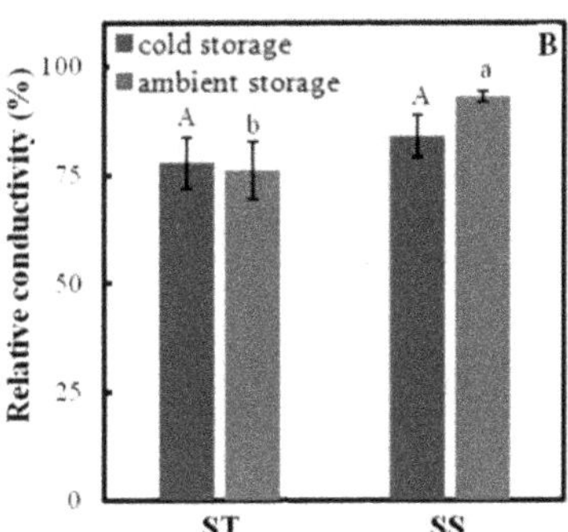

Fig. 2. Malondialdehyde (MDA) content and relative conductivity in storage-tolerant (ST) and storage-sensitive (SS) seeds stored in cold (−18 °C) and ambient (20 °C) conditions for 10 years. The results were expressed as a mean ± SD of three independent experiments ($n = 3$). Different alphabets indicate statistically significant difference between ST and SS genotypes according to Student's *t*-test ($p < 0.05$).

storage in ST, but lower in SS seeds. *CAT2* transcription was stable under two storage conditions in ST, but lower under ambient than under cold storage in SS seeds. The abundance of *ascorbate peroxidase 1* (*APX1*), *APX2*, *dehydroascorbate reductase 1* (*DHAR1*), and *DHAR2* transcripts remained constant in ST, but was significantly lower in SS seeds. In contrast, the expression of *glutathione reductase 1* (*GR1*) was significantly higher under ambient storage in ST and SS than under cold storage. The level of *monodehydroascorbate reductase 1* (*MDHAR1*) and *MDHAR2* transcription was significantly higher in ST, whereas in SS seeds, *MDHAR1* transcription showed no difference, but that of *MDHAR2* was significantly higher under ambient storage. These results indicated that ST seeds had a greater ability to maintain the expression of genes coding for antioxidant enzymes during storage.

To investigate the roles of AsA and GSH in storage tolerance of seeds, AsA, dehydroascorbate (DHA), GSH, and glutathione disulfide were measured in ST and SS seeds. All substances were significantly lower in seeds stored at ambient conditions for 10 years (Fig. 4). The contents of the AsA pool (AsA + DHA), AsA, GSH pool (GSH and glutathione disulfide), and GSH were significantly higher in ST than in SS seeds. These results indicated that the ST seeds had a greater ability to maintain the activities of AsA and GSH during storage.

3.3. Proteomic analysis of ST and SS seeds

Proteomic profile analysis was used to investigate the proteins associated with seed storability. Imaging of the proteins separated by two-dimensional gel electrophoresis revealed more than 600 spots that were reproducibly detected by the PDQuest 8.0 software (Fig. 5). Comparing

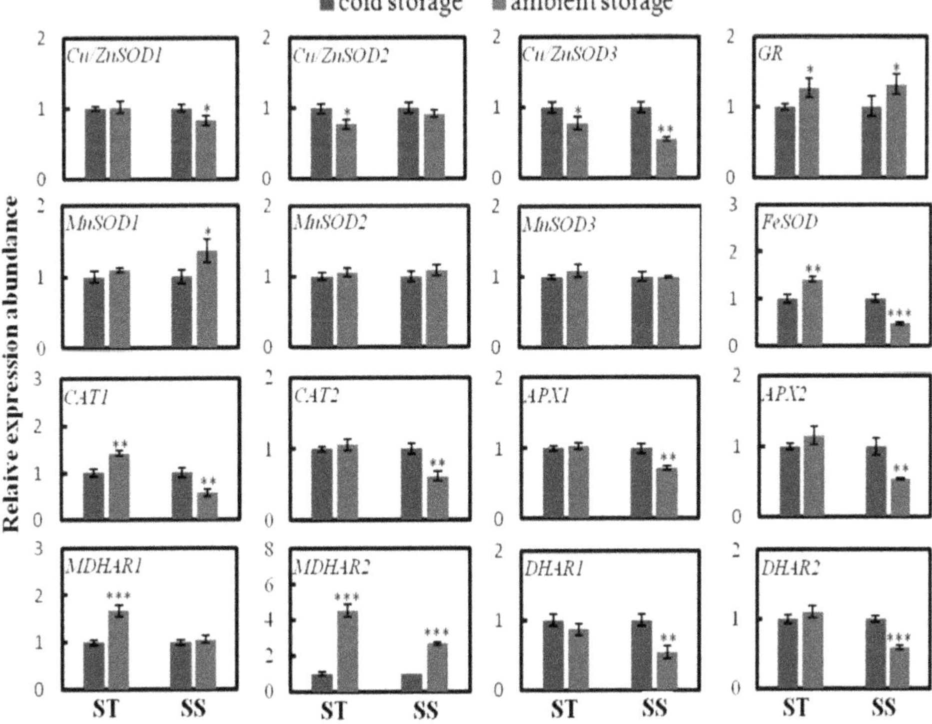

Fig. 3. Gene expression of major antioxidant enzymes in storage-tolerant (ST) and storage-sensitive (SS) seeds stored in cold (−18 °C) and ambient (20 °C) conditions for 10 years. The results were expressed as a mean ± SD of three independent experiments ($n = 3$). Asterisk indicates statistically significant difference between ambient and cold storage conditions, determined by Student's *t*-test (*, $p < 0.05$ **, $p < 0.01$ and ***, $p < 0.001$).

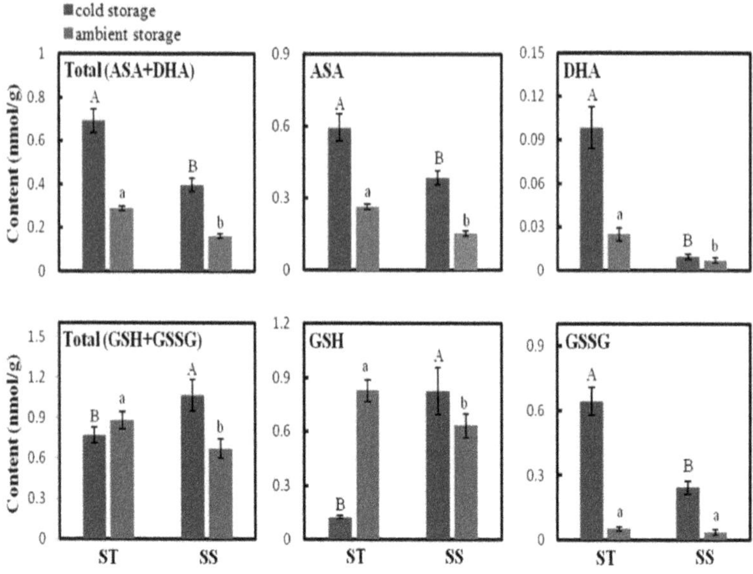

Fig. 4. Contents of glutathione and ascorbate in storage-tolerant (ST) and storage-sensitive (SS) seeds stored in cold (−18 °C) and ambient (20 °C) conditions for 10 years. The results were expressed as a mean ± SD of three independent experiments ($n = 3$). Different letters indicate statistically significant difference between ST and SS genotypes according to Student's t-test ($p < 0.05$). GSH, glutathione reduced form; GSSG, glutathione oxidized form; ASA, ascorbate reduced form; DHA, ascorbate oxidized form.

the difference in the each genotype seeds which were stored in − 18 °C and 20 °C conditions for 10 years, 93 spots in ST and 105 spots in SS seeds, exhibited a more than 1.5-fold change in abundance and their analysis by MALDI-TOF/TOF MS/MS revealed MOWSE scores significantly higher than 65 (Supplemental Table 2). Proteins at spot 1 to spot 93 (in ST) and at spot 33 to spot 137 (in SS seeds) were differentially expressed. Sixty-one spots (spots 33–93) were found in both ST and SS seeds. Among the detected differentially expressed proteins (spots 1–137), 19 were downregulated and 74 upregulated in ST seeds, and 65 were downregulated and 40 upregulated in SS seeds.

The proteins in ST and SS seeds were classified into nine and ten groups, respectively, following the classification by Bevan et al. (1998) (Fig. 6). The 93 differentially expressed proteins in ST seeds were related to disease or defense (29%), protein destination and storage

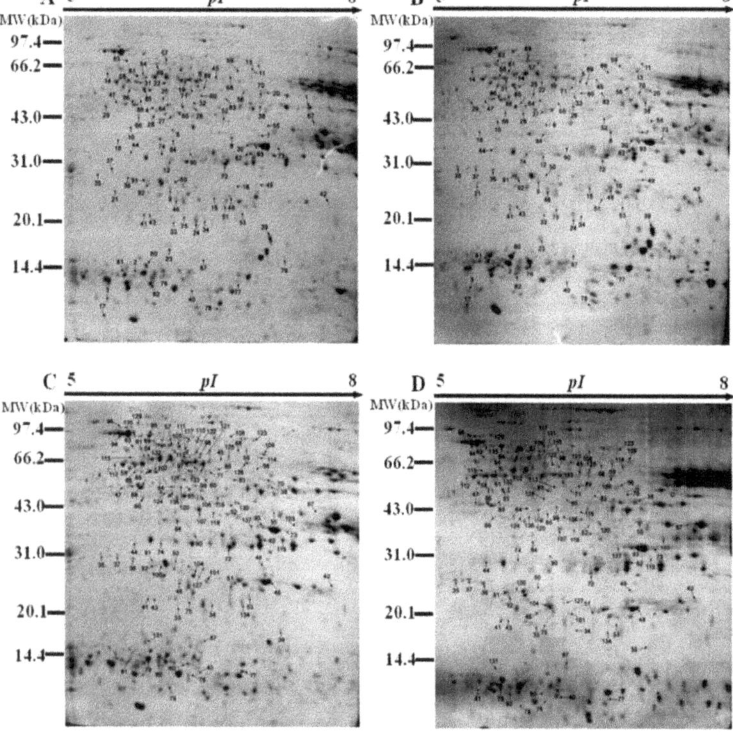

Fig. 5. Representative isoelectric focusing (IEF)/dodecyl sulfate polyacrylamide gel electrophoresis separation gels of proteins from storage tolerant and storage sensitive seeds stored for 10 years. An equal amount (500 µg) of total proteins was loaded onto 17 cm gels trip (pH5–8, linear). The pI and molecular mass standards are indicated on the top and left side of each gel image. A, ST seed stored in cold (− 18 °C) condition; B, ST seed stored in ambient condition; C, SS seed stored in cold (− 18 °C) condition; D, SS seed stored in ambient condition.

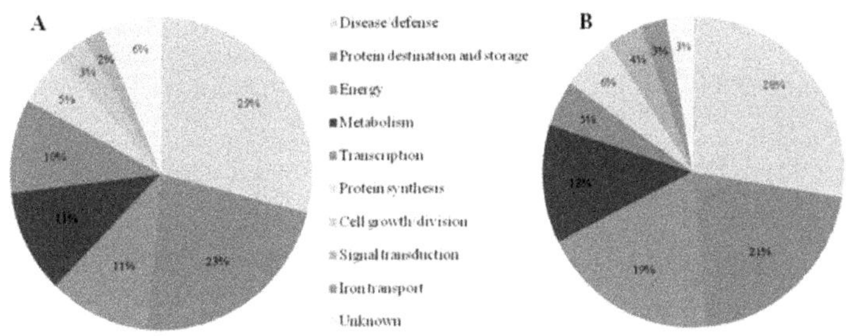

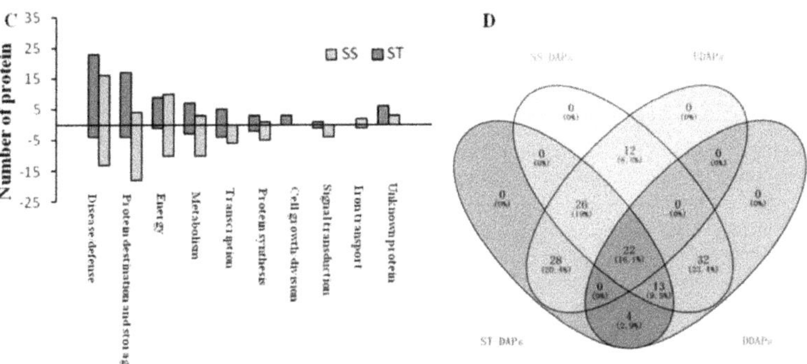

Fig. 6. Distribution of differentially expressed proteins in storage-tolerant (ST) and storage-sensitive (SS) seed stored in cold (−18 °C) and ambient conditions (20 °C) for 10 years. Pie chart showing function classification of 137 differentially expressed proteins in ST (A) and SS (B) seeds. Histogram showing protein expression model of ten categories in ST and SS (C) seeds. Venn diagram showing the distribution of differentially expressed proteins in ST and SS seeds (D). Abbreviation: UDAPs, up–regulated differential abundance proteins and DDAPs, down–regulated differential abundance proteins.

(23%), energy (11%), metabolism (11%), transcription (10%), protein synthesis (5%), cell growth and division (3%), signal transduction (2%) and unknown (6%). The 105 differentially expressed proteins in the SS seeds were related to disease or defense (28%), protein destination and storage (21%), energy (19%), metabolism (12%), transcription (5%), protein synthesis (6%), signal transduction (4%), iron transport (3%), and unknown (3%). The expression of those groups showed different patterns in the two genotypes—they were upregulated in the ST and downregulated in the SS genotype (Fig. 5). The major functional categories in the biological processes were response to disease or defense, protein destination and storage, and energy; 23 proteins in ST seeds and 13 proteins in SS seeds related to disease or defense metabolism were upregulated and downregulated, respectively. Among the proteins related to protein destination and storage metabolism, 17 proteins in ST seeds were upregulated and 18 proteins in SS seeds were downregulated; of the proteins related to the energy metabolism, 9 were upregulated in ST seeds and 10 were downregulated in SS seeds. These results indicated that the ST seeds had greater ability to maintain the metabolism during storage as compared with the SS seeds.

4. Discussion

Seed conservation in Genebanks is a general and ideal method for preserving global plant resources (FAO, 2010). However, seeds continue to slowly age, even when stored at Genebank (Lu et al., 2005; Walters et al., 2005). The factors affecting seed longevity are storage conditions and seed storability. Genebank Standards (FAO, 2014) have strict guidelines for seeds of orthodox species about seed collection, drying, packaging, storage, and other steps. Given that the storage conditions are similar for all seeds, the most plausible reason for differences in seed vigor loss during their storage in Genebank is their storability. In the present study, we compared the seeds of two wheat genotypes (TRI_23248 and TRI_10230) that were stored at −18 °C and 20 °C for 10 years in the IPK Genebank. The seed viability monitoring results showed a difference in their resistance to storage, TRI_23248 was substantially more tolerant than TRI_10230 (Fig. 1). Therefore, we categorized TRI_23248 as storage–tolerant (ST) and TRI_10230 as storage-sensitive (SS). The storage tolerance might be an essential characteristic in preventing vigor loss during storage. Furthermore, a proteomic approach was used to investigate the proteins responsible for the difference in storability between the ST and SS genotypes. Through MALDI-TOF/TOF analysis, 137 differentially expressed proteins were identified to be potentially involved in wheat seed storability (Supplementary Table 2).

4.1. Storage–tolerant seeds possess a stronger antioxidant defense system

ROS and reactive carbonyl species (RCS) are produced at an elevated rate during seed storage, which might cause oxidative damage in seeds (Rajjou et al., 2008; Yin et al., 2017). In the present study, the MDA content and relative conductivity were higher in SS than in ST seeds stored at −18 °C and 20 °C for 10 years (Fig. 2), indicating that serious oxidative damage occurred in the SS seeds. Many proteomic analyses of the mechanism of seed ageing showed that defense- or stress-related proteins were an extremely important component group in differentially expressed proteins (Rajjou et al., 2008; Yin et al., 2016). These studies indicated that the oxidative damage occurred during seed storage. Seeds possess defense systems to scavenge ROS and RCS and prevent oxidative damage. Thus, storability of seeds is believed to be associated with the capacity to detoxify ROS and RCS. In the present study, defense- or stress-related proteins were the largest component that displayed a different pattern in ST and SS seeds. These proteins were assigned to detoxification and stress response function. A noteworthy observation was that the detoxification-related proteins in ST seeds were mostly upregulated after storage at ambient temperature for 10 years, suggesting that the oxidative damage in seeds occurred after the storage period. Furthermore, the ST seeds possessed a stronger ability to activate defense or stress response systems than did SS seeds.

Antioxidases, including SOD, CAT, APX, GR, MDHAR and DHAR play an important role in detoxification of ROS. Proteomic analysis revealed that the level of GR (spot 45) decreased in ST and increased in SS seeds. Two spots, spot 46 and spot 101, were identified as DHAR; the first one increased in ST and SS seeds, and the second one showed no change in ST but decreased in SS seeds. The DHAR levels in ST seeds might be higher than those in SS seeds. Spots 47 and 97 were identified as MDHAR. MDHAR (spot 97) levels showed no change in ST but they decreased in SS seeds. GR, DHAR, and MDHAR play key roles in AsA and GSH regeneration (Noctor et al., 2012). Furthermore, we analyzed the expression patterns of 16 genes from the AsA–GSH cycle in ST and SS seeds (Fig. 3). *Cu/Zn SOD1, CAT1, APX1, GR1, MDHAR1*, and *DHAR1* play major roles in their isoenzyme genes, and overexpression of these genes results in significant stress tolerance (Feki et al., 2015). Most of those genes were upregulated in ST and downregulated in SS seeds. Besides, the total AsA and GSH contents were significantly higher in ST than in SS seeds (Fig. 4). These results indicated that the activity of the AsA–GSH cycle in ST seeds might be higher than that in SS seeds. Thus, compared with the SS seeds, the ST seeds could maintain higher activity of the AsA–GSH cycle after storage for 10 years at ambient temperature, which could prevent oxidative damage.

We identified the GSH S-transferase family of proteins (GST, spot 49, 50, 51, 100) and 1-Cys peroxiredoxin (PER1, spot 48), which are associated with detoxification of ROS. GSTs are a large family of multifunctional enzymes that detoxify xenobiotic and endobiotic compounds by conjugating GSH and protect against ROS-induced oxidative damage (Hayes and McLellan, 1999). GSTs were significantly upregulated in ST and downregulated in SS seeds. Compared with cold storage, the content of GSH in ST significantly increased after storage in ambient conditions for 10 years, but decreased significantly in SS seeds (Fig. 4). These results indicated that the ST seeds could utilize GSTs for ROS detoxification and oxidative stress tolerance. In addition, PER1 protects against ROS by catalyzing the reduction of H_2O_2 and alkyl hydroperoxides (Rhee et al., 2005). In the present study, similar to GSTs, PER1 increased significantly in ST and decreased in SS seeds. Our results are in agreement with previous reports that overexpression of PER1 could enhance seed tolerance with aging (Chen et al., 2016). We propose that storage tolerance of ST seeds is enhanced by ROS detoxification using GSTs and PER1.

RCS are downstream products of ROS, which oxidize lipids to form various aldehydes and ketones (Farmer and Mueller, 2013). The RCS levels are increased in response to various environmental stresses, which in turn may activate the RCS scavenging system (Mano, 2012). We identified three proteins (spot 60, 98 and 106) that are involved in the RCS scavenging system. Aldehyde dehydrogenase (spot 98) plays a key role in detoxification of RCS under stress in plants (Stiti et al., 2011). Alcohol dehydrogenase ADH1D (spot 60) and alcohol dehydrogenase-like 2 (spot 106) are the members of the alcohol dehydrogenase superfamily involved in the interconversion between alcohols and aldehydes or ketones (Thompson et al., 2007). The levels of these three proteins were decreased significantly in SS seeds, but, in ST seeds, spot 98 and 106 showed no change. This result suggested that the activity of the RCS scavenging system might be reduced in SS seeds, leading to RCS accumulation. The elevated RCS levels might cause protein carbonylation, which affects enzyme activity and even accelerates rice seed aging process (Yin et al., 2017). Thus, the ST seed might possess a stronger ability to scavenge RCS during storage, which might prevent protein carbonylation.

4.2. Storage-tolerant seeds utilize storage proteins

In the present study, destination-and storage-related proteins were the second largest component of differentially expressed proteins in both ST and SS seeds. During seed germination, storage proteins are degraded by proteasomes into smaller polypeptides or amino acids, which are then utilized in new protein synthesis (Bewley, 1997). Many studies reported a differential expression of these cluster proteins in aged seeds (Nguyen et al., 2015; Yin et al., 2017), indicating their possible involvement in the regulation of seed longevity. In the present study, we found five globulin-related spots (spot 76, 77, 78, 79, 80) in both ST and SS seeds. Globulin is a major storage protein family in wheat (Song and Zheng, 2007). The upregulation of storage proteins in wheat seeds could improve heat, salt, and drought stress tolerance (Giuliani et al., 2015). Consistent with those studies, five spots (76, 77, 78, 79, 80) showed a significant increase in ST seeds. In contrast, all the five spots decreased significantly in SS seeds. Storage proteins are the first target of ROS (Davies, 2005). Oxidative storage proteins more easily degrade to smaller polypeptides or amino acids needed for seed germination. Besides, the accumulation of storage proteins could inhibit ROS to attack other proteins. Eight spots were identified that belong to the 26S proteasome family. This supercomplex contains a core subcomplex (20S proteasome) and one or two regulatory subcomplexes (19S proteasome). Two beta subunits (spot 15, 17) and alpha subunits were identified; their levels were increased significantly in ST and showed no change in SS seeds. The alpha and beta subunits belong to the 20S proteasome, which is involved in proteolysis. The other five subunits (spot 70, 71, 72, 124, 125) belong to the 19S proteasome, which targets proteins and transports them to the 20S proteasome. The levels of those subunits were significantly increased or not modified in ST and decreased in SS seeds. These results indicated that the activity of protein proteolysis was greater in ST seeds than in SS seeds. Taken together, ST seeds possess greater ability for accumulation and metabolism of storage proteins, which might improve storage tolerance.

4.3. Storage-tolerant seeds maintain energy metabolism

The proteins related to energy metabolism were the third largest component of differentially expressed proteins in the ST and SS seeds. Among those, 10 proteins were associated with the glycolytic pathway (EMP), and 12 proteins were associated with the tricarboxylic acid (TCA) cycle pathway. The EMP–TCA provides energy and carbon skeletons for other metabolic processes during seed imbibition. A general picture of some of these proteins is shown in Fig. 5. Dysfunctional energy metabolism has been shown to contribute to seed ageing during storage (Rajjou et al., 2008; Xin et al., 2011; Yin et al., 2016, 2017). The EMP comprises 10 key steps during which glucose is converted to pyruvate (Kim and Dang, 2005). In the present study, five proteins were identified that belong to the key enzymes of the EMP. Interestingly, most of the proteins related to the EMP showed a significant increase in ST and SS seeds; the fold increase in SS seeds was significantly higher than that in ST seeds. These results showed that the EMP was upregulated in SS as compared with that in ST. The enzyme 6-phosphofructokinase 2 (PFK2, spot 56) catalyzes the conversion of fructose 6-phosphate and ATP to fructose 2,6-bisphosphate (F2,6BP) and ADP. In the EMP, the fructose 6-phosphate is catalyzed by 6-phosphofructokinase 1 to fructose1,6-bisphosphate, which is then catalyzed by fructose 1,6–bisphosphatase to glyceraldehyde-3-phosphate. The elevated activity of PFK2 could increase F2,6BP levels, further promote glycolysis, leading to the raise of C3 metabolites, especially phosphoenolpyruvate, to facilitate seed germination. Compared with the SS seeds, the higher expression of PFK2 in the ST seeds indicated higher levels of F2,6BP and the potential for the higher survival ability and glycolysis activity. Glyceraldehyde-3-phosphate dehydrogenase (GAPDH, spot 55, 103) catalyzes the conversion of glyceraldehyde-3-phosphate to 1,3-bisphosphoglycerate. In wheat, 22 *GAPDH* genes were identified that were classified into four types of plant *GAPDHs* (*gapA/B, gapC, gapCp*, and *gapN*) (Zeng et al., 2016). According to their protein sequences, spot 55 and spot 103 were distributed into gapC subfamily, which could be induced under stress (Guo et al., 2012). Enolase (spot 54, 102) catalyzes the conversion of 2-phosphoglycerate to phosphoenolpyruvate. Proteomic analysis reported the upregulation of enolase in wheat under drought stress (Cheng et al., 2015). Consistent with these studies, GAPDH and enolase were upregulated after storage for 10 years in both ST and SS seeds, with the fold increase being significantly higher in SS seeds than in ST seeds. In addition, the levels of the NADP-dependent malic enzyme (ME, spot 108) showed no change in ST but it

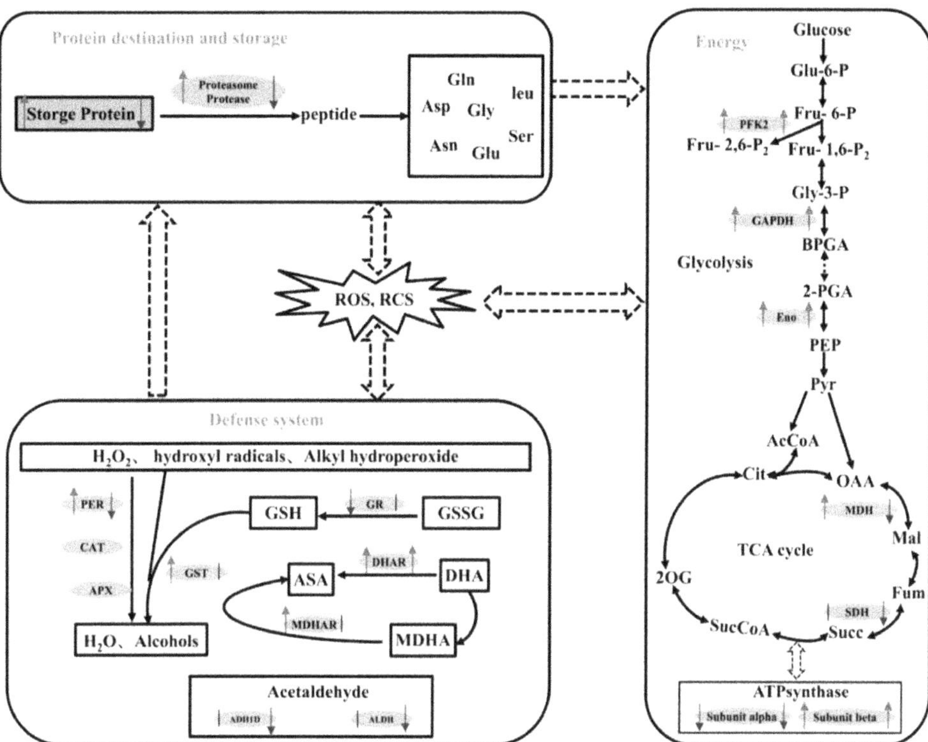

Fig. 7. Main metabolic pathways involved in seed storage tolerance. Protein abundance is marked in green ellipse and rectangle. Changes in protein abundance of storage tolerant and sensitive seed are presented in the left and right side, respectively. Arrow indicated up-regulation (red) and down-regulation (purple). Black vertical lines represented no change. (For interpretation of the references to color in this figure legend, the reader is referred to the Web version of this article.)

increased significantly in SS seeds. ME converts malate to pyruvate and CO_2 (Chang and Tong, 2003). As ME plays an important role in supplying pyruvate, the results presented herein indicated that pyruvate content in SS might be higher than that in ST seeds.

Pyruvate is converted to acetyl-coenzyme A by the pyruvate dehydrogenase complex, which has three key enzymes, namely pyruvate dehydrogenase, dihydrolipoyl transacetylase, and dihydrolipoyl dehydrogenase (DLD) (Izard et al., 1999). In the present study, we found that the DLD (spot 59, 109) levels were significantly higher in ST seeds than in SS seeds. DLD is a part of three mitochondrial enzyme complexes, i.e., the pyruvate dehydrogenase complex, α-ketoglutarate dehydrogenase complex, and branched chain alpha-ketoacid dehydrogenase complex (Babady et al., 2007). DLD deficiency shows variable phenotypes based on its function. The downregulation of DLD suggested that acetyl-coenzyme A, which is a key member of the TCA cycle, might be reduced from pyruvate in SS seeds. In addition, pyruvate could also be converted into acetaldehyde and CO_2 by pyruvate decarboxylase (PDC), and the acetaldehyde is then converted to ethanol by alcohol dehydrogenase. Four *PDC* (*PDC1, PDC2, PDC3, PDC4*) genes had been identified in rice, of which *PDC2* was upregulated under hypoxic stress (Rivoal et al., 1997). Two PDC2 (spot 57, 105) were identified in this study. . The downregulation of spot 105 in SS seeds might indicated a high tendency for pyruvate conversion to acetaldehyde in SS seeds after the 10-year storage at ambient temperature, leading to serious oxidative damage. Besides, ST seeds might maintain the ability to convert pyruvate to acetyl–coenzyme A, which is involved in the TCA cycle.

We also identified two proteins related to the TCA cycle, succinate dehydrogenase flavoprotein subunit (SDHA, spot 110) and mitochondrial malate dehydrogenase (mMDH, spot 7). SDHA is one of the four members of the succinate dehydrogenase complex (SDH), which oxidizes succinate to fumarate and transfers electrons to ubiquinone (Huang and Millar, 2013). In our previous studies, succinate-dependent O_2 uptake was decreased, indicating that SDH activity was affected in aged rice and soybean seeds (Xin et al., 2014; Yin et al., 2016). In the present study, the SDHA levels remained constant in ST seeds, but decreased significantly in SS seeds, indicating that ST seeds were able to preserve their ability to produce fumarate. MDH, which is located in the cytosol, mitochondria, and chloroplasts, reversibly catalyzes the oxidation of malate to oxaloacetate. The mMDH activity was reduced in aged rice and soybean seeds (Yin et al., 2016), and its expression was downregulated in soybean seeds under chilling stress (Yin et al., 2009). In the present study, the mMDH level was increased significantly in ST, but remained constant in SS seeds, indicating that the mitochondrial activity after 10-year storage at ambient temperature was higher in the ST genotype than in the SS genotype. Furthermore, we also detected the cytosolic MDH (cyMDH, spot 8 and 107). Spot 8 increased in abundance in ST seeds and had no change in SS seeds, whereas spot 107 showed no change in ST seeds but decreased significantly in SS seeds. MDH level in mitochondria and the cytosol might be higher in the ST genotype, indicating that ST seeds maintain the ability to produce malate and oxaloacetate. Downstream from the TCA is the electron transport chain in the mitochondria. The ATP synthase complex, comprising F_0 and F_1 subcomplexes, is the key component of the electron transport chain, and it combines ATP synthesis with the transport of protons across the mitochondrial membrane (Rühle and Leister, 2015). We found four subunits of the ATP synthase complex, including an alpha subunit (spot 113) and two beta subunits (spot 58,

112). The levels of the beta subunits of the ATP synthase decreased significantly in aged rice seeds (Yin et al., 2016), and the abundance of the two subunits were decreased significantly under oxidative stress in *Arabidopsis* (Sweetlove et al., 2002). These foundations would affect the assembly of ATP synthase, leading to a decrease in ATP production. In the present study, the alpha subunit level decreased in SS seeds, while showed no change in ST seeds. Besides, the beta subunit levels in ST seeds were significantly higher when compared with those in SS seeds. These changes might have a significant effect on ATP production capacity, especially in SS seeds. Taken together, the ST seeds might possess higher TCA activity and ATP production after a 10-year storage at ambient temperature in comparison with the SS seeds.

5. Conclusions

In conclusion, this work compared the storability of two wheat genotypes (accessions TRI_23248 and TRI_10230) that had been stored at −18 °C and 20 °C for 10 years in the IPK-Gatersleben Genebank. A comparison of the two accessions in terms of seed viability revealed that TRI_23248 was a storage-tolerant genotype. We analyzed the protein expression patterns of the two genotypes using proteomics; a model for the regulatory effect of protein expression on seed storage tolerance is proposed in Fig. 7. Based on our findings, seed storage tolerance is enhanced if the seeds could possess a stronger ability to: (1) activate defense system for preventing oxidative damage; (2) utilize storage proteins for germination; (3) maintain energy metabolism for supplying ATP. However, the mechanism of storage tolerance needs to be further investigated.

Author contributions

All authors read and commented on the manuscript.

XXL and XL designed the study and guided the research. XLC and GKY performed the experiments and wrote the main manuscript text. XX and JJH prepared all the figures and performed some of the experiments. AB and MN modified this manuscript.

Acknowledgments

This work was supported by the National Key Technology R&D Program (2013BAD01B01), the Agricultural Science and Technology Innovation Program/Crop Germplasm Resources Preservation and Sharing Innovation Team, the National Natural Science Foundation of China Program (31371713 and 31401470).

Appendix A. Supplementary data

Supplementary data related to this article can be found at https://doi.org/10.1016/j.plaphy.2018.07.022.

References

Babady, N.E., Pang, Y.P., Elpeleg, O., Isaya, G., 2007. Cryptic proteolytic activity of dihydrolipoamide dehydrogenase. Proc. Natl. Acad. Sci. U.S.A. 104, 6158–6163.
Bevan, M., Bancroft, I., Bent, E., Love, K., Goodman, H., Dean, C., Bergkamp, R., Dirkse, W., Staveren, M.V., Stiekema, W., et al., 1998. Analysis of 1.9 Mb of contiguous sequence from chromosome 4 of Arabidopsis thaliana. Nature 391, 485–488.
Bewley, J.D., 1997. Seed germination and dormancy. Plant Cell 9, 1055–1066.
Chang, G.G., Tong, L., 2003. Structure and function of malic enzymes, a new class of oxidative decarboxylases. Biochem 42, 12721–12733.
Chen, H.H., Chu, P., Zhou, Y.L., Ding, Y., Li, Y., Liu, J., Jiang, L.W., Huang, S.Z., 2016. Ectopic expression of NnPER1, a Nelumbo nucifera 1–cysteine peroxiredoxin antioxidant, enhances seed longevity and stress tolerance in Arabidopsis. Plant J. 88, 608–619.
Cheng, Z.W., Dong, K., Ge, P., Bian, Y.W., Dong, L.W., Deng, X., Li, X.H., Yan, Y.M., 2015. Identification of leaf proteins differentially accumulated between wheat cultivars distinct in their levels of drought tolerance. PLoS One 10, e0125302.
Davies, M.J., 2005. The oxidative environment and protein damage. BBA–Proteins Proteom 1703, 93–109.
FAO, 2014. Genbank Stardards. Food and Agriculture Organization of the United Nations, Rome. pp. 36.
FAO, 2010. Second report on the state of the world's plants genetic resources for food and agriculture. In: Commission on Genetic Resources for Food and Agriculture, Food and Agriculture Organization of the United Nations, Rome, pp. 47.
Farmer, E.E., Mueller, M.J., 2013. ROS–mediated lipid peroxidation and RES–activated signaling. Annu. Rev. Plant Biol. 64, 429–450.
Feki, K., Kamoun, Y., Mahmoud, R.B., Farhat–Khemakhem, A., Gargouri, A., Brini, F., 2015. Multiple abiotic stress tolerance of the transformants yeast cells and the transgenic Arabidopsis plants expressing a novel durum wheat catalase. Plant Physiol. Biochem. 97, 420–431.
Giuliani, M.M., Palermo, C., De Santis, M.A., Mentana, A., Pompa, M., Giuzio, L., Masci, S., Centonze, D., Flagella, Z., 2015. Differential expression of durum wheat gluten proteome under water stress during grain filling. J. Agric. Food Chem. 63, 6501–6512.
Guo, L., Devaiah, S.P., Narasimhan, R., Pan, X.Q., Zhang, Y.Y., Zhang, W.H., Wang, X.M., 2012. Cytosolic glyceraldehyde–3–phosphate dehydrogenases interact with phospholipase Dδ to transduce hydrogen peroxide signals in the Arabidopsis response to stress. Plant Cell 24, 2200–2212.
Hayes, J.D., McLellan, L., 1999. Glutathione and glutathione–dependent enzymes represent a co–ordinately regulated defence against oxidative stress. Free Radic. Res. 31, 273–300.
Heath, R.L., Packer, L., 1968. Photoperoxidation in isolated chloroplasts I. Kinetics and stoichiometry of fatty acid peroxidation. Arch. Biochem. Biophys. 125, 189–198.
Huang, S., Millar, A.H., 2013. Succinate dehydrogenase: the complex roles of a simple enzyme. Curr. Opin. Plant Biol. 16, 344–349.
Izard, T., Aevarsson, A., Allen, M.D., Westphal, A.H., Perham, R.N., Kok, A.D., 1999. Principles of quasi–equivalence and Euclidean geometry govern the assembly of cubic and dodecahedral cores of pyruvate dehydrogenase complexes. Proc. Natl. Acad. Sci. U.S.A. 96, 1240–1245.
Jiang, W.Z., Lee, J., Jin, Y.M., Qiao, Y.L., Piao, R., Jang, S.M., Woo, M.O., Kwon, S.W., Liu, X., Pan, H.Y., Du, X., Koh, H.J., 2011. Identification of QTLs for seed germination capability after various storage periods using two RIL populations in rice. Mol. Cell. 31, 385–392.
Kim, J.W., Dang, C.V., 2005. Multifaceted roles of glycolytic enzymes. Trends Biochem. Sci. 30, 142–150.
Landjeva, S., Lohwasser, U., Börner, A., 2010. Genetic mapping within the wheat D genome reveals QTL for germination, seed vigour and longevity, and early seedling growth. Euphytica 171, 129–143.
Lu, X.X., Chen, X.L., Guo, Y.H., 2005. Seed germinability of 23 crop species after a decade of storage in the national genebank of China. Agric. Sci. China 4, 408–412.
Mano, J., 2012. Reactive carbonyl species: their production from lipid peroxides, action in environmental stress, and the detoxification mechanism. Plant Physiol. Biochem. 59, 90–97.
Nguyen, T.P., Cueff, G., Hegedus, D.D., Rajjou, L., Bentsink, L., 2015. A role for seed storage proteins in Arabidopsis seed longevity. J. Exp. Bot. 66, 6399–6413.
Noctor, G., Mhamdi, A., Chaouch, S., Han, Y., Neukermans, J., Marquez–Garcia, B., Queval, G., Foyer, C.H., 2012. Glutathione in plants: an integrated overview. Plant Cell Environ. 35, 454–484.
Rajjou, L., Lovigny, Y., Groot, S.P.C., Belghazi, M., Job, C., Job, D., 2008. Proteome–Wide characterization of seed aging in Arabidopsis: a comparison between artificial and natural aging protocols. Plant Physiol. 148, 620–641.
Rhee, S.G., Kang, S.W., Jeong, W., Chang, T.S., Yang, K.S., Woo, H.A., 2005. Intracellular messenger function of hydrogen peroxide and its regulation by peroxiredoxins. Curr. Opin. Cell Biol. 17, 183–189.
Rivoal, J., Thind, S., Pradet, A., Ricard, B., 1997. Differential induction of pyruvate decarboxylase subunits and transcripts in anoxic rice seedlings. Plant Physiol. 114, 1021–1029.
Rühle, T., Leister, D., 2015. Assembly of F1F0–ATP synthases. Biochim. Biophys. Acta 1847, 849–860.
Song, Y., Zheng, Q., 2007. Dynamic rheological properties of wheat flour dough and proteins. Trends Food Sci. Technol. 18, 132–138.
Stiti, N., Missihoun, T.D., Kotchoni, S.O., Kirch, H.H., Bartels, D., 2011. Aldehyde dehydrogenases in Arabidopsis thaliana: biochemical requirements, metabolic pathways, and functional analysis. Front. Plant Sci. 2, 1–11.
Sweetlove, L.J., Heazlewood, J.L., Herald, V., Holtzapffel, R., Day, D., Leaver, C.J., Millar, A.H., 2002. The impact of oxidative stress on Arabidopsis mitochondria. Plant J. 32, 891–904.
Thompson, C.E., Salzano, F.M., De Souza, O.N., Freitas, L.B., 2007. Sequence and structural aspects of the functional diversification of plant alcohol dehydrogenases. Gene 396, 108–115.
Walters, C., Wheeler, L.M., Grotenhuis, J.M., 2005. Longevity of seeds stored in a genebank: species characteristics. Seed Sci. Res. 15, 1–20.
Wang, W.Q., Liu, S.J., Song, S.Q., Møller, I.M., 2015. Proteomics of seed development, desiccation tolerance, germination and vigor. Plant Physiol. Biochem. 86, 1–15.
Xin, X., Lin, X.H., Zhou, Y.C., Chen, X.L., Liu, X., Lu, X.X., 2011. Proteome analysis of maize seeds: the effect of artificial ageing. Physiol. Plantarum 143, 126–138.
Xin, X., Tian, Q., Yin, G.K., Chen, X.L., Zhang, J.M., Ng, S., 2014. Reduced mitochondrial and ascorbate–glutathione activity after artificial ageing in soybean seed. J. Plant Physiol. 171, 140–147.
Yin, G.K., Xin, X., Fu, S.Z., An, M.N., Wu, S.H., Chen, X.L., Zhang, J.M., He, J.J., Whelan, J., Lu, X.X., 2017. Proteomic and carbonylation profile analysis at the critical node of seed ageing in oryza sativa. Sci. Rep. 7, 1–12.
Yin, G.K., Whelan, J., Wu, S.H., Zhou, J., Chen, B.Y., Chen, X.L., Zhang, J.M., He, J.J., Xin, X., Lu, X.X., 2016. Comprehensive mitochondrial metabolic shift during the critical node of seed ageing in rice. PLoS One 11, e0148013.
Yin, G.K., Sun, H.M., Xin, X., Qin, G.Z., Liang, Z., Jing, X.M., 2009. Mitochondrial damage in the soybean seed axis during imbibition at chilling temperatures. Plant Cell Physiol. 50, 1305–1318.
Zeng, L.F., Deng, R., Guo, Z.P., Yang, S.S., Deng, X.P., 2016. Genome–wide identification and characterization of Glyceraldehyde–3–phosphate dehydrogenase genes family in wheat (Triticum aestivum). BMC Genom. 17, 240.

3.6 PAPER 6:

Age-dependent loss of seed viability is associated with increased lipid oxidation and hydrolysis

by

Janine Wiebach, **Manuela Nagel**, Andreas Börner, Thomas Altmann and David Riewe

Published in

Plant, Cell & Environment (2020) 43, 303-314

https://doi.org/10.1111/pce.13651

To view supplementary material for this article, please visit
https://onlinelibrary.wiley.com/doi/10.1111/pce.13651

Received: 12 June 2019 | Revised: 26 August 2019 | Accepted: 28 August 2019
DOI: 10.1111/pce.13651

ORIGINAL ARTICLE

WILEY 

Age-dependent loss of seed viability is associated with increased lipid oxidation and hydrolysis

Janine Wiebach[1,2] | Manuela Nagel[3] | Andreas Börner[3] | Thomas Altmann[1] | David Riewe[1,4]

[1] Department of Molecular Genetics, Leibniz Institute of Plant Genetics and Crop Plant Research (IPK), Seeland 06466, Germany

[2] Charité - Universitaetsmedizin Berlin, corporate member of Freie Universitaet Berlin, Humboldt-Universitaet zu Berlin, and Berlin Institute of Health, Institute of Biometry and Clinical Epidemiology, Berlin 10117, Germany

[3] Genebank Department, Leibniz Institute of Plant Genetics and Crop Plant Research (IPK), Seeland 06466, Germany

[4] Julius Kuehn-Institute (JKI), Federal Research Centre for Cultivated Plants, Institute for Ecological Chemistry, Plant Analysis and Stored Product Protection, Berlin 14195, Germany

Correspondence
D. Riewe, Julius Kuehn-Institute, Federal Research Centre for Cultivated Plants, Institute for Ecological Chemistry, Plant Analysis and Stored Product Protection, Koenigin-Luise-Strasse 19, 14195 Berlin, Germany.
Email: david.riewe@julius-kuehn.de; dariewe@googlemail.com

Abstract

The accumulation of reactive oxygen species has been associated with a loss of seed viability. Therefore, we have investigated the germination ability of a range of seed stocks, including two wheat collections and one barley collection that had been dry-aged for 5–40 years. Metabolite profiling analysis revealed that the accumulation of glycerol was negatively correlated with the ability to germinate in all seed sets. Furthermore, lipid degradation products such as glycerol phosphates and galactose were accumulated in some seed sets. A quantitative analysis of nonoxidized and oxidized lipids was performed in the wheat seed set that showed the greatest variation in germination. This analysis revealed that the levels of fully acylated and nonoxidized storage lipids like triacylglycerols and structural lipids like phospho- and galactolipids were decreasing. Moreover, the abundance of oxidized variants and hydrolysed products such as mono-/diacylglycerols, lysophospholipids, and fatty acids accumulated as viability decreased. The proportional formation of oxidized and nonoxidized fatty acids provides evidence for an enzymatic hydrolysis of specifically oxidized lipids in dry seeds. The results link reactive oxygen species with lipid oxidation, structural damage, and death in long-term aged seeds.

KEYWORDS

ageing, germination, hydrolysis, lipid, oxidation, seed, seed longevity, storage, *Triticum aestivum* (wheat), viability

1 | INTRODUCTION

Ageing is the process of becoming older and, unless immortal, leads to the death of an organism. The life span of plants in the form of dried seeds is highest among beings without clonal reproduction cycle. For example, seeds aged 1,300 (Shen-Miller, Mudgett, Schopf, Clarke, & Berger, 1995) and 2,000 years (Sallon et al., 2008) were still able to germinate. The placental tissue of 32,000-year-old seeds found below the permafrost was still viable and formed plants (Yashina et al., 2012).

Though seeds are a resilient form of life, they cannot completely defy ageing. Seed viability and longevity are affected by interspecific and intraspecific genetic variation (Bentsink et al., 2000; Nagel et al., 2016; Probert, Daws, & Hay, 2009; Sasaki et al., 2015; Christina Walters, Wheeler, & Grotenhuis, 2005) and environmental influences during seed development and storage (Ellis, Osei-Bonsu, & Roberts, 1982; Roberts, 1961). Survival during long-term storage of dry seeds is important for the conservation of valuable genetic resources stored in gene bank collections (Li & Pritchard, 2009). Furthermore, the frequency and uniformity of seed germination are important determinants of crop yields and quality and, thus, are of high-economic relevance. High temperature, high humidity (Harrington, 1963; Roberts, 1960), high oxygen vapour pressure (Groot, Surki, de Vos, & Kodde, 2012), and long storage intervals (Steiner & Ruckenbauer, 1995; Telewski & Zeevaart, 2002) are known to reduce the germination of orthodox (desiccation tolerant) seeds.

Seeds exposed to artificial ageing regimes (typically high temperature and humidity for a short period) produce reactive oxygen species (ROS), malondialdehyde, and 4-hydroxynonenal, the latter products of lipid oxidation, as shown for Arabidopsis, rice, or legumes (Chen et al., 2016a; Colville et al., 2012; Petla et al., 2016). Yet biochemical changes in response to these artificial conditions may not reflect the situation in naturally or long-term dry aged seeds (Galleschi et al., 2002; Roach, Nagel, Börner, Eberle, & Kranner, 2018). Oenel et al. (2017) have found that singly oxidized triacylglycerols and fatty acids (FAs) were elevated in two long-term aged Arabidopsis seed batches whose germinability was lower than the control seeds. Oxidized lipids from other major lipid classes were not reported in this study. However, hundreds of different singly and multiply oxidized lipids of various classes are formed in long-term stored wheat seeds (Riewe, Wiebach, & Altmann, 2017).

To investigate the occurrence of lipid oxidation and degradation during long-term dry storage (seed water content <10%) and its relation to seed viability, we analysed 90 differently aged wheat seed stocks that vary in genotype, storage period, and storage conditions. Central metabolites (e.g., sugars, amino acids, organic acids and other polar intermediates from essential plant metabolic pathways) and lipids correlated with germinability were identified using gas chromatography-mass spectrometry (GC-MS) and a recently developed liquid chromatography-mass spectrometry (LC-MS) method for the large-scale quantitative detection of oxidized and nonoxidized lipids (Riewe et al., 2017). We have found that germination frequencies were positively correlated with concentrations of nonoxidized lipids, such as triacylglycerols (TAGs), phospho-, and galactolipids, and were negatively correlated with oxidized lipids and lipid hydrolysis products such as oxidized diacylglycerols (DAGs), lysophospholipids (lysoPLs), FAs, and glycerol. These results suggest that the occurrence of lipid oxidation triggers the hydrolysis of storage and structural lipids and results in the loss of seed viability. Moreover, the GC-MS analyses of further sets of long-term cold stored wheat (176 genotypes) and barley (170 genotypes) seed stocks confirmed the aforementioned observations.

2 | MATERIAL AND METHODS

2.1 | Seed storage conditions

Seeds (Table S1) were obtained from plants grown in the field in Gatersleben, Germany, and stored in the Federal *Ex situ* Gene Bank, according to contemporary Genebank standards (FAO, 2014). For collection "wheat 1," seeds were harvested in 1998, 2000–2006, and 2008. Every year, five wheat accessions were harvested and stored. The seeds were dried at 20°C/20% relative humidity (RH) for 4 months and then threshed, cleaned, and further dried at 20°C/13% RH for 2 weeks. After drying, the moisture content was approximately 6%. Then, stocks were split: One batch of seeds was transferred into glass jars covered with silica gel and stored at 0°C until 2008 and then at −18°C thereafter. Following the cold storage, the equilibrate relative humidity was 17.4%, which corresponds to a moisture content of approximately 6%. The other batch was stored in paper bags at ambient conditions (20.3 ± 1.3°C/50.5 ± 6.3% RH), corresponding to moisture contents of around 10%, according to the Kew Seed Information Database (Royal Botanic Gardens Kew, 2019). In collections "wheat 2" and "barley," the seeds of 176 wheat and 170 barley accessions were harvested in 1974 and then threshed and dried in the same manner as the wheat 1 set. Seeds were initially stored in paper bags at ambient conditions. In 1978, the seeds were transferred to glass jars with silica gel and kept at 0°C, and from 2008 onwards, they were kept at −18°C. In 2013, aliquots of wheat 1 seeds were subjected to germination assays, and other aliquots were shock frozen for metabolite analysis. In 2014, aliquots of wheat 2 and barley seeds were subjected to the same procedure.

2.2 | Germination assay

Four replicates of 50 seeds were placed on moistened filter paper and kept in an incubator at 20°C, 60% RH for 8 hr of light. Physiological germination was assessed as radicle emergence of at least 2 mm. In accordance with the International Rules for Seed Testing (ISTA, 2013), the percentage of germination was recorded for wheat after 8 days and for barley after 7 days.

2.3 | GC-MS analysis

Central metabolites were extracted and analysed as described previously (Riewe et al., 2012; Riewe et al., 2016). Fifty deep frozen seeds per stock were pooled in scintillation vials containing three steel balls with 8 mm diameters and ground in a ball mill (5 × 2 min at 50 Hz at −70°C) using a cryogenic robot (Labman, North Yorkshire, UK). Fifteen milligrams deep frozen material was extracted with 1 ml ice-cold methanol: chloroform: water (2.5:1:1, containing 2 μg internal standard) to inhibit enzymatic conversion of metabolites. After an addition of 400 μl water and centrifugation, 100 μl of the polar phase were dried and stored under Ar. The dried metabolites were in-line derivatized and analysed using an Agilent 7890 gas chromatograph (Agilent, Santa Clara, CA, USA) coupled to a Pegasus HT mass spectrometer (LECO, St. Joseph, MI, USA). The data were normalized using the internal standard L-[2,3,3,3-d_4] alanine (www.sigmaaldrich.com) and exact weight (Table S2).

2.4 | LC-MS analysis

The LC-MS analysis of the wheat 1 set was described in detail recently (Riewe et al., 2017). Twenty-five milligrams of homogenized deep frozen seed material was extracted with 750 μl ice-cold methanol:methyl *tert*-butyl ether:water (1:3:0.5) to inhibit the enzymatic conversion of lipids. After an addition of 487.5 μl methanol: water (1:3) and centrifugation, 80 µl of the apolar phase were dried and stored under Ar. The samples were resolubilized prior to analysis in 50 µl acetonitrile: isopropanol (7:3). All 90 samples were quantitatively analysed in positive and negative modes using high-resolution LC-MS and were structurally analysed in the positive mode using LC-MS/MS. The LC-MS spectra were deconvoluted to identify pseudospectra/base peaks of nonredundant analytes (Kuhl, Tautenhahn, Bottcher, Larson, & Neumann, 2012). In the positive mode, oxidized lipids were annotated by accurate mass (1.5 ppm tolerance), and lipids of major classes were cross-validated by MS/MS spectra. In total, 624 annotations with this level of confidence were made for nonoxidized and oxidized TAGs, DAGs, monoacylglycerols (MAGs), phosphatidylcholines (PCs), phosphatidylethanolamines (PEs), phosphatidylglycerols (PGs), phosphatidylinositols, lysoPCs, lysoPEs, lysoPGs, lysophosphatidylinositols, monogalactosyldiacylglycerols, and digalactosyl-diacylglycerols (DGDGs). In the positive mode data, 154 (oxidized) FAs were annotated based on accurate mass (1.5 ppm) and filtering procedures (Riewe et al., 2017). Six (oxidized) phosphatidic acids (PAs) and four lysophosphatidic acids were quantified in the negative mode as [M-2H + NH_4] adduct with 5 ppm tolerance if they were detected in the positive and negative modes ("Bothmode," "Adduct," and "Retention filter" = TRUE, Table S3).

2.5 | Statistical information

Data processing and analysis were performed using RStudio (R Development Core Team, 2018; RStudio Team, 2016). All *P* values were false discovery rate (FDR) adjusted according to Benjamini and Hochberg (1995) and referred to as FDR. Significant means FDR < 0.05.

2.6 | Accession numbers

Accession numbers of the 436 wheat and barley accessions are provided in the first column ("ACCENUMB") of Table S1. Of the 436 accessions, 416 are currently available at the IPK gene bank (http://www.ipk-gatersleben.de/en/gbisipk-gaterslebendegbis-i/informationordering/).

3 | RESULTS

3.1 | Germination variation in a genetically diverse wheat population after long-term dry storage at different conditions

To study molecular processes associated with seed deterioration, we compiled long-term stored seed stocks from globally collected wheat accessions with large variations in germination. The set comprised 45 wheat genotypes stored at ambient or cold storage conditions for 5–15 years and formed 90 individual seed stocks (wheat 1, Table S1). The range of germination was 0–99%, and more stocks had a germination over 50% (Figure 1a). Germination was typically more varied between seeds from the same genotype that were harvested in the same year but stored at different temperatures than those within different genotypes harvested in the same year and stored at the same temperature (Table S2). This finding suggests that the storage environment has a large effect on seed viability. The observation that seeds harvested in 2008 display a lower germination than expected indicates that seed production, not storage, may have additionally lowered germination in this case.

3.2 | Glycerol accumulates in seeds with lower germinability

Central metabolism is hub, donor, and acceptor of matter for all specialized pathways, such as cell wall synthesis and specialized or lipid metabolism (Buchanan, Gruissem, & Jones, 2000). Changes in distant pathways with effects on seed viability are likely to be reflected in central metabolism as well. In wheat 1 seeds using GC-MS, 89 known and 277 unknown metabolites were quantified (Table S2). We identified 7 and 67 metabolites with positive or negative correlations to germination (Figure 1b), ranging from $R = -.82$ to.55. All correlation coefficients of known and unknown central metabolites are provided in Table S2. The top 12 negatively correlated metabolites of known chemical identity are displayed in Table 1. The three most strongly correlated metabolites are glycerol ($R = -.82$, Figure 1c), glycerol-3-phosphate ($R = -.80$), and glycerol-2-phosphate ($R = -.78$), and they are not topped by other compounds of unknown chemical structure. The only significantly positively correlated identified metabolite was glutamine, and this correlation was weak ($R = .30$, FDR = 0.023).

To validate the negative correlation between glycerol content and germination, we investigated another set of 176 wheat accessions (wheat 2) stored in the cold from 1974–2014 (Table S1). Although the seeds were older than those of wheat 1 set, 85% had a germination higher than 50% (Table S2). Furthermore, 85 known and 254 unknown metabolites were determined. Similar to wheat 1 set, more metabolites were negatively correlated to germination (83) than positively correlated to germination (23). Glycerol was ranked fourth among known germination-correlated metabolites ($R = -.58$, Table 3), again supporting a close link to seed longevity and the deterioration processes that occur during long-term dry storage.

3.3 | Wheat and barley show similar germination-associated central metabolite profiles

Barley seeds are similar in structure and composition to wheat seeds, and their ageing may lead to similar metabolic changes. To test this hypothesis, we assayed seed germination and central metabolites in a set of 170 barley accessions stored at cold storage for 40 years (Table S1) by following the same process conducted for wheat 2. In this case, 85 known and 253 unknown metabolites were quantified, and the correlations of their concentrations with germination were calculated (Table S2). As observed in the wheat 2 set, germination was over 50% for more than 85% of all accessions. Again, more negative (141) than positive (16) metabolite-to-germination correlations were detected. Among all detected known metabolites, glycerol was the second strongest germination-correlated metabolite ($R = -.57$,

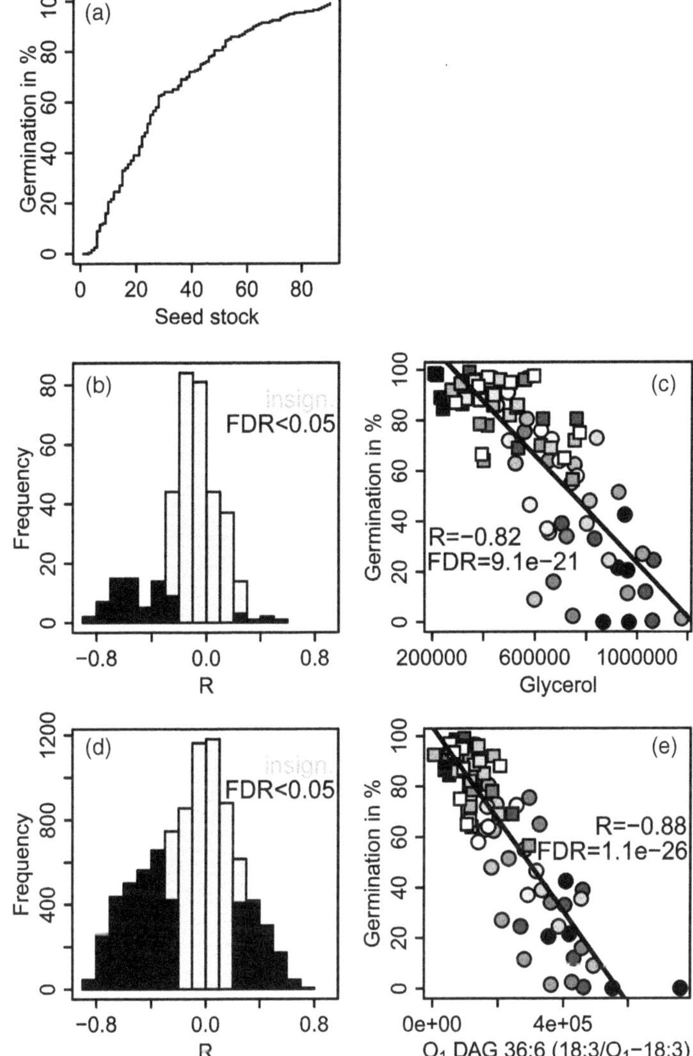

FIGURE 1 Correlation between germination and central/lipid metabolites in aged wheat seeds (wheat 1). (a) Germination frequency (ascending from left to right). (b) Frequency of measure of central metabolite-to-germination correlations; 67 negative, 292 insignificant (grey), and seven positive correlations (FDR < 0.05). (c) Glycerol-to-germination correlation (darkest symbols = 15 years of storage, lightest symbols = 5 years of storage). (d) Frequency of measure of lipid-to-germination correlations, 2,878 negative, 4,611 insignificant (grey), and 1,344 positive correlations (FDR < 0.05). (e) Negative correlation of O_1 DAG 36:6 (FA composition = 18:3/O_1-18:3, Formula ID = 7379) to germination. Glycerol and O_1 DAG 36:6 levels are displayed in mass spectral ion count (*n* = 90)

Table 2) in the set. Of the 12 top-ranked germination-correlated metabolites in the wheat 2 and barley sets, six were in common (Table 2), which suggests that similar metabolic processes coincide with losses in the viability of wheat and barley seeds.

3.4 | Seed viability is largely correlated to lipids

Glycerol and its phosphorylated derivatives are products of TAG and phospholipid (PL) catabolism. Galactose, which was high ranking among the germination-correlated metabolites in the wheat 2 and barley sets (third and fifth positions, respectively, in Table 2), can result from galactolipid (GL) degradation. Using LC-MS, we recorded untargeted lipid profiles of the wheat 1 set and identified nonoxidized and oxidized lipids by accurate mass and MS/MS fragmentation patterns (Table S3). For details, see Riewe et al. (2017). Pseudospectra (composed by insource fragments, adducts, and isotopes from single analytes; Kuhl et al., 2012) for 8,836 nonredundant analytes were constructed by a deconvolution of all 18,556 quantitative *m/z* features of the dataset. Each pseudospectrum contains one or more features. The most abundant feature within each pseudospectrum was defined as base peak and used for the untargeted nonredundant quantification of analytes. Approximately 50% (4,222) of all lipophilic analytes were correlated to germination, and, as with central metabolites, the number of negative correlations (2,878) was higher than the number of positive correlations (1,344), although a wider range (R = −.89 to.81) was observed (Figure 1d, all correlation coefficients of known and unknown lipids are provided in Table S3). Of the 8,836 detected analytes, 771 were previously annotated as nonoxidized and oxidized

TABLE 1 Top 12 negative central metabolite-to-germination correlations in wheat 1 ($n = 90$, FDRs < 6.4e-7)

Central metabolite	R^a
Glycerol	-.82
Glycerol-3-phosphate	-.80
Glycerol-2-phosphate	-.78
Secologanin	-.73
Vanillic acid	-.70
Inositol-1-phosphate	-.68
Azelaic acid	-.66
Ferulic acid, trans-	-.66
Urea	-.64
Pyruvic acid	-.58
Ribonic acid	-.57
Adenine	-.53

[a]Pearson correlation coefficient.

TABLE 2 Top 12 negative identified central metabolite-to-germination correlations in wheat 2 and barley

Wheat 2	R^a	Barley	R^a
Ribonic acid	-.63	Butanoic acid, 4-amino-	-.60
Butanoic acid, 4-amino-	-.63	**Glycerol**	-.57
Galactose	-.60	Gluconic acid	-.56
Glycerol	-.58	Adenine	-.55
Adenine	-.53	Galactose	-.49
Arabinonic acid	-.48	Proline	-.48
Glyceric acid	-.46	Galactitol	-.42
Xylitol	-.46	Alanine, beta-	-.42
Lyxonic acid	-.45	Ribonic acid	-.42
Aconitic acid, cis-	-.43	Fructose	-.42
Glucose	-.42	Aconitic acid, cis-	-.40
Inositol, myo-	-.41	Valine	-.39

Note. The six highest ranking metabolites of both lists are underlined, and glycerol is bold ($n = 176$ for wheat 2, $n = 170$ for barley, FDRs < 2.4e-6).

[a]Pearson correlation coefficient.

TABLE 3 Top 12 negative/positive identified lipid-to-germination correlations in wheat 1 as well as the most abundant lipid species of TAG, DAG, PC, MGDG, and DGDG in bold ($n = 90$, FDRs < 2.6e-5)

Lipid	Acyl composition[a]	ID[b]	R^c
O_1 DAG 36:6	(18:3/O_1-18:3)	7379	-.88
O_1 DAG 36:5	(18:2/O_1-18:3)	7380	-.88
O_1 DAG 38:4	(18:2/O_1-20:2)	7394	-.87
O_2 DAG 36:6	(O_1-18:3/O_1-18:3)	8017	-.87
O_1 DAG 36:5	(18:2/O_1-18:3)	7381	-.87
O_1 DAG 38:3	(18:3/O_1-20:0)	7397	-.85
O_1 DAG 38:5	(18:3/O_1-20:2)	7391	-.84
O_1 DAG 34:3	(16:0/O_1-18:3)	7370	-.84
O_1 DAG 34:4	(16:0/O_1-18:4)	7369	-.84
DAG 34:4	(16:1/18:3)	6591	-.84
O_1 DAG 36:4	(18:2/O_1-18:2)	7382	-.81
O_1 DAG 38:3	(18:2/O_1-20:1)	7396	-.81
DAG 36:4	**(18:2/18:2)**	**12058**	**-.62**
TAG 54:6	**(18:1/18:2/18:3)**	**19764**	**.43**
MGDG 36:4	**(18:2/18:2)**	**16888**	**.51**
PC 36:4	**(18:2/18:2)**	**16461**	**.59**
DGDG 36:2	(18:0/18:2)	6714	.62
DGDG 32:1	(16:0/16:1)	6705	.62
DGDG 32:2	(14:0/18:2)	6706	.62
DGDG 38:2	(18:1/20:1)	6719	.63
DGDG 36:3	(18:1/18:2)	6715	.64
PC 34:1	NA	630	.65
DGDG 36:5	(18:2/18:3)	6717	.65
DGDG 34:1	(16:0/18:1)	6709	.66
DGDG 34:2	(16:0/18:2)	6710	.67
O_1 DGDG 36:4	(18:2/O_1-18:2)	7495	.67
PC 36:3	NA	643	.67
DGDG 36:4	**(18:2/18:2)**	**6716**	**.69**

Abbreviations: DAG, diacylglycerols; DGDG, digalactosyldiacylglycerols; MGDG, monogalactosyldiacylglycerols; PC, phosphatidylcholines; TAG, triacylglycerols.

[a]"Putative Composition" in Table S3.

[b]"Formula ID" in Table S3.

[c]Pearson correlation coefficient.

(having one to four additional oxygen atoms) major lipids (Riewe et al., 2017). Ten annotations of (oxidized) (lyso)phosphatidic acids with lower annotation confidence were added from the negative mode recordings (Table S3) to complement this study. Of these 781 annotated lipids, 513 were correlated to germination.

3.5 | The extremal lipid-to-germination correlations are made up by known major lipids

Unlike unknown analytes, identified lipids were quantified using the dominant adduct formed per lipid class to allow for comparison within lipid classes (Collins, Edwards, Fredricks, & Van Mooy, 2016). But this *m/z* feature is not necessarily the base peak of a pseudospectrum. The 20 highest positively germination-correlated quantitative features belong to six pseudospectra (Table S3 "pcgroup" = 99, 89, 22, 960, 92, and 1103), and three of these pseudospectra were identified as PCs ("Formula ID" = 613, 643, and 625). A similar observation was noted for the 20 highest negatively correlated features in the dataset. They belong to eight pseudospectra ("pcgroup" = 893, 415, 283, 8,261, 134, 6,582, 216, and 129), of which six are annotated as singly oxidized DAGs ("Formula ID" = 8019, 7379, 7397, 7380, 7394, and 7381). The closest correlations observed in this untargeted dataset are not attributed to unknown analytes (which represent >90% of all analytes). Instead, they are formed by known major lipids and their oxides, which underline the special role of these lipids in seed integrity within the entirety of the lipidome.

3.6 | Nonoxidized membrane lipids have the highest positive correlations with seed viability

The top 12 positive lipid-to-germination correlations of identified lipids range from R = .62 to.69 (Table 3). Nonoxidized lipids are overrepresented given their frequency in the dataset. In addition, lipids typically found in membranes are overrepresented as the list only contains PCs and DGDGs with 32–38 carbon atoms in the FA moieties. The lipid with the highest positive correlation to germination, DGDG 36:4 ("Formula ID" = 6716), is also the most abundant cellular DGDG. Other high ranking positively correlated lipids also belong to the higher abundant representatives of their subclasses (Table S3 "Median intensity"). With few exceptions, the lipids consisted of FAs with 16 or 18 carbon atoms, which are typical of wheat/plant seed lipids (González-Thuillier et al., 2015).

3.7 | Singly oxidized DAGs have the highest negative correlations with seed viability

Of the top 12 negatively correlated annotated lipids (R = −.81 to −.88), 11 were oxidized (Table 3). Of these, 10 were singly oxidized, and all 12 were DAGs. The DAGs consisted mainly of FAs with 18 C atoms, such as O_1 DAG 36:6 (18:3/O_1-18:3), the lipid with the highest correlation to germination (Figure 1e), although they also contained FAs with 16 or 20 C atoms. Another lipid class with a high number of high negative lipid-to-germination correlations was oxidized and nonoxidized free FAs, typically with 18 but possibly 16 or 20 C atoms. Twelve of such FAs were among the next 20 highest negatively correlated identified lipids (Table S3).

3.8 | Lipid-to-germination correlation distribution within major lipid classes and oxidized subclasses

An analysis of enrichment of lipids that were positively or negatively correlated to germination was conducted within individual lipid subclasses to support the interpretation of the results in a broader context. A two-sided binomial test was applicable to 24 (having six or more significant positive or negative correlations to germination) of the 43 detected lipid subclasses. Of the 24 subclasses, 20 were significantly enriched in positive (7) or negative (13) correlations between individual lipids and germination frequency (Figure 2). These clear enrichments (e.g., for TAGs, O_1 TAGs, DAG, PCs, and lysoPCs) reveal that in most cases, the individual lipids of subclasses are collectively altered with progressing loss of seed viability.

3.9 | Nonoxidized TAGs are exclusively positively correlated to seed germination, and oxidized TAGs are predominantly negatively correlated to seed germination

Of the 87 nonoxidized TAGs, 52 are positively correlated to germination, and none are negatively correlated (FDR = 5.3e − 15, Figure 2a). The situation has been reversed in the detected 82 singly oxidized TAGs, where the number of negatively correlated lipids is 61, and 2 is positively correlated. The same trend was observed for twofold, threefold, and fourfold oxidized TAGs, which, in all cases, contained significantly more negative than positive correlations.

3.10 | DAGs, MAGs, and FAs are predominantly negatively correlated to germination

Nonoxidized or oxidized TAGs can be hydrolysed to corresponding forms of DAGs, MAGs, and FAs (Figure 3). Independent of the degree of oxidation, more negative (165) than positive (6) correlations were found for these glycerides/FAs (Figure 2a). The trend was significant for DAGs and FAs with up to three added oxygen atoms.

3.11 | PLs are positively correlated to germination

Of 33 nonoxidized PCs, 28 are positively correlated to germination and, similar to the nonoxidized TAGs, negative correlations were not detected (Figure 2b). In addition, the PEs and PGs display more positive correlations to germination than negative correlations to germination, and all four identified PAs were also positively correlated. The numbers of identified/correlated oxidized PLs were typically too low for enrichment analyses. Only O_2 PCs demonstrated a significant enrichment in positive correlations.

3.12 | LysoPLs are negatively correlated to germination

While 30 of 42 detected nonoxidized and oxidized lysoPLs were negatively correlated to germination, none indicated a positive correlation (Figure 2c). On the individual subclass level, correlation enrichments were significant for the lysoPCs and lysoPLs.

3.13 | GLs are predominantly positively correlated to germination

Except for two subclasses, GLs were more often positively correlated to germination than negatively correlated to germination. However, statistical testing was restricted to monogalactosylglycerols, DGDGs, and O_2 DGDGs due to low numbers of identifications/correlations in the other subclasses (Figure 2d). Nonetheless, all 21 identified DGDGs were significantly positively correlated to germination, which renders this finding a clear result. Lastly, the O_2 DGDGs are significantly enriched for positive correlations.

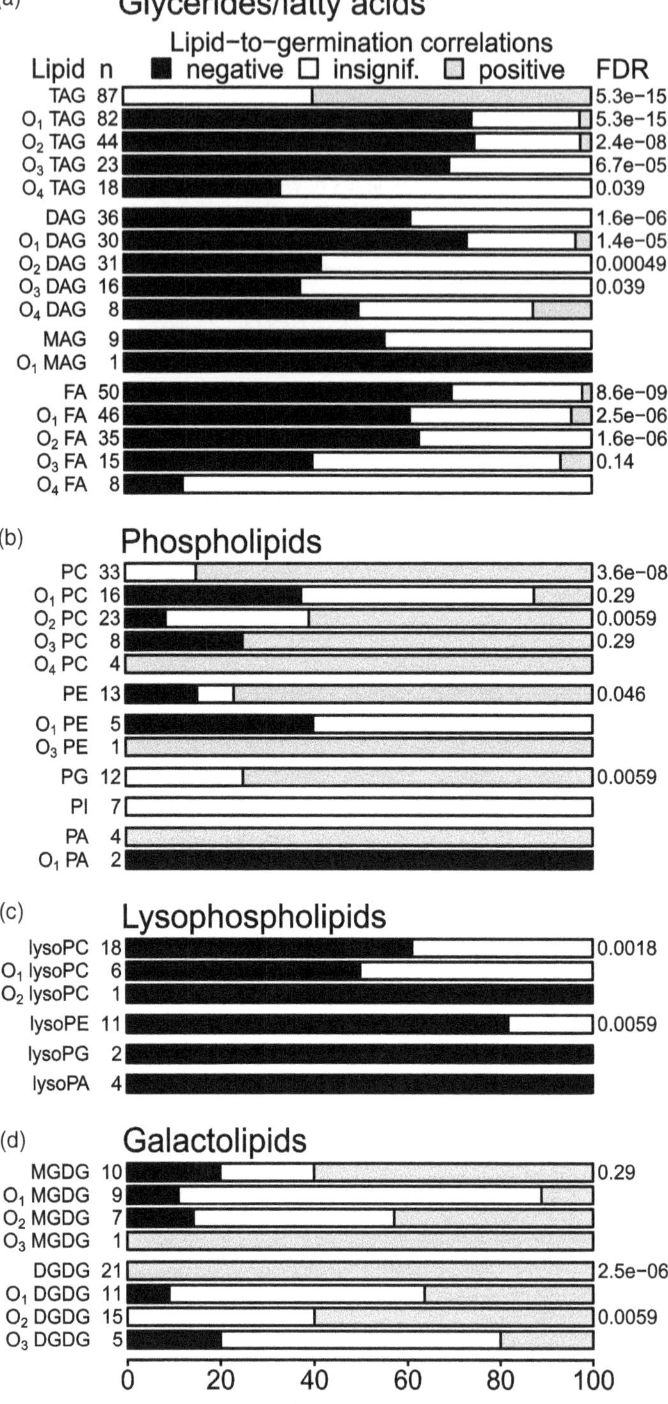

FIGURE 2 Correlation enrichment analysis. Lipid-to-germination correlations were counted for identified (oxidized) lipids belonging to (a) glycerides/fatty acids, (b) phospholipids, (c) lysophospholipids, and (d) galactolipids, and the number of lipids (*n*) was normalized to 100% (horizontal bars). Black sections represent negative lipid-to-germination correlations, white sections represent insignificant correlations, and grey sections refer to positive correlations (FDR < 0.05). FDR-corrected *P* values of a two-sided binomial test (if applicable) on enrichment of positive or negative correlations within lipid subclasses are provided on the right side. DAG, diacylglycerols; DGDG, digalactosyldiacylglycerols; FA, fatty acids; FDR, false recovery rate; MAG, monoacylglycerols; MGDG, monogalactosyldiacylglycerols; PA, phosphatidic acids; PC, phosphatidylcholines; PE, phosphatidylethanolamines; PG, phosphatidylglycerols; TAG, triacylglycerols

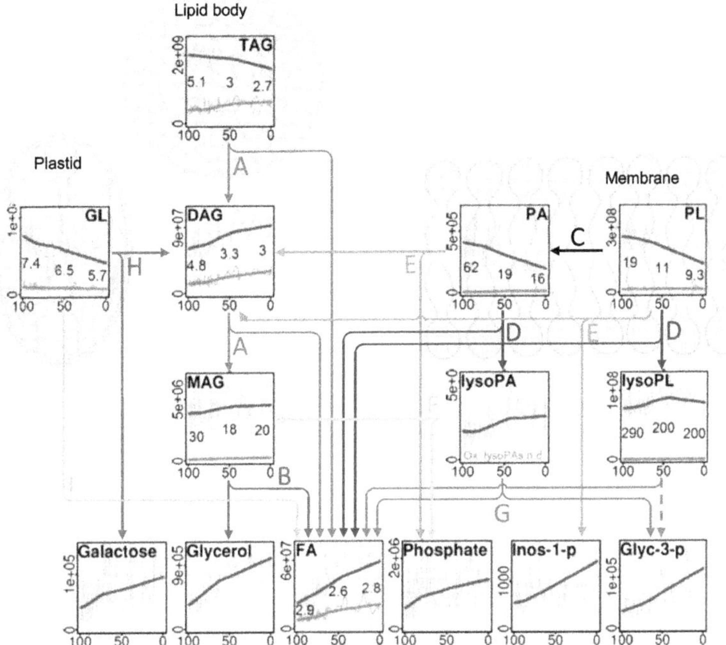

FIGURE 3 Biochemical/cellular model of lipid oxidation and degradation. For central metabolites and phosphate levels, mass spectral ion counts (Y axis) of the individual metabolites are displayed as line diagram (light blue) with moving average (dark blue) in descending order of germination (in %, X axis) of the 90 wheat 1 seed stocks. For lipid classes, the sum abundance of all nonoxidized and oxidized lipid species of the corresponding classes is displayed as the sum ion count of all nonoxidized (blue) and oxidized (red) species per class. The three numbers in each diagram containing oxidized lipid data are the ratios in abundance of nonoxidized to oxidized lipids for seeds with more than 90% (left), less than 10% (right), or 44–55% germination (middle). Known enzymatic interconversion pathways are indicated by coloured lines for (a) TAG lipase, (b) MAG lipase, (c) phospholipase D, (d) phospholipase A, (e) phospholipase C, (f) lysophosphatidic acid lipase, (g) lysophosphorylase, (h) α-galactosidase, and (i) galactolipase. Inos-1-p = inositol-1-phosphate, Glyc-3-p = glycerol-3-phophate. DAG, diacylglycerols; FA, fatty acids; GL, galactolipid; MAG, monoacylglycerols; TAG, triacylglycerols

3.14 | Cross-validation of lipid analysis

We have not used the negative mode data (with the exception of oxidized and nonoxidized lysophosphatidic acids and PAs) due to the lower degree of annotation confidence (lower mass accuracy and no validation by MS/MS spectra), lower analyte coverage, and redundancy with the positive mode data. However, the above presented results, such as the extreme correlations of major lipids to germination within the lipidomic dataset, the outstanding negative association of oxidized DAGs and FAs with germination, and the distribution of lipid-germination correlations within lipid classes, are also found in the negative mode dataset (Table S3).

4 | DISCUSSION

The ROS-caused lipid oxidation during seed ageing has been postulated to cause seed deterioration and death (Chen et al., 2016a; Hu et al., 2012; Ratajczak, Małecka, Bagniewska-Zadworna, & Kalemba, 2015). Here, we present a detailed analysis of the contents of central metabolites, lipids, oxidized lipids, and products of lipid hydrolysis of differently aged wheat and barley seeds and their relation to germination.

Several factors, including genotype, seed production year, storage time, moisture content, and storage temperature, have influenced variation in germination of seed stocks (Roberts, 1961). Factors such as storage temperature and moisture content change over time within a set. Although we cannot clearly disentangle the relative effects of these factors on ageing, our results suggest that higher temperatures and moisture content led to faster ageing, and such conditions also favour chemical/enzymatic turnover of lipids. In addition, seed viability may be reduced by storing seeds together with silica gel. More specifically, silica gel may lead to storage conditions below the critical water content, which is detrimental for seeds (C. Walters & Engels, 1998; Christina Walters, Hill, & Wheeler, 2005). However, for wheat seeds, the critical water content is estimated below 5% at temperatures of 65°C (Yan, 2017). As ERH was measured at 17.4 ± 10.0%, which corresponds to a 6% moisture content; we assume that seeds were not affected by an overdrying induced by silica gel.

Figure 3 integrates the findings of this study into a biochemical pathway of (oxidized) lipid degradation in a cellular context. Except for the production of glycerol-3-phosphate (Glyc-3-P), the complete pathway chart is composed of known single enzymatic hydrolytic reactions and otherwise contains no gaps regarding biochemical resolution. Any cleavage/hydrolysis of headgroup or FA of fully acylated lipid classes (TAGs, GLs, and PLs/PAs) results in the formation of lipids associated with seed deterioration. Lipid degradation is potentially executed by several hydrolytic enzymes known as lipases, which hydrolyse phospho or acyl esters. Although flux information would be needed to draw conclusions on the directions of lipid conversions in the ageing wheat seed, our data does not support a substantial role for phospholipase D (conversion of PLs to PAs) in seed viability, as was the case for Arabidopsis (Devaiah et al., 2007), soybean (J. Lee et al., 2012), and recalcitrant seeds of different species (Chen et al., 2017). In this study, (oxidized) PAs were detected only in small amounts (Table S3), and ethanolamine, a product of PE to PA conversion and typically well measurable using GC-MS, was not detectable. Phospholipase A (PLA, PLs to lysoPLs and FAs) is likely to be involved in seed deterioration. LysoPLs were quantified in notable amounts, and while the substrates (PLs) of this enzymatic reaction were overall positively correlated to germination, the products (lysoPLs and FAs) displayed the opposite (Figures 2 and 3). Nonetheless, one of the two enzymes, phospholipase D or PLA, or both, must have been active for the production of Glyc-3-P from PLs. Phospholipase C (PLs to DAGs) may also contribute to the reduction of PL by removing the complete headgroup. Substrates/products show inverse correlations, supporting the course of this reaction. Independent and in support of this conversion, we found the PI specific degradation product Inositol-phosphate to be the sixth strongest negatively correlated metabolite detected in the wheat 1 GC-MS analysis (Table 1). While reductions in the levels of these phospholipids may lead to damages of cellular membranes, hydrolytic reductions of galactolipids by galactolipase (GLs to FAs) or galactosidase (GLs to DAGs) would specifically alter plastidial membranes (Dörmann, Hoffmann-Benning, Balbo, & Benning, 1995; Jarvis et al., 2000; R.-H. Lee, Hsu, Huang, Lo, & Grace Chen, 2009). The TAGs form the largest lipid pool and are concentrated in lipid bodies. Conversion of oxidized TAGs and DAGs by TAG lipase (TAGs to DAGs to MAGs and FAs) likely results from the exclusive conversion of TAGs by this enzyme and is supported by reduced TAG and elevated DAG and FA levels. Given that TAG is the principle lipid in wheat seed (Morrison, 1998), which has a lipid content of 2% (http://www.fao.org), this reaction could deliver the largest fraction of (oxidized) FAs produced in connection with the loss in seed viability.

The contribution of each of these enzymes to general lipid hydrolysis cannot be reliably specified. Moreover, other enzymes may be involved. However, the low ratio of 2.9 for nonoxidized to oxidized FAs in seeds with 90–100% germination (Figure 3, FA) suggests that lipases with higher specificity for oxidized than nonoxidized lipids as substrates are involved in lipid degradation because a higher ratio would be expected if the specificity was equal or higher for nonoxidized lipids due to their higher availability in such seeds. If these lipases specifically released oxidized or nonoxidized FA from oxidized lipids, either should accumulate relative to the other with reducing degrees of viability. Yet, they accumulate both in a proportional manner, and the ratio of nonoxidized to oxidized FAs remains constant between 2.6–2.9 in seeds with high, medium, or low germination. This constant ratio suggests that the same lipases have little product specificity. The theoretical ratio of nonoxidized to oxidized FAs obtained from the random hydrolysis of FA moieties from TAGs with a single oxidation, which is the most abundant oxidized lipid subclass in the seeds of this study (Riewe et al., 2017), would be two.

A higher specificity for oxidized over nonoxidized lipids could also explain why in many lipid classes (GL, MAG, (lyso)PA, (lyso)PL) oxidized lipids are low in abundance or are undetectable. The specific conversion of oxidized lipids due to higher affinity/turnover may also account for the highest correlations of oxidized DAGs and oxidized FAs with germination because these species would be formed at a ratio of two from singly oxidized TAGs. The twofold greater theoretical production of oxidized DAGs compared with oxidized FAs from TAGs may contribute to the observed slightly higher precision in predicting germination by oxidized DAGs compared with oxidized FAs. As a novel aspect, however, preferred conversion of oxidized lipids implies that lipid oxidation precedes lipid hydrolysis.

The seeds used in this study had moisture contents below 0.1 g H_2O g^{-1} DW^{-1} and were in a so-called glassy state wherein molecular mobility and enzyme activity were reduced severely or completely (Buitink & Leprince, 2008; Fernández-Marín et al., 2013). Our findings provide evidence for enzymatic catalysis for the reactions leading to the proportional production of oxidized and nonoxidized FAs (see above) and potentially for other reactions depicted in Figure 3. As such, diffusion limitations may be less severe for lipases. The seed-specific Arabidopsis TAG lipase gene SDP1 is expressed during late seed development. The SDP1 protein is associated with oil bodies and is enzymatically active during seed imbibitions (Eastmond, 2006). If seed lipases are in close proximity to massively concentrated amounts of their substrates and are able to diffuse in a lipophilic microenvironment, catalysis could occur in a glassy state. Lipases could lose their activity by alterations in the 3D structure under the low moisture conditions. However, lipases including wheat germ lipase are active in organic solvents with low additions of water (Caro et al., 2002; Yang & Russell, 1995), which suggests that they might remain stable and active in seeds with low moisture content.

Although we do not have direct evidence that lipid oxidation/hydrolysis precedes and is causal for the losses in viability, the high abundance of oxidized lipids and FAs in seeds with approximately 100% germination implies that these processes happen before death and not after. Additional posthumous lipid oxidation would require a mechanism to prevent/revert these processes in the dry and metabolically inert seed that would become inactive when the seed loses its ability to germinate. Lipid oxidation/hydrolysis could lead to death in many ways. For example, membrane damage may lead to uncontrolled permeability of small molecules or proteins across cells or organelles, causing cellular damage/death. The formation of pH-gradients important for energy production or transport may be impaired. Oxidized lipids/FA may inhibit enzymes or trigger fatal signal transduction. Further research is required to identify the causative ROS-induced molecular processes that lead to the termination of life.

Our present knowledge on metabolic processes in ageing seeds has been limited due to the indirect or incomprehensive analysis of oxidized lipids and other metabolites or suboptimal model systems such as controlled deterioration treatments. Here, we consider the associations of hundreds of metabolites, mostly lipids, with longevity of long-term aged cereal seeds at single metabolite, lipid class, and pathway level. Our results link ROS with losses in viability caused by lipid oxidation and hydrolysis in the longest living of all mortal life forms, the plant seed.

ACKNOWLEDGEMENTS

Sibylle Pistrick, Gabrielle Matzig, Michael Grau, and Andrea Apelt are gratefully acknowledged for technical support in viability testing and GC-/LC-MS analyses and provision of the material.

AUTHOR CONTRIBUTIONS

A.B. and D.R. initiated the work. M.N. assayed germination of wheat 1 and barley, and J.W. assayed germination of wheat 2 and measured central metabolites in wheat 2 and barley. D.R. measured central metabolites in wheat 1 and designed the research. J.W. measured lipids with advice of D.R., and both analysed all data. D.R. wrote the manuscript with the support of J.W. M.N., T.A., and A.B. provided edits.

ORCID

Janine Wiebach https://orcid.org/0000-0002-7416-9859
Manuela Nagel https://orcid.org/0000-0003-0396-0333
Andreas Börner https://orcid.org/0000-0003-3301-9026
Thomas Altmann https://orcid.org/0000-0002-3759-360X
David Riewe https://orcid.org/0000-0002-9095-5518

REFERENCES

Benjamini, Y., & Hochberg, Y. (1995). Controlling the false discovery rate—A practical and powerful approach to multiple testing. *Journal of the Royal Statistical Society: Series B (Methodological)*, *57*, 289–300.

Bentsink, L., Alonso-Blanco, C., Vreugdenhil, D., Tesnier, K., Groot, S. P., & Koornneef, M. (2000). Genetic analysis of seed-soluble oligosaccharides in relation to seed storability of Arabidopsis. *Plant Physiology*, *124*(4), 1595–1604. https://doi.org/10.1104/pp.124.4.1595

Buchanan, B. B., Gruissem, W., & Jones, R. L. (2000). *Biochemistry & molecular biology of plants*. New York, NY: John Wiley & Sons.

Buitink, J., & Leprince, O. (2008). Intracellular glasses and seed survival in the dry state. *Comptes Rendus Biologies*, *331*(10), 788–795. https://doi.org/10.1016/j.crvi.2008.08.002

Caro, Y., Pina, M., Turon, F., Guilbert, S., Mougeot, E., Fetsch, D. V., … Graille, J. (2002). Plant lipases: Biocatalyst aqueous environment in relation to optimal catalytic activity in lipase-catalyzed synthesis reactions. *Biotechnology and Bioengineering*, *77*(6), 693–703. https://doi.org/10.1002/bit.10155

Chen, H., Yu, X., Zhang, X., Yang, L., Huang, X., Zhang, J., … Li, W. (2017). Phospholipase Dα1-mediated phosphatidic acid change is a key determinant of desiccation-induced viability loss in seeds. *Plant, Cell & Environment*, *41*, 50–63. https://doi.org/10.1111/pce.12925

Chen, H.-h., Chu, P., Zhou, Y.-l., Ding, Y., Li, Y., Liu, J., … Huang, S.-z. (2016a). Ectopic expression of NnPER1, a Nelumbo nucifera 1-cysteine peroxiredoxin antioxidant, enhances seed longevity and stress tolerance in Arabidopsis. *The Plant Journal*, *88*(4), 608–619. https://doi.org/10.1111/tpj.13286

Collins, J. R., Edwards, B. R., Fredricks, H. F., & Van Mooy, B. A. (2016). LOBSTAHS: An adduct-based lipidomics strategy for discovery and identification of oxidative stress biomarkers. *Analytical Chemistry*, *88*(14), 7154–7162. https://doi.org/10.1021/acs.analchem.6b01260

Colville, L., Bradley, E. L., Lloyd, A. S., Pritchard, H. W., Castle, L., & Kranner, I. (2012). Volatile fingerprints of seeds of four species indicate the involvement of alcoholic fermentation, lipid peroxidation, and Maillard reactions in seed deterioration during ageing and desiccation stress. *Journal of Experimental Botany*, *63*(18), 6519–6530. https://doi.org/10.1093/jxb/ers307

Devaiah, S. P., Pan, X., Hong, Y., Roth, M., Welti, R., & Wang, X. (2007). Enhancing seed quality and viability by suppressing phospholipase D in Arabidopsis. *The Plant Journal*, *50*(6), 950–957. https://doi.org/10.1111/j.1365-313X.2007.03103.x

Dörmann, P., Hoffmann-Benning, S., Balbo, I., & Benning, C. (1995). Isolation and characterization of an Arabidopsis mutant deficient in the thylakoid lipid digalactosyl diacylglycerol. *The Plant Cell*, *7*(11), 1801–1810. https://doi.org/10.1105/tpc.7.11.1801

Eastmond, P. J. (2006). Sugar-dependent1 encodes a patatin domain triacylglycerol lipase that initiates storage oil breakdown in germinating Arabidopsis seeds. *The Plant Cell*, *18*(3), 665–675. https://doi.org/10.1105/tpc.105.040543

Ellis, R. H., Osei-Bonsu, K., & Roberts, E. H. (1982). The influence of genotype, temperature and moisture on seed longevity in chickpea, cowpea and soya bean. *Annals of Botany*, *50*(1), 69–82. https://doi.org/10.1093/oxfordjournals.aob.a086347

Fernández-Marín, B., Kranner, I., Sebastián, M. S., Artetxe, U., Laza, J. M., Vilas, J. L., … García-Plazaola, J. I. (2013). Evidence for the absence of enzymatic reactions in the glassy state. A case study of xanthophyll cycle pigments in the desiccation-tolerant moss *Syntrichia ruralis*. *Journal of Experimental Botany*, *64*(10), 3033–3043. https://doi.org/10.1093/jxb/ert145

Galleschi, L., Capocchi, A., Ghiringhelli, S., Saviozzi, F., Calucci, L., Pinzino, C., & Zandomeneghi, M. (2002). Antioxidants, free radicals, storage proteins, and proteolytic activities in wheat (*Triticum durum*) seeds during accelerated aging. *Journal of Agricultural and Food Chemistry*, *50*(19), 5450–5457. https://doi.org/10.1021/jf0201430

González-Thuillier, I., Salt, L., Chope, G., Penson, S., Skeggs, P., Tosi, P., … Haslam, R. P. (2015). Distribution of lipids in the grain of wheat (cv. Hereward) determined by lipidomic analysis of milling and pearling fractions. *Journal of Agricultural and Food Chemistry*, *63*(49), 10705–10716. https://doi.org/10.1021/acs.jafc.5b05289

Groot, S. P. C., Surki, A. A., de Vos, R. C. H., & Kodde, J. (2012). Seed storage at elevated partial pressure of oxygen, a fast method for analysing seed ageing under dry conditions. *Annals of Botany*, *110*(6), 1149–1159. https://doi.org/10.1093/aob/mcs198

Harrington, J. F. (1963). Practical instructions and advice on seed storage. *Proceedings of the International Seed Testing Association*, *28*, 989–994.

Hu, D., Ma, G., Wang, Q., Yao, J., Wang, Y., Pritchard, H. W., & Wang, X. (2012). Spatial and temporal nature of reactive oxygen species production and programmed cell death in elm (*Ulmus pumila* L.) seeds during controlled deterioration. *Plant, Cell & Environment*, *35*(11), 2045–2059. https://doi.org/10.1111/j.1365-3040.2012.02535.x

International Seed Testing Association (2013). *International rules for seed testing*. Bassersdorf, Switzerland: ISTA.

Jarvis, P., Dörmann, P., Peto, C. A., Lutes, J., Benning, C., & Chory, J. (2000). Galactolipid deficiency and abnormal chloroplast development in the Arabidopsis MGD synthase 1 mutant. *Proceedings of the National*

Academy of Sciences of the United States of America, *97*(14), 8175–8179. https://doi.org/10.1073/pnas.100132197

Kuhl, C., Tautenhahn, R., Bottcher, C., Larson, T. R., & Neumann, S. (2012). CAMERA: An integrated strategy for compound spectra extraction and annotation of liquid chromatography/mass spectrometry data sets. *Analytical Chemistry*, *84*(1), 283–289. https://doi.org/10.1021/ac202450g

Lee, J., Welti, R., Roth, M., Schapaugh, W. T., Li, J., & Trick, H. N. (2012). Enhanced seed viability and lipid compositional changes during natural aging by suppressing phospholipase Dα in soybean seed. *Plant Biotechnology Journal*, *10*(2), 164–173. https://doi.org/10.1111/j.1467-7652.2011.00650.x

Lee, R.-H., Hsu, J.-H., Huang, H.-J., Lo, S.-F., & Grace Chen, S.-C. (2009). Alkaline α-galactosidase degrades thylakoid membranes in the chloroplast during leaf senescence in rice. *New Phytologist*, *184*(3), 596–606. https://doi.org/10.1111/j.1469-8137.2009.02999.x

Li, D.-Z., & Pritchard, H. W. (2009). The science and economics of ex situ plant conservation. *Trends in Plant Science*, *14*(11), 614–621. https://doi.org/10.1016/j.tplants.2009.09.005

Morrison, W. R. (1998). Wheat lipid composition. *Cereal Chemistry*, *55*, 548–558.

Nagel, M., Kodde, J., Pistrick, S., Mascher, M., Börner, A., & Groot, S. P. C. (2016). Barley seed aging: Genetics behind the dry elevated pressure of oxygen aging and moist controlled deterioration. *Frontiers in Plant Science*, *7*, 388. https://doi.org/10.3389/fpls.2016.00388

Oenel, A., Fekete, A., Krischke, M., Faul, S. C., Gresser, G., Havaux, M., ... Berger, S. (2017). Enzymatic and non-enzymatic mechanisms contribute to lipid oxidation during seed aging. *Plant and Cell Physiology*, *58*(5), 925–933. https://doi.org/10.1093/pcp/pcx036

Petla, B. P., Kamble, N. U., Kumar, M., Verma, P., Ghosh, S., Singh, A., ... Majee, M. (2016). Rice protein L-isoaspartyl methyltransferase isoforms differentially accumulate during seed maturation to restrict deleterious isoAsp and reactive oxygen species accumulation and are implicated in seed vigor and longevity. *New Phytologist*, *211*(2), 627–645. https://doi.org/10.1111/nph.13923

Probert, R. J., Daws, M. I., & Hay, F. R. (2009). Ecological correlates of ex situ seed longevity: A comparative study on 195 species. *Annals of Botany*, *104*(1), 57–69. https://doi.org/10.1093/aob/mcp082

R Development Core Team (2018). *R: A language and environment for statistical computing*. Vienna, Austria: R Foundation for Statistical Computing. Retrieved from. http://www.R-project.org/

Ratajczak, E., Małecka, A., Bagniewska-Zadworna, A., & Kalemba, E. M (2015). The production, localization and spreading of reactive oxygen species contributes to the low vitality of long-term stored common beech (Fagus sylvatica L.) seeds. *Journal of Plant Physiology*, *174*, 147–156. https://doi.org/10.1016/j.jplph.2014.08.021

Riewe, D., Jeon, H. J., Lisec, J., Heuermann, M. C., Schmeichel, J., Seyfarth, M., ... Altmann, T. (2016). A naturally occurring promoter polymorphism of the Arabidopsis FUM2 gene causes expression variation, and is associated with metabolic and growth traits. *The Plant Journal*, *88*(5), 826–838. https://doi.org/10.1111/tpj.13303

Riewe, D., Koohi, M., Lisec, J., Pfeiffer, M., Lippmann, R., Schmeichel, J., ... Altmann, T. (2012). A tyrosine aminotransferase involved in tocopherol synthesis in Arabidopsis. *The Plant Journal*, *71*(5), 850–859. https://doi.org/10.1111/j.1365-313X.2012.05035.x

Riewe, D., Wiebach, J., & Altmann, T. (2017). Structure annotation and quantification of wheat seed oxidized lipids by high-resolution LC-MS/MS. *Plant Physiology*, *175*(2), 600–618. https://doi.org/10.1104/pp.17.00470

Roach, T., Nagel, M., Börner, A., Eberle, C., & Kranner, I. (2018). *Changes in tocochromanols and glutathione reveal differences in the mechanisms of seed ageing under seedbank conditions and controlled deterioration in barley* (Vol. 156).

Roberts, E. H. (1960). The viability of cereal seed in relation to temperature and moisture: With eight figures in the text. *Annals of Botany*, *24*(1), 12–31. https://doi.org/10.1093/oxfordjournals.aob.a083684

Roberts, E. H. (1961). The viability of rice seed in relation to temperature, moisture content, and gaseous environment. *Annals of Botany*, *25*(3), 381–390. https://doi.org/10.1093/oxfordjournals.aob.a083759

Royal Botanic Gardens Kew. (2019). Seed information database (SID). Version 7.1. Available from: http://data.kew.org/sid/ (August 2019).

RStudio Team. (2016). RStudio: Integrated development environment for R. Boston, MA. Retrieved from http://www.rstudio.com/

Sallon, S., Solowey, E., Cohen, Y., Korchinsky, R., Egli, M., Woodhatch, I., ... Kislev, M. (2008). Germination, genetics, and growth of an ancient date seed. *Science*, *320*(5882), 1464–1464. https://doi.org/10.1126/science.1153600

Sasaki, K., Takeuchi, Y., Miura, K., Yamaguchi, T., Ando, T., Ebitani, T., ... Sato, T. (2015). Fine mapping of a major quantitative trait locus, qLG-9, that controls seed longevity in rice (Oryza sativa L.). *Theoretical and Applied Genetics*, *128*(4), 769–778. https://doi.org/10.1007/s00122-015-2471-7

Shen-Miller, J., Mudgett, M. B., Schopf, J. W., Clarke, S., & Berger, R. (1995). Exceptional seed longevity and robust growth: Ancient sacred lotus from China. *American Journal of Botany*, *82*(11), 1367–1380. https://doi.org/10.2307/2445863

Steiner, A. M., & Ruckenbauer, P. (1995). *Germination of 110-year-old cereal and weed seeds, the Vienna Sample of 1877. Verification of effective ultra-dry storage at ambient temperature* (Vol. 5).

Telewski, F. W., & Zeevaart, J. A. D. (2002). The 120-yr period for Dr. Beal's seed viability experiment. *American Journal of Botany*, *89*(8), 1285–1288. https://doi.org/10.3732/ajb.89.8.1285

The Food and Agriculture Organization of the United Nations (2014). Genebank standards for plant genetic resources for food and agriculture. In Rev (Ed.), *Commission on Genetic Resources for Food and Agriculture*. Rome: FAO.

Walters, C., & Engels, J. (1998). The effects of storing seeds under extremely dry conditions. *Seed Science Research*, *8*, 3–8.

Walters, C., Hill, L. M., & Wheeler, L. J. (2005). Dying while dry: Kinetics and mechanisms of deterioration in desiccated organisms. *Integrative and Comparative Biology*, *45*(5), 751–758. https://doi.org/10.1093/icb/45.5.751

Walters, C., Wheeler, L. M., & Grotenhuis, J. M. (2005). Longevity of seeds stored in a genebank: Species characteristics. *Seed Science Research*, *15*(1), 1–20. https://doi.org/10.1079/SSR2004195

Yan, M. (2017). The preliminary study on the optimum moisture content of ultra-dry storage and its related chemicals in seeds from six crop species. *Plant Genetic Resources*, *15*(6), 506–514. https://doi.org/10.1017/S1479262116000216

Yang, F., & Russell, A. J. (1995). A comparison of lipase-catalyzed ester hydrolysis in reverse micelles, organic solvents, and biphasic systems. *Biotechnology and Bioengineering*, *47*(1), 60–70. https://doi.org/10.1002/bit.260470108

Yashina, S., Gubin, S., Maksimovich, S., Yashina, A., Gakhova, E., & Gilichinsky, D. (2012). Regeneration of whole fertile plants from 30,000-y-old fruit tissue buried in Siberian permafrost. *Proceedings of the National Academy of Sciences*, *109*(10), 4008–4013. https://doi.org/10.1073/pnas.1118386109

SUPPORTING INFORMATION

Additional supporting information may be found online in the Supporting Information section at the end of the article.

Table S1. Wheat and barley accessions passport information.

Table S2. Germination, central metabolite profiles (GC-MS), and central metabolite-to-germination correlations for wheat 1, wheat 2, and barley.

Table S3. Lipid profiles (LC-MS) and lipid-to-germination correlations for wheat 1.

How to cite this article: Wiebach J, Nagel M, Börner A, Altmann T, Riewe D. Age-dependent loss of seed viability is associated with increased lipid oxidation and hydrolysis. *Plant Cell Environ*. 2020;43:303–314. https://doi.org/10.1111/pce.3651

3.7 PAPER 7:

Barley seed ageing: genetics behind the dry elevated pressure of oxygen ageing and moist controlled deterioration

by

Manuela Nagel, Jan Kodde, Sibylle Pistrick, Martin Mascher, Andreas Börner and Steven P. C. Groot

Published in

Frontiers in Plant Science (2016) 7, 388

https://doi.org/10.3389/fpls.2016.00388

To view supplementary material for this article, please visit
https://www.frontiersin.org/article/10.3389/fpls.2016.00388

frontiers in Plant Science

ORIGINAL RESEARCH
published: 31 March 2016
doi: 10.3389/fpls.2016.00388

Barley Seed Aging: Genetics behind the Dry Elevated Pressure of Oxygen Aging and Moist Controlled Deterioration

Manuela Nagel[1], Jan Kodde[2], Sibylle Pistrick[1], Martin Mascher[1], Andreas Börner[1] and Steven P. C. Groot[2]*

[1] Genebank Department, Leibniz Institute of Plant Genetics and Crop Plant Research (IPK Gatersleben), Stadt Seeland, Germany, [2] Wageningen UR, Plant Research International B.V., Wageningen, Netherlands

Experimental seed aging approaches intend to mimic seed deterioration processes to achieve a storage interval reduction. Common methods apply higher seed moisture levels and temperatures. In contrast, the "elevated partial pressure of oxygen" (EPPO) approach treats dry seed stored at ambient temperatures with high oxygen pressure. To analyse the genetic background of seed longevity and the effects of seed aging under dry conditions, the EPPO approach was applied to the progeny of the Oregon Wolfe Barley (OWB) mapping population. In comparison to a non-treated control and a control high-pressure nitrogen treatment, EPPO stored seeds showed typical symptoms of aging with a significant reduction of normal seedlings, slower germination, and less total germination. Thereby, the parent Dom ("OWB-D"), carrying dominant alleles, is more sensitive to aging in comparison to the population mean and in most cases to the parent Rec ("OWB-R"), carrying recessive alleles. Quantitative trait locus (QTL) analyses using 2832 markers revealed 65 QTLs, including two major loci for seed vigor on 2H and 7H. QTLs for EPPO tolerance were detected on 3H, 4H, and 5H. An applied controlled deterioration (CD) treatment (aged at higher moisture level and temperature) revealed a tolerance QTL on 5H, indicating that the mechanism of seed deterioration differs in part between EPPO or CD conditions.

Keywords: caryopsis, seed conservation, seed storage, genotype, germination, linkage mapping

OPEN ACCESS

Edited by:
Changbin Chen,
University of Minnesota, USA

Reviewed by:
Penny M. A. Kianian,
University of Minnesota, USA
Chengdao Li,
Murdoch University, Australia

****Correspondence:***
Manuela Nagel
Nagel@ipk-gatersleben.de

Specialty section:
This article was submitted to Plant Genetics and Genomics, a section of the journal Frontiers in Plant Science

Received: *13 December 2015*
Accepted: *14 March 2016*
Published: *31 March 2016*

Citation:
Nagel M, Kodde J, Pistrick S, Mascher M, Börner A and Groot SPC (2016) Barley Seed Aging: Genetics behind the Dry Elevated Pressure of Oxygen Aging and Moist Controlled Deterioration. Front. Plant Sci. 7:388. doi: 10.3389/fpls.2016.00388

INTRODUCTION

Knowledge of the mechanism of seed longevity has become important since international companies have started to ship seeds around the world and gene banks initiated the first seed collections at the beginning of the twentieth century. Thereby, long-term survival is greatly influenced by plant breeders who continuously work on seed quality performance.

Desiccation-tolerant (orthodox) seeds extend their life span when they reach maturity, dry, and enter the glassy state (Walters et al., 2005a; Buitink and Leprince, 2008). A glass is an amorphous, solid state with high viscosity (Buitink and Leprince, 2004) where metabolic processes are reduced to a minimum or can be even excluded (Kranner et al., 2010; Fernández-Marín et al., 2013). In this state, reactive oxygen species (ROS; Bailly, 2004; El-Maarouf-Bouteau et al., 2011), non-enzymatic Amadori and Maillard reactions (Sun and Leopold, 1995), and lipid peroxidations can damage macromolecules such as DNA, proteins, and lipids, and lead to seed deterioration. Further extrinsic factors can be multifaceted and include abiotic and biotic stress during seed development, seed maturity, harvest, drying, and cleaning. The rate of decay depends largely on the storage environment, as temperatures (Roberts, 1973), atmosphere (Groot et al., 2014) and relative humidity (RH) interact with the genotype (Walters et al., 2005b; Nagel and Börner, 2010). Consequently, a temperature or RH increase cause a viscosity decrease and thus a glass to rubber state change with higher molecular mobility and reaction kinetics (Walters, 1998).

The investigation of seed deterioration under natural conditions is protracted, as optimally stored orthodox seeds can survive for decades without any effect on the germination. To study seed deterioration more quickly, experimental approaches were invented by Roberts (1960) and refined by Hampton and Tekrony (1995) and Hay et al. (2008). Basically, temperatures are adjusted to above 30°C and seed moisture levels are increased, which converts the glassy state into a rubber state, with dramatic effects on chemical reactions (Walters, 1998). These methods have been widely applied for investigations on ecological (Probert et al., 2009), molecular (Waterworth et al., 2010), biochemical (Lehner et al., 2008), and genetic (Sasaki et al., 2015) levels. However, biochemical processes of artificially aged seeds in the rubber state can deviate from those in industrial dry storage at ambient temperatures or in cold, dry storage in gene banks, such as the half-reduction potential of the antioxidant glutathione and an assumed shift in pH (Nagel et al., 2015). Therefore, Groot et al. (2012) developed the "elevated partial pressure of oxygen" (EPPO) approach, which avoids a glass–rubber transition while increasing oxidation by exposing the dry seeds to high oxygen pressure (up to 18 MPa). This treatment affects germination in such a way that the visual appearance of seed deterioration is consistent with that observed after long term dry storage.

So far, the genetic backgrounds of seed longevity have been investigated by exposing seeds to relatively moist experimental aging conditions. Quantitative trait locus (QTL) analysis and association mapping revealed that longevity QTLs often co-located with QTLs for other traits, including morphological traits such as spike compactness (Rehman Arif et al., 2012). In barley, whose seeds are categorized as having medium-long storage potential (Walters et al., 2005b), QTLs for longevity co-located with QTLs for plant height and naked caryopsis on two chromosomes (Nagel et al., 2009).

The aim of this study was to investigate the genetic background of barley seed longevity by applying the EPPO treatment to dry-stored seeds. Double haploid (DH) lines of the Oregon Wolfe Barley (OWB) mapping population, genotyped with 2383 markers (Chutimanitsakun et al., 2011), were equilibrated to 40% RH at 20°C and exposed to high oxygen pressure at 18 MPa. Control and deterioration effects after treatment were assessed by germination speed and compared with previous experiments using CD (Nagel et al., 2009). Both data sets were analyzed and reanalyzed with the currently available marker sets and differences after QTL mapping are discussed.

Abbreviations: AUC, Area under the curve; BLASTX, Basic Local Alignment Tool; C08, Control of CD experiment in 2008; CD, Controlled deterioration; CD08, Controlled deterioration experiment in 2008; DArT, Diversity Array Technology; DH, Double haploid; E, Expect Value; EPPO, Elevated partial pressure of oxygen; EST, Expressed sequence tag; LOD, Logarithm of odds; OWB, Oregon Wolfe Barley; "OWB-D," Dominant parent Dom; "OWB-R," Recessive parent Rec; P50, Half-viability period; QTL, Quantitative trait locus; R^2, Explained phenotypic variation; RH, Relative humidity; ROS, Reactive oxygen species; TC, Treatment Control of EPPO experiment in 2011; T50, Time to 50% germination; %NS, Percentage of normal seedlings; %TG, Percentage of total germination.

MATERIAL AND METHODS

Experimental Material

Caryopses, henceforth termed seeds, of the DH OWB mapping population were studied on their ability to cope with stress during storage. The set of 94 spring barley lines was developed by pollinating the *Hordeum vulgare* F1 hybrid produced by the dominant parent Dom ("OWB-D") × recessive parent Rec ("OWB-R") with *H. bulbosum* (Costa et al., 2001). Characteristics of "OWB-D" and "OWB-R" were selected as dominant and recessive morphological marker stocks (Wolfe and Franckowiak, 1991), whereby more information can be found at http://barleyworld.org/oregonwolfe.php. A marker set of 2832 markers including 463 Restriction site Associated DNA (RAD) markers was developed by Chutimanitsakun et al. (2011) and is available at http://wheat.pw.usda.gov/ggpages/maps/OWB/. In 2005, a large number of seeds were produced under optimum conditions to maximize seed yield and quality in Gatersleben, Germany, and stored in conditions of 10% seed moisture content and −18°C.

Storage Treatment

The EPPO treatment was applied at Wageningen UR in 2011 according to Groot et al. (2012). Seeds of the OWB population were equilibrated to 40% RH at 20°C for 2 weeks. Thereafter, the seeds were stored under ambient atmospheric pressure in 1 l rubber-sealed glass jars (TC, treatment control) or in 12 l steel tanks under an elevated partial pressure of oxygen (EPPO) or nitrogen (EPPN) gas. The latter were used to control for potential damage induced by the high pressure and/or pressure release from the high pressure storage. The seed samples were placed within the tanks in 13 ml polystyrene tubes (Sarstedt, Germany) perforated with holes of ~1 mm diameter and closed with low-density polyethylene push caps. The tanks were filled slowly (~6 bar/min) with oxygen or nitrogen from large buffer tanks, until the tank pressure reached ~18 MPa. To minimize temperature changes during filling, the tanks were placed in water at ambient temperature. Filling of the tanks with oxygen was performed by skilled staff at a local scuba diving shop (4Divers, Veenendaal, The Netherlands). As tanks were not flushed before filling, they retained the initial atmospheric amounts of oxygen (0.021 MPa partial pressure), nitrogen and other minor gases present in air. Silica gel, equilibrated to 40% RH, was also added to the tanks and glass jars to buffer RH in the containers. The tank pressure was released with an average relative pressure decline of 0.5% per min (maximum 3.0%) using a computer-controlled relative flow rate. Between experiments and prior to germination testing, seeds were stored in a cabinet with air circulating above a saturated $CaCl_2$ salt solution (RH 35%) at a temperature of 20°C. Seeds were packaged in sealed laminated foil bags and sent to IPK Gatersleben. As an initial test, 10 lines of the "W766" barley population, multiplied and stored together with seeds of the OWB population, were treated for 4, 6, 8, and 12 weeks in oxygen at 18 MPa (atmospheric pressure is ~0.1 MPa) and showed a significant reduction in germination and germination speed between 8 and 12 weeks (Supplementary Figure 1). For the main experiment, the seeds were stored for 9 weeks under EPPO, EPPN, or ambient pressure conditions (TC).

Germination Test

At IPK Gatersleben, seeds were subjected to germination tests. Four replicates of 50 seeds (for a few lines, only 10–40 seeds per replicate were available) were placed on moistened filter paper and kept in a germination chamber at 20°C, 60% RH for 8 h of light. Germinating seeds were determined daily by examining radicle protrusion. After a period of 10 days, seeds were classified into normal seedlings, which show the potential to develop into satisfactory plants; abnormal seedlings, which are damaged, deformed, or decayed, and do not show potential to develop into a normal plant; and non-germinating seeds according to (ISTA, 2012). The percentage of total germinated seeds (%TG) and normal seedlings (%NS), the time to reach 50% of the maximum germination (T50), and the area under the curve (AUC) after 100 h were calculated using the curve-fitting module of the Germinator (Joosen et al., 2010).

Controlled Deterioration

In 2008, DH lines of the OWB population were experimentally aged at 44°C and 18% seed moisture content, following the procedure of CD by Hampton and Tekrony (1995). Four replicates of 50 seeds [control of 2008 tests (C08) and controlled deteriorated seeds of 2008 tests (CD08)] were germinated according to ISTA (2008) and published in Nagel et al. (2009). Here, %NS and %TG were recalculated and, on the basis of viability equation $v = K_i–\sigma^{-1}$ p of Roberts (1973), the half-viability period (P50) for CD and EPPO aging was analyzed using %NS for initial seed germination (K_i) and 63 days and 3 days to calculate sigma (σ) for EPPO and CD, respectively. Further, %NS were transformed to probit units and used to subtract control and EPPN from the EPPO and CD treatments and to overcome the effect of the different initial germinations of the genotypes. This resulted in additional traits for QTL mapping: 3-year storage (C08-TC), TC-EPPN, EPPO-TC, EPPO-EPPN, CD08–TC08.

The Shapiro–Wilk test indicated that, except for T50, all data did not match the pattern produced by a population with normal distribution. Therefore, Friedman Repeated Measures Analysis of Variance on Ranks, followed by Tukey's and Dunn's tests, was used to analyse differences ($P < 0.05$) between treatments.

QTL Mapping

QTL analyses for each of the phenotypic characters were conducted for the 93 OWB lines using the composite interval mapping procedure Zeng (1994) implemented in Windows QTL Cartographer 2.5 (Wang et al., 2011). Experiment-wise significance likelihood ratio test (LR) statistic thresholds ($P \leq 0.05$) for QTL identification were determined with 1000 permutations, expressed as logarithm of odds (LOD = 0.217LR), and reinforced the set LOD threshold of 3. Significant QTL were characterized by the position of significant flanking markers, the marker with highest LOD, the proportion of phenotypic variance explained by the individual QTL (R^2), the additive effect (expressed as one-half of the difference between the two allelic classes), and the LOD threshold. Negative values indicate that alleles were contributed by the parent "OWB-R" and positive values by the parent "OWB-D."

Gene Content in QTL Regions

Sequences of flanking markers of QTL regions were mapped to the whole-genome shotgun assembly of barley cv. "Morex" (The International Barley Genome Sequencing Consortium, 2012) with BWA mem version 0.7.12 (Li, 2013). Primary alignments with mapping quality ≥ 30 were extracted with SAMtools (Li et al., 2009), converted to BED format with BEDTools (Quinlan and Hall, 2010) and imported into R (R Core Team, 2015). Genetic positions of QTL regions in the POPSEQ genetic map (Mascher et al., 2013) were determined from the alignments of marker sequences to the Morex WGS contigs anchored by POPSEQ. If both flanking markers of a QTL regions were positioned in the POPSEQ map, the identifiers and functional annotation of all genes (The International Barley Genome Sequencing Consortium, 2012) between flanking markers were exported into a spreadsheet using functionalities of the R packages "openxlsx" (https://cran.r-project.org/web/packages/openxlsx/) and "data.table" (https://cran.r-project.org/web/packages/data.table/).

RESULTS

Phenotypic Data

Population lines were selected as dominant and recessive morphological marker stocks and show high phenotypic diversity. After 6 years of storage at 10% seed moisture content and −18°C, the seed material had on average 86.2 ± 16.3%NS and differed significantly between "OWB-D" (54.0%) and "OWB-R" (92.5%; Supplementary Figure 2). These differences were also reflected in the AUC and T50, which showed lower area and longer time to reach 50% germination for "OWB-D" (AUC = 35.8; T50 = 49.2 h) in comparison to "OWB-R" (AUC = 72.3; T50 = 27.6 h; **Figure 1** and Supplementary Table 1). Except for T50 after EPPO and for %TG and P50 after CD08, the parent "OWB-D" showed worse performance than the mean of the population. "OWB-R" was only inferior to the mean of population after EPPO treatment.

To compensate for potential damage induced by high pressure and pressure release, high nitrogen pressure (EPPN) was applied to seeds of the OWB population. The %NS was the only trait that showed a significant ($P < 0.05$) effect between both controls, storage at ambient pressure and EPPN treatment (Supplementary Table 1). In contrast, a high oxygen concentration, as applied in the EPPO treatment, affected germination significantly ($P < 0.05$), resulting in a reduced %NS, %TG, and AUC and extended

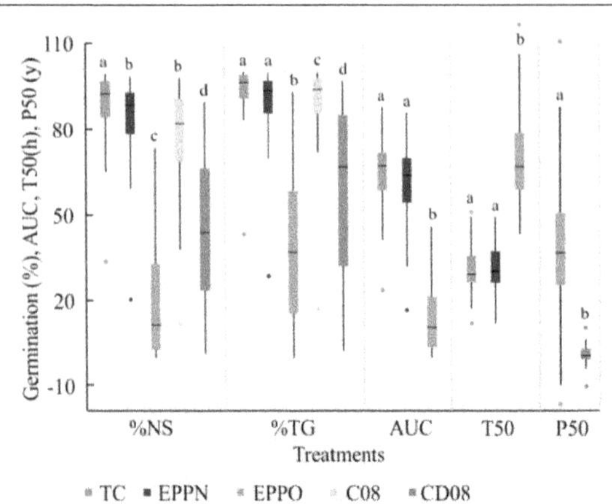

FIGURE 1 | Germination performance of 94 Oregon Wolfe Barley (OWB) lines after different storage treatments tested in 2008 and 2011. Germination performance is expressed as the percentage of normal seedlings (NS%), total germination (TG%), area under the curve (***AUC***), time to 50% germination (T50 in hours), and half-viability period (P50 in days). Boxplots show minimum and maximum values (spots), 25% and 75% quartiles, and whiskers of the 1.5 interquartile range. Color code is given in the figure. C08, control performed in 2008; CD08, controlled deterioration performed in 2008; EPPN, elevated partial pressure of nitrogen storage; EPPO, elevated partial pressure of oxygen storage; TC, treatment control at ambient air pressure. a, b, c, and d symbolize significant differences between treatments at *P < 0.05.*

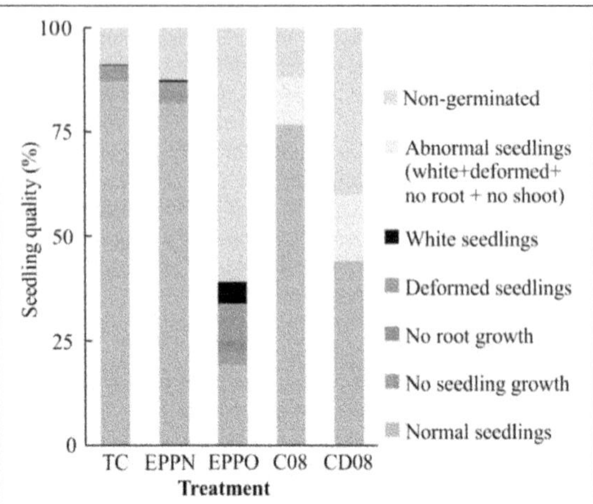

FIGURE 2 | Seedling quality of Oregon Wolfe Barley (OWB) lines after storage under atmospheric pressure (TC, treatment control), high-pressure nitrogen (EPPN), and high-pressure oxygen (EPPO). For comparisons, normal, abnormal (not discriminated into white, deformed, no root, no seedling growth), and non-germinated seedlings of C08 (control performed in 2008) and CD08 (controlled deterioration performed in 2008) are shown.

T50 (**Figure 1**). Seedling quality differed between the treatments and was best in the control. After the high-pressure nitrogen treatment, there was a minor relative increase in abnormal seedlings with no shoot, white shoots, or deformations, and of non-germinated seeds, resulting in an ~5% decline in normal seedlings in comparison to the control seeds stored under ambient pressure conditions (**Figure 2**). The EPPO treatment caused extensive damage which resulted in a relative increase in seedlings with no shoot, no root, white shoots, deformation, and no germination of 190, 1026, 1600, 235, and 587%, respectively.

Comparing the 9 weeks EPPO treatment with the CD08 treatment, less damage was observed after the CD08 treatment. Only 38% more abnormal seedlings (not differentiated between white, deformed, no root and no shoot seedlings) were Counted (**Figure 2**). After 3 days of CD, CD08 was significantly different ($P < 0.05$) to control in 2008 and to EPPO treatment in 2011, and revealed an average decline in %NS and %TG of OWB population to 44.3 and 60.3%, respectively. Again, with some exceptions, the parent "OWB-D" performs less well than the population mean in all traits whereas "OWB-R" revealed higher initial germination but comparable germinations after CD.

On the basis of %NS in the ambient control, EPPN, and after EPPO, the P50 was calculated and resulted in a mean for the whole population of 38.5 ± 21.8 days after the EPPO treatment and of 2.3 ± 3.5 days after CD08 treatment. As demonstrated by the previous data, the parent "OWB-D" has a shorter half-viability period in comparison with "OWB-R" and the mean half-viability period of the DH population was higher than that of either parent.

Correlation analysis revealed that EPPN treatment showed a higher correlation with control treatments than did EPPO treatment. Here, the highest coefficients ($r = -0.62$; $P < 0.001$) were found between %TG of EPPO and T50 of controls, and indicate a weak relationship between both. All the germination parameters measured showed highly significant correlations ($P < 0.001$) with controls following EPPN treatment but this was not so following EPPO treatment, when not all parameters correlated significantly and those that did had lower r values (**Table 1**).

QTL Mapping

In total, 2832 markers were used to analyse the EPPO experiments and to reanalyse the results of the 2008 experiments (**Figure 3**; **Table 2**). In total, 65 significant QTLs (LOD > 3) were found across all chromosomes. Eight QTLs were detected under control, 9 for EPPN, 13 for EPPO conditions, 3 for C08, and 12 for CD08. In addition, 13 QTLs and 7 QTLs appeared for NS (probit) and TG (probit), respectively, after subtraction of control and EPPN treatment from EPPO, CD08, and C08. Most QTLs were found on chromosome 2H between 109 and 182 cM and on chromosome 7H between 63 and 122 cM. QTLs with the highest LOD scores were found for control (T50; $R^2 = 42.6$; LOD = 17.7) on 7H, for EPPN (T50; $R^2 = 55.0$; LOD = 25.1) on 7H, for EPPO treatment (AUC; $R^2 = 25.5$; LOD = 11.5) on 3H, for C08 (%TG; $R^2 = 17.9$; LOD = 7.3) on 2H, and for CD08 (%TG; $R^2 = 42.9$; LOD = 16.3) on 2H.

Subtraction of control or EPPN treatment from EPPO, CD, and C08 is assumed to compensate for the effect of control

TABLE 1 | Correlation coefficients between germination parameters of different treatments.

	TC_ %TG	TC_ T50	TC_ AUC	EPPN_ %NS	EPPN_ %TG	EPPN_ T50	EPPN_ AUC	EPPO_ %NS	EPPO_ %TG	EPPO_ T50	EPPO_ AUC	C08_ %NS	C08_ %TG	CD08_ %NS	CD08_ %TG	P50_ EPPO	P50_ CD08
TC_%NS	0.93	−0.61	0.80	0.79	0.75	−0.54	0.70	0.38	0.52	−0.24	0.45	0.31	0.35		0.25	0.58	0.23
TC_%TG		−0.59	0.81	0.72	0.71	−0.54	0.67	0.33	0.48	−0.21	0.40	0.30	0.34		0.22	0.55	0.21
TC_T50			−0.93	−0.58	−0.58	0.84	−0.79	−0.34	−0.62	0.32	−0.56					−0.66	
TC_AUC				0.70	0.68	−0.81	0.84	0.33	0.60	−0.26	0.52		0.22			0.66	0.03
EPPN_%NS					0.95	−0.62	0.85	0.49	0.68		0.58	0.33	0.35			0.69	0.22
EPPN_%TG						−0.59	0.84	0.49	0.67	−0.23	0.59	0.29	0.31			0.70	0.24
EPPN_T50							−0.90	−0.24	−0.56	0.21	−0.46	−0.15	−0.19	0.22		−0.59	
EPPN_AUC								0.39	0.68	−0.23	0.58	0.23	0.26			0.71	
EPPO_%NS									0.85	−0.71	0.90	0.39	0.46	0.29	0.37	0.81	0.37
EPPO_%TG										−0.57	0.96	0.29	0.38			0.97	
EPPO_T50											−0.75	−0.21	−0.25		−0.24	−0.56	
EPPO_AUC												0.27	0.35			0.93	
C08_%NS													0.93	0.51	0.56	0.33	0.51
C08_%TG														0.59	0.66	0.40	0.57
CD08_%NS															0.97	0.07	0.83
CD08_TG%																	0.83
P50_TC																	
P50_EPPN																	
P50_EPPO																	
P50_C08																	

Spearmen correlation was performed on the basis of 93 Oregon Wolfe Barley (OWB) lines. C08, control performed in 2008; CD08, controlled deterioration performed in 2008; EPPN, elevated partial pressure of nitrogen storage; EPPO, elevated partial pressure of oxygen storage; P50, half-viability period; TC, treatment control at ambient air pressure; %NS, percentage of normal seedlings; %TG, percentage of total germination; T50, time to reach 50% of germinated seeds (hours); AUC, area under the curve; $P \le 0.001$; $P \le 0.01$; $P < 0.05$.

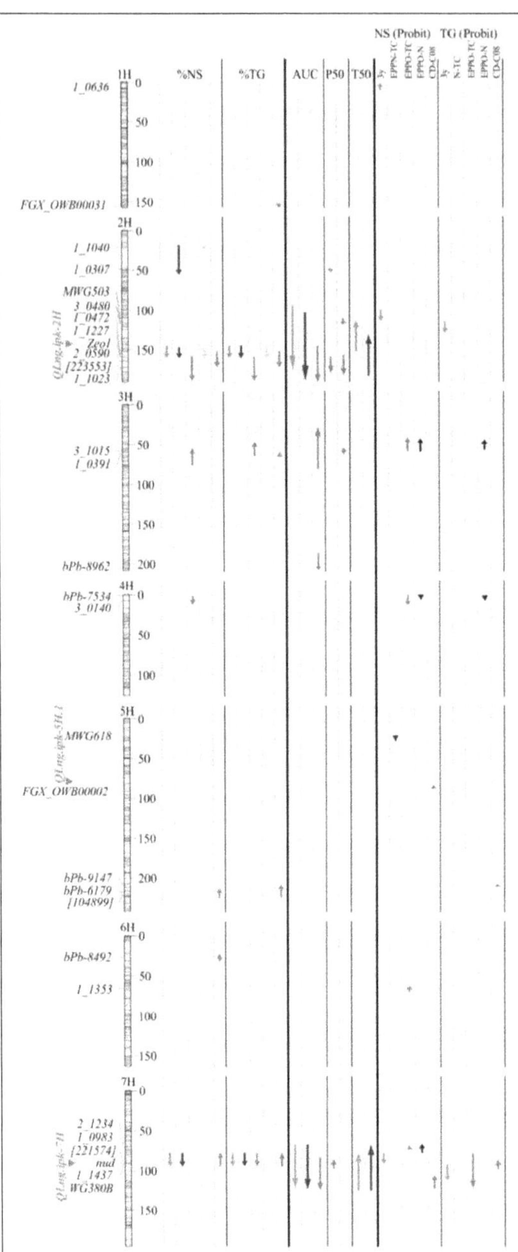

FIGURE 3 | Linkage map of Oregon Wolfe Barley (OWB) mapping population and quantitative trait locus (QTL) locations. QTLs detected under control conditions are marked in green (TC, treatment control, dark green; C08, control prepared in 2008, light green), those under nitrogen pressure (EPPN) in brown, under oxygen pressure (EPPO) in blue, and those after CD in 2008 red. Map distances are shown in cM. QTLs declared above a logarithm of odds (LOD) threshold of 3. Arrow lengths indicate QTL intervals having LOD > 3 and arrows up indicate that alleles came from the "OWB-D" parent and those down from the "OWB-R" parent.

and high-pressure treatment and reveal treatment-specific loci. QTLs specific for EPPO were found on 3H, 4H, and 6H. On 5H, two QTLs were found for CD (94.9 and 207.7 cM) and one for EPPN (21.4 cM) and, on 2H, one QTL accounted for the differences after a 3-year storage period. Between 67 and 122 cM on chromosome 7H, QTLs were detected under all treatments.

Interestingly, alleles with the greatest effect on most QTLs (39 out of 65) were contributed by the parent "OWB-R" showing the highest additive effects in CD08 and EPPO treatment with −2236.5 and −1290.1 for %TG. The dominant parent "OWB-D" contributed the most alleles for QTLs on chromosome 3H after EPPO treatment and on 5H and 7H after CD08 treatment.

DISCUSSION

High Nitrogen Pressure Induce White Seedling Appearance

Anoxic conditions (produced by storage under 100% nitrogen gas) are used in the food industry to preserve organic material from microbial growth (Van Campenhout et al., 2013). In the current experiment, EPPN was used as a second control to compensate for the potential effects of high pressure and pressure release. With respect to %TG, AUC, and T50, high nitrogen pressure did not induce a significant effect compared to the control under ambient air pressure, but %NS was significantly reduced, as a higher number of seedlings with white shoots and no roots were produced. A release of pressure from the tanks at the end of the storage period can create an increase in the volume of gases trapped in internal spaces, thereby causing physical disruption of tissues and producing seedlings with deformations. However, in our experiments with barley, the frequency of seedlings with deformations was not increased. A possible explanation for the white seedlings is nitrogen conversion to nitrogen oxides, which affect metabolism, oxidation state, or signaling (Bethke et al., 2007). This might also explain the observed dormancy alleviation for some lines, as indicated by an increase in the %NS.

Detrimental Effects of EPPO on Seed Germination

Oxygen is essential for the survival of most living organisms under moist conditions, as it is required for aerobic respiration. With desiccation tolerant organisms such as seeds, respiration activity ceases under dry conditions and they can survive without oxygen. Molecular oxygen can have a detrimental effect through the formation of super oxide which reacts with almost all organic components and damages DNA, proteins and lipids, resulting in membrane leakage and cell death (Van Breusegem and Dat, 2006). The effect of high oxygen pressure was investigated by Caldwell (1964), who found that dry bean and pea seeds are able to tolerate high atmospheric oxygen pressure whereas imbibed beans and peas cannot. In our experiments, the dry barley seeds under investigation responded to high-pressure oxygen treatment with a considerable reduction in germination performance. After 9 weeks of dry storage under ∼18 MPa oxygen, the %NS and %TG were reduced by 67 and 52% and

TABLE 2 | Quantitative trait locus (QTL) analysis of the germination performance of the Oregon Wolfe Barley (OWB) mapping population after experimental aging methods.

Trait	Treat	Unit	C	Flanking markers	Marker	Position	LOD	R^2	Additive effect
NS	TC	%	2	146.7-164.3	1_0472	155.5	12.1	24.8	−893.1
	TC	%	7	82.6-94.7	1_1437	94.7	6.9	14.3	−614.3
	EPPN	%	2	21.9-55.2	1_0307	49.7	3.7	5.9	−488.5
	EPPN	%	2	148.9-164.3	Zeo1	159.9	12.9	26.8	−1096.1
	EPPN	%	7	82.6-94.7	nud	93.7	11.5	22.8	−914.7
	EPPO	%	2	159.9-182.2	1_1023	172.3	9.1	22.9	−946.6
	EPPO	%	3	55.2-69.4	1_0391	60.6	9.7	27.2	1027.7
	EPPO	%	4	1.1-9	3_0140	9.0	3.7	8.4	−586.1
	C08	%	2	-	1_1040	21.9	3.5	10.1	−855.7
	C08	%	2	148.9-164.3	1_1227	157.7	7.3	19.0	−1438.2
	CD08	%	2	152.2-174.5	2_0590	164.3	12.0	31.0	−1599.1
	CD08	%	5	212.1-228.6	bPb-6179	218.7	4.6	9.7	851.8
	CD08	%	6	20-30	bPb-8492	25.5	3.7	7.7	779.4
	CD08	%	7	82.6-94.7	1_1437	94.7	6.0	13.2	1033.7
NS	TC	Probit	2	86.9-105.6	1_0859	96.8	4.9	9.0	−23.1
	TC	Probit	2	146.7-164.3	1_0472	155.5	14.2	31.2	−42.7
	TC	Probit	7	87.1-94.7	1_1437	94.7	6.3	13.0	−26.1
	EPPN	Probit	2	12.1-17.5	3_0155	15.3	3.4	5.5	−16.3
	EPPN	Probit	2	148.9-164.3	Zeo1	159.9	17.1	40.0	−45.0
	EPPN	Probit	7	82.6-94.7	1_1437	94.7	8.9	16.7	−29.0
	EPPO	Probit	2	159.9-182.2	1_1023	172.3	9.7	22.1	−46.8
	EPPO	Probit	3	46.2-66.1	3_1015	57.3	10.2	23.8	46.5
	EPPO	Probit	4	0-9	bPb-7534	1.1	4.3	8.5	−29.0
	C08	Probit	2	146.7-164.3	3_0396	158.8	10.5	32.2	−40.2
	C08	Probit	5	211-220.9	[104899]	219.8	4.4	11.5	23.1
	CD08	Probit	2	152.2-174.5	2_0590	164.3	6.3	17.6	−33.1
	CD08	Probit	5	207.7-228.6	[104899]	219.8	7.4	20.5	34.6
NS	TC-C08	Probit	1	0-8.9	1_0636	8.9	3.2	9.2	24.4
	TC-C08	Probit	2	99-114.4	MWG503	113.3	4.8	13.5	−29.7
	TC-C08	Probit	7	82.6-94.7	[221574]	91.5	5.1	15.2	−31.0
	CD-C08	Probit	5	91.6-94.9	FGX_OWB00002	94.9	3.6	12.4	−38.6
	CD-C08	Probit	7	105.5-122	WG380B	122.0	6.5	25.7	54.3
	EPPO-TC	Probit	3	48.4-66.1	3_1015	57.3	7.3	17.4	44.6
	EPPO-TC	Probit	4	9-22.7	3_0140	9.0	3.4	7.4	−26.8
	EPPO-TC	Probit	6	59.7-69.5	1_1353	66.2	4.1	8.9	28.9
	FPPO-TC	Probit	7	73.8-74.9	2_1234	73.8	3.5	7.4	26.3
	EPPO-EPPN	Probit	3	46.2-66.1	3_1015	57.3	14.1	35.9	53.7
	EPPO-EPPN	Probit	4	0-1.1	bPb-7534	1.1	3.7	7.0	−25.0
	EPPO-EPPN	Probit	7	67.3-82.6	1_0983	74.9	4.9	10.1	27.8
	EPPN-TC	Probit	5	21.4-23.6	MWG618	21.4	3.6	13.8	−23.9
TG	TC	%	2	146.7-164.3	1_0472	155.5	12.1	30.3	−821.9
	TC	%	7	82.6-94.7	1_1437	94.7	6.1	13.8	−508.0
	EPPN	%	2	146.7-164.3	1_0472	155.5	12.3	28.5	−977.6
	EPPN	%	7	82.6-94.7	1_1437	94.7	9.4	19.9	−760.4
	EPPO	%	2	159.9-182.2	1_1023	172.3	8.6	18.5	−1226.8
	EPPO	%	3	46.2-65	3_1015	57.3	6.8	13.9	1031.1
	EPPO	%	7	82.6-94.7	1_1437	94.7	9.2	20.8	−1290.1
	C08	%	2	148.9-164.3	1_1227	157.7	7.3	17.9	−1420.9
	CD08	%	1	-	FGX_OWB00031	155.1	3.6	7.0	−886.0
	CD08	%	2	152.2-174.5	2_0590	164.3	16.3	43.0	−2236.5

(Continued)

TABLE 2 | Continued

Trait	Treat	Unit	C	Flanking markers	Marker	Position	LOD	R^2	Additive effect
	CD08	%	3	56.2-58.4	3_1015	57.3	3.8	7.1	882.4
	CD08	%	5	207.7-228.6	[104899]	219.8	8.2	17.3	1356.4
	CD08	%	7	82.6-94.7	1_1437	94.7	3.9	6.1	852.9
TG	TC	Probit	2	91.3-105.6	1_0859	96.8	4.2	8.4	−24.5
	TC	Probit	2	146.7-164.3	1_0472	155.5	11.7	27.3	−44.2
	TC	Probit	6	-	3_1308	45.4	3.0	5.9	19.9
	TC	Probit	7	94.7-108.8	1_1437	94.7	5.2	11.4	−26.9
	EPPN	Probit	2	93.5-113.3	MWG503	113.3	4.4	7.2	−22.0
	EPPN	Probit	2	148.9-164.3	Zeo1	159.9	13.5	28.5	−41.9
	EPPN	Probit	6	64-69.5	bPb-9082	69.5	4.1	6.8	21.8
	EPPN	Probit	7	82.6-94.7	nud	93.7	11.4	22.6	−35.7
	EPPO	Probit	2	159.9-173.4	3_0248	163.3	12.7	34.3	−56.1
	EPPO	Probit	3	35.1-55.2	FGX_OWB00221	46.2	4.5	9.5	29.4
	EPPO	Probit	7	82.6-94.7	nud	93.7	8.1	18.6	−41.2
	C08	Probit	2	146.7-182.4	3_0396	158.8	11.9	35.4	−50.6
	C08	Probit	5	181.3-182.4	[105067]	182.4	3.3	7.9	23.0
	CD08	Probit	2	152.2-174.5	2_0590	164.3	12.6	32.2	−55.1
	CD08	Probit	5	207.7-228.6	bPb-6179	218.7	6.7	14.3	35.2
	CD08	Probit	6	20-31.1	bPb-8492	25.5	4.0	8.3	27.0
	CD08	Probit	7	91.5-94.7	1_1437	94.7	3.7	7.5	25.8
TG	3y	Probit	2	113.3-127.6	3_0480	121.0	5.5	16.6	−37.2
	3y	Probit	7	94.7-108.8	1_1437	94.7	4.4	13.8	−33.5
	CD-C08	Probit	5	-	bPb-9147	207.7	3.2	11.6	25.9
	CD-C08	Probit	7	94.7-122	1_1437	94.7	3.5	13.9	27.2
	EPPO-TC	Probit	3	47.3-66.1	3_1015	57.3	5.6	17.7	35.9
	EPPO-EPPN	Probit	3	48.4-66.1	3_1015	57.3	8.0	24.5	53.9
	EPPO-EPPN	Probit	4	0-9	bPb-7534	1.1	3.9	11.0	−23.3
AUC	TC	mm^2	2	91.3-173.4	1_0472	155.5	7.5	12.8	−5.6
	TC	mm^2	7	63.9-119.8	1_1437	94.7	13.0	26.5	−7.4
	EPPN	mm^2	2	100.1-186.6	Zeo1	159.9	14.3	21.3	−8.2
	EPPN	mm^2	7	63.9-120.9	nud	93.7	20.3	36.8	−9.7
	EPPO	mm^2	2	142.3-188.8	Zeo1	159.9	10.2	21.6	−5.7
	EPPO	mm^2	3	27.5-80.3	1_0391	60.6	11.5	25.5	6.2
	EPPO	mm^2	3	185.9-206.7	bPb-0902	206.7	3.7	6.8	−3.3
	EPPO	mm^2	7	80.4-122.0	nud	93.7	5.6	10.4	−4.0
T50	TC	h	2	109.9-148.9	FGX_OWB00464	127.6	6.5	13.9	3.0
	TC	h	7	63.9-118.7	nud	93.7	17.7	42.6	5.0
	EPPN	h	2	127.6-180.0	Zeo1	159.9	7.0	9.3	2.5
	EPPN	h	7	63.9-122.0	1_1437	94.7	25.1	55.0	5.6
P50	EPPO	d	2	109.9-115.4	FGX_OWB00273	109.9	4.1	9.7	−7.9
	EPPO	d	2	159.9 – 180.0	1_1023	172.3	9.2	26.1	−12.7
	EPPO	d	3		1_0391	60.6	3.2	7.2	6.4
	CD	d	2	49.7-50.8	1_0307	164.3	3.1	10.4	−1.0
	CD	d	2	164.3-176.7	[223553]	171.3	4.2	12.4	−1.2
	CD	d	7	80.4-94.8	[221574]	91.5	7.2	23.3	1.7

QTLs declared above a logarithm of odds (LOD) of 3. Positive additive effects indicate a contribution by "OWB-D." C08, control performed in 2008; CD08, controlled deterioration performed in 2008; EPPN, elevated partial pressure of nitrogen storage; EPPO, elevated partial pressure of oxygen storage; %NS, percentage of normal seedlings; %TG, percentage of total germination; TC, treatment control at ambient air pressure; T50, time to reach 50% of germinated seeds (hours); AUC, area under the curve; P50, half-viability period. Chr, chromosome; R^2, proportion of explained phenotypic variance (%); cM, centiMorgan. −, before the additive effect of the allele indicates that the larger value allele came from the "OWB-R"; +, before the additive effect of the allele indicates that the larger value allele came from the "OWB-D" parent.

the time to 50% germination almost doubled. This increased deterioration confirms the deleterious effects of oxygen on seeds.

Aging Treatment Affects Seed Deterioration

The reduction in total number of germinating seeds indicates that barley seeds deteriorate more after 3 days of CD treatment than after 63 days of EPPO treatment. At the higher moisture content used in the CD, the fluidity of intercellular glasses increases (Walters et al., 2010) and biochemical processes are accelerated (Lehner et al., 2008). The EPPO approach intends to use seed equilibrated to ambient moisture and temperature conditions (here 20°C and 40% RH) and are assumed, according to Walters et al. (2005a), in the glassy state. As the glass-phase transition temperature is assumed to be hardly affected by higher gas pressure (Groot et al., 2012), seeds might remain in the glassy, metabolically inactive (Fernández-Marín et al., 2013) state during EPPO storage and deterioration is caused by oxygen. Groot et al. (2012) demonstrated that the EPPO approach might simulate ambient storage better than CD, as tocopherol content was linearly reduced with EPPO storage time, but not with CD, when seed quality declined.

Linked Markers and Genes Indicate Relationship between Morphology, Stress Response Mechanism, and Seed Deterioration

In general, seed deterioration is affected by many factors, which pop up during seed development, ripening, threshing, cleaning and storage. This activates a plethora of mechanisms on cellular level as exemplified by the functional annotation of genes (Supplementary Table 2). However, most QTLs were found with two previously defined areas termed *QLng.ipk-2H* and *QLng.ipk-7H* (Nagel et al., 2009; **Figure 3**) and confirmed an association with the genetic panel of 175 diverse barley genotypes (Nagel et al., 2015). The QTL area on 2H estimated between 110 and 172 cM influences 21 traits, including plant height, spike length, grain number and grain yield. Thereby, the parent "OWB-D" provides dominant alleles to the progeny, is characterized by shorter plants (Chutimanitsakun et al., 2011) and is more sensitive to seed deterioration as "OWB-R". In agreement with this the detected *Zeo1* gene determines plant height, spike compactness and kernel width (Franckowiak et al., 1997) caused by polymorphisms in the microRNA172 (miR172)-binding site of the barley ortholog of an APELATA2 transcription factor (HvAP2). It is indicated that changes result in altered timing of developmental events (Houston et al., 2013) which might affect seed quality and deterioration behavior. Further, in the centromeric region of 7H from 73 to 95 cM, 18 QTLs were identified. The *nud* gene, an ethylene response family transcription factor gene responsible for naked/covered caryopsis (Franckowiak and Konishi, 1997) is also in this area. It further controls the lipid biosynthesis pathway in the inner side of the hull, generating the adhesion between lemma and palea (Taketa et al., 2008) and β-glucan content (Capo-Chichi et al., 2012).

Further annotation of gene functions in the QTL areas revealed the appearance of harpin-induced protein 1 family (Gechev et al., 2004) and glutaredoxins (Rouhier et al., 2008) which indicate a relationship to abiotic and especially oxidative stress response mechanism and programmed cell death. Similar, the regulation of plant defense response against pathogens are demonstrated by the protein cysteine proteinase RD21a (Shindo et al., 2012), proline extensin-like receptor kinase 1 (PERK1; Silva and Goring, 2002), the disease resistance protein NBS-LRR (Belkhadir et al., 2004) and the MAP kinase substrate (MKS1; Andreasson et al., 2005). Due to metabolic inactivation during storage we speculate that proteins, kinases and factors are activated during seed development or germination.

Specialized QTLs for Tolerance to the Two Aging Treatments

Germination after aging/storage is influenced by the initial germination which seeds show after harvest (Roberts, 1973). This effect is due to differences in susceptibility to pathogens, attraction for insects, and abiotic stress tolerances during seed production. Here, the effect of initial germination was mathematically omitted by the subtraction of control or EPPN from EPPO. This mathematical procedure elucidated and/or confirmed QTLs for EPPO tolerance on 3H, 4H, and 6H, for CD08 on 5H, and dormancy release on 2H calculated by differences between control in 2008 and the treatment control in 2011 (3 years). The QTL for CD08 shows co-linearity (Stein et al., 2007) to strong and highly significant QTLs found for rice seed deterioration (Sasaki et al., 2005, 2015). This region contains a gene, annotated as encoding trehalose-6-phosphate phosphatase, which can be used to manipulate abiotic stress tolerance in rice (Sasaki et al., 2005).

Biochemical pathways and deterioration processes are influenced by water activity (seed moisture level), temperature and atmosphere. In agreement with this, we observed QTLs in response to both EPPO and CD storage, as well QTLs expressed as a results of genetic variation in longevity under either EPPO or CD storage conditions. This indicates the causes of deterioration under CD and storage under EPPO are partly similar and partly distinct controlled.

AUTHOR CONTRIBUTIONS

MN, AB, and SG conceived and designed research. MN, JK, SP, and SG conducted experiments. MN and SG analyzed data. MN and SG wrote the manuscript. All authors read and approved the manuscript.

FUNDING

Financial support was provided by the EU "Seventh Framework Programme" grant EcoSeed (No. 311840) and the Netherlands Ministry of Economic Affairs in the research project Improved seed storage, in the frame of the "Topconsortium Kennis en Innovatie Tuinbouw & Uitgangsmaterialen program Meer met minder." The publication of this article was funded by the Open Access fund of the Leibniz Association.

ACKNOWLEDGMENTS

We wish to acknowledge Annette Marlow, Stephanie Thumm and Peter Schreiber for experimental support and seed multiplication.

SUPPLEMENTARY MATERIAL

The Supplementary Material for this article can be found online at: http://journal.frontiersin.org/article/10.3389/fpls.2016.00388

REFERENCES

Andreasson, E., Jenkins, T., Brodersen, P., Thorgrimsen, S., Petersen, N. H. T., Zhu, S., et al. (2005). The MAP kinase substrate MKS1 is a regulator of plant defense responses. *EMBO J.* 24, 2579–2589. doi: 10.1038/sj.emboj.7600737

Bailly, C. (2004). Active oxygen species and antioxidants in seed biology. *Seed Sci. Res.* 14, 93–107. doi: 10.1079/SSR2004159

Belkhadir, Y., Subramaniam, R., and Dangl, J. L. (2004). Plant disease resistance protein signaling: NBS-LRR proteins and their partners. *Curr. Opin. Plant Biol.* 7, 391–399. doi: 10.1016/j.pbi.2004.05.009

Bethke, P. C., Libourel, I. G. L., and Jones, R. L. (2007). "Nitric oxide in seed dormancy and germination," in *Annual Plant Reviews: Vol. 27, Seed Development, Dormancy and Germination*, eds K. J. Bradford and H. Nonogaki (Oxford: Blackwell Publishing Ltd), 153–175.

Buitink, J., and Leprince, O. (2004). Glass formation in plant anhydrobiotes: survival in the dry state. *Cryobiology* 48, 215–228. doi: 10.1016/j.cryobiol.2004.02.011

Buitink, J., and Leprince, O. (2008). Intracellular glasses and seed survival in the dry state. *C. R. Biol.* 331, 788–795. doi: 10.1016/j.crvi.2008.08.002

Caldwell, J. (1964). Effect of high pressures of pure oxygen on tissues. *Nature* 201, 514–515. doi: 10.1038/201514a0

Capo-Chichi, L., Kenward, K., Nyachiro, J., and Anyia, A. (2012). *Nud* locus and the effects on seedling vigour related traits for genetic improvement of hulless barley. *J. Plant Sci. Mol. Breed.* 1, 1–9. doi: 10.7243/2050-2389-1-2

Chutimanitsakun, Y., Nipper, R., Cuesta-Marcos, A., Cistué, L., Corey, A., Filichkina, T., et al. (2011). Construction and application for QTL analysis of a Restriction Site Associated DNA (RAD) linkage map in barley. *BMC Genomics* 12:4. doi: 10.1186/1471-2164-12-4

Costa, J. M., Corey, A., Hayes, P. M., Jobet, C., Kleinhofs, A., Kopisch-Obusch, A., et al. (2001). Molecular mapping of the Oregon Wolfe Barleys: a phenotypically polymorphic doubled-haploid population. *Theoret. Appl. Genet.* 103, 415–424. doi: 10.1007/s001220100622

El-Maarouf-Bouteau, H., Mazuy, C., Corbineau, F., and Bailly, C. (2011). DNA alteration and programmed cell death during ageing of sunflower seed. *J. Exp. Bot.* 62, 5003–5011. doi: 10.1093/jxb/err198

Fernández-Marín, B., Kranner, I., Sebastián, M. S., Artetxe, U., Laza, J. M., Vilas, J. L., et al. (2013). Evidence for the absence of enzymatic reactions in the glassy state. A case study of xanthophyll cycle pigments in the desiccation-tolerant moss Syntrichia ruralis. *J. Exp. Bot.* 64, 3033–3043. doi: 10.1093/jxb/ert145

Franckowiak, D., and Konishi, T. (1997). Naked caryopsis. *Barley Genet. Newsl.* 26, 51–52.

Franckowiak, D., Lundqvist, U., and Konishi, T. (1997). New and revised names for barley genes. *Barley Genet. Newsl.* 26, 22–516.

Gechev, T. S., Gadjev, I. Z., and Hille, J. (2004). An extensive microarray analysis of AAL-toxin-induced cell death in *Arabidopsis thaliana* brings new insights into the complexity of programmed cell death in plants. *Cell. Mol. Life Sci.* 61, 1185–1197. doi: 10.1007/s00018-004-4067-2

Groot, S. P. C., De Groot, L., Kodde, J., and Van Treuren, R. (2014). Prolonging the longevity of *ex situ* conserved seeds by storage under anoxia. *Plant Genet. Resour.* 13, 18–26. doi: 10.1017/S1479262114000586

Groot, S. P. C., Surki, A. A., De Vos, R. C. H., and Kodde, J. (2012). Seed storage at elevated partial pressure of oxygen, a fast method for analysing seed ageing under dry conditions. *Ann. Bot.* 110, 1149–1159. doi: 10.1093/aob/mcs198

Hampton, J. G., and Tekrony, D. M. (1995). *Handbook of Vigour Test Methods.* Zürich: International Seed Testing Association.

Hay, F. R., Adams, J., Manger, K., and Probert, R. (2008). The use of non-saturated lithium chloride solutions for experimental control of seed water content. *Seed Sci. Technol.* 36, 737–746. doi: 10.15258/sst.2008.36.3.23

Houston, K., Mckim, S. M., Comadran, J., Bonar, N., Druka, I., Uzrek, N., et al. (2013). Variation in the interaction between alleles of HvAPETALA2 and microRNA172 determines the density of grains on the barley inflorescence. *Proc. Natl. Acad. Sci. U.S.A.* 110, 16675–16680. doi: 10.1073/pnas.1311681110

ISTA (2008). *International Rules for Seed Testing.* Bassersdorf: International Seed Testing Association.

ISTA (2012). *International Rules for Seed Testing.* Bassersdorf: International Seed Testing Association.

Joosen, R. V. L., Kodde, J., Willems, L. A. J., Ligterink, W., Van Der Plas, L. H. W., and Hilhorst, H. W. M. (2010). GERMINATOR: a software package for high-throughput scoring and curve fitting of Arabidopsis seed germination. *Plant J.* 62, 148–159. doi: 10.1111/j.1365-313X.2009.04116.x

Kranner, I., Minibayeva, F. V., Beckett, R. P., and Seal, C. E. (2010). What is stress? *Concepts, definitions and applications in seed science. New Phytol.* 188, 655–673. doi: 10.1111/j.1469-8137.2010.03461.x

Lehner, A., Mamadou, N., Poels, P., Come, D., Bailly, C., and Corbineau, F. (2008). Changes in soluble carbohydrates, lipid peroxidation and antioxidant enzyme activities in the embryo during ageing in wheat grains. *J. Cereal Sci.* 47, 555–565. doi: 10.1016/j.jcs.2007.06.017

Li, H. (2013). Aligning sequence reads, clone sequences and assembly contigs with BWA-MEM. *arXiv:1303.3997 [q-bio.GN]* [Online].

Li, H., Handsaker, B., Wysoker, A., Fennell, T., Ruan, J., Homer, N., et al. (2009). The Sequence Alignment/Map format and SAMtools. *Bioinformatics* 25, 2078–2079. doi: 10.1093/bioinformatics/btp352

Mascher, M., Muehlbauer, G. J., Rokhsar, D. S., Chapman, J., Schmutz, J., Barry, K., et al. (2013). Anchoring and ordering NGS contig assemblies by population sequencing (POPSEQ). *Plant J.* 76, 718–727. doi: 10.1111/tpj.12319

Nagel, M., and Börner, A. (2010). The longevity of crop seeds stored under ambient conditions. *Seed Sci. Res.* 20, 1–12. doi: 10.1017/S0960258509990213

Nagel, M., Kranner, I., Neumann, K., Rolletschek, H., Seal, C. E., Colville, L., et al. (2015). Genome-wide association mapping and biochemical markers reveal that seed ageing and longevity are intricately affected by genetic background and developmental and environmental conditions in barley. *Plant Cell Environ.* 38, 1011–1022. doi: 10.1111/pce.12474

Nagel, M., Vogel, H., Landjeva, S., Buck-Sorlin, G., Lohwasser, U., Scholz, U., et al. (2009). Seed conservation in *ex-situ* genebanks - genetic studies on longevity in barley. *Euphytica* 170, 1–10. doi: 10.1007/s10681-009-9975-7

Probert, R. J., Daws, M. I., and Hay, F. R. (2009). Ecological correlates of *ex situ* seed longevity: a comparative study on 195 species. *Ann. Bot.* 104, 57–69. doi: 10.1093/aob/mcp082

Quinlan, A. R., and Hall, I. M. (2010). BEDTools: a flexible suite of utilities for comparing genomic features. *Bioinformatics* 26, 841–842. doi: 10.1093/bioinformatics/btq.033

R Core Team (2015). *R Foundation for Statistical Computing.* Göttingen: R Core Team.

Rehman Arif, M. A., Nagel, M., Neumann, K., Kobiljski, B., Lohwasser, U., and Börner, A. (2012). Genetic studies of seed longevity in hexaploid wheat using segregation and association mapping approaches. *Euphytica* 186, 1–13. doi: 10.1007/s10681-011-0471-5

Roberts, E. H. (1960). The viability of cereal seed in relation to temperature and moisture: with eight figures in the text. *Ann. Bot.* 24, 12–31.

Roberts, E. H. (1973). Predicting the storage life of seeds. *Seed Sci. Technol.* 1, 499–514.

Rouhier, N., Lemaire, S. D., and Jacquot, J.-P. (2008). The role of glutathione in photosynthetic organisms: emerging functions for glutaredoxins and glutathionylation. *Annu. Rev. Plant Biol.* 59, 143–166. doi: 10.1146/annurev.arplant.59.032607.092811

Sasaki, K., Fukuta, Y., and Sato, T. (2005). Mapping of quantitative trait loci controlling seed longevity of rice (Oryza sativa L.) after various periods of seed storage. *Plant Breed.* 124, 361–366. doi: 10.1111/j.1439-0523.2005.01109.x

Sasaki, K., Takeuchi, Y., Miura, K., Yamaguchi, T., Ando, T., Ebitani, T., et al. (2015). Fine mapping of a major quantitative trait locus, qLG-9, that controls seed longevity in rice (Oryza sativa L.). *Theoret. Appl. Genet.* 128, 769–778. doi: 10.1007/s00122-015-2471-7

The International Barley Genome Sequencing Consortium. (2012). A physical, genetic and functional sequence assembly of the barley genome. *Nature* 491, 711–716. doi: 10.1038/nature11543

Shindo, T., Misas-Villamil, J. C., Horger, A. C., Song, J., and Van Der Hoorn, R. A. L. (2012). A role in immunity for Arabidopsis cysteine protease RD21, the ortholog of the tomato immune protease C14. *PLoS ONE* 7:e29317. doi: 10.1371/journal.pone.0029317

Silva, N. F., and Goring, D. R. (2002). The proline-rich, extensin-like receptor kinase-1 (PERK1) gene is rapidly induced by wounding. *Plant Mol. Biol.* 50, 667–685. doi: 10.1023/A:1019951120788

Stein, N., Prasad, M., Scholz, U., Thiel, T., Zhang, H. N., Wolf, M., et al. (2007). A 1,000-loci transcript map of the barley genome: new anchoring points for integrative grass genomics. *Theoret. Appl. Genet.* 114, 823–839. doi: 10.1007/s00122-006-0480-2

Sun, W. Q., and Leopold, A. C. (1995). The Maillard reaction and oxidative stress during aging of soybean seeds. *Physiol. Plant.* 94, 94–104. doi: 10.1111/j.1399-3054.1995.tb00789.x

Taketa, S., Amano, S., Tsujino, Y., Sato, T., Saisho, D., Kakeda, K., et al. (2008). Barley grain with adhering hulls is controlled by an ERF family transcription factor gene regulating a lipid biosynthesis pathway. *Proc. Natl. Acad. Sci. U.S.A.* 105, 4062–4067. doi: 10.1073/pnas.0711034105

Van Breusegem, F., and Dat, J. F. (2006). Reactive oxygen species in plant cell death. *Plant Physiol.* 141, 384–390. doi: 10.1104/pp.106.078295

Van Campenhout, L., Maes, P., and Claes, J. (2013). Modified atmosphere packaging of Tofu: headspace gas profiles and microflora during storage. *J. Food Process. Preserv.* 37, 46–56. doi: 10.1111/j.1745-4549.2011.00612.x

Walters, C. (1998). Understanding the mechanisms and kinetics of seed aging. *Seed Sci. Res.* 8, 223–244. doi: 10.1017/S096025850000 0413X

Walters, C., Ballesteros, D., and Vertucci, V. A. (2010). Structural mechanics of seed deterioration: standing the test of time. *Plant Sci.* 179, 565–573. doi: 10.1016/j.plantsci.2010.06.016

Walters, C., Hill, L. M., and Wheeler, L. J. (2005a). Dying while dry: kinetics and mechanisms of deterioration in desiccated organisms. *Integr. Comp. Biol.* 45, 751–758. doi: 10.1093/icb/45.5.751

Walters, C., Wheeler, L. M., and Grotenhuis, J. M. (2005b). Longevity of seeds stored in a genebank: species characteristics. *Seed Sci. Res.* 15, 1–20. doi: 10.1079/SSR2004195

Wang, S., Basten, C. J., and Zeng, Z.-B. (2011). *Windows QTL Cartographer 2.5*. Department of Statistics, North Carolina State University, Raleigh, NC. Available online at: http://statgen.ncsu.edu/qtlcart/WQTLCart.htm.

Waterworth, W. M., Masnavi, G., Bhardwaj, R. M., Jiang, Q., Bray, C. M., and West, C. E. (2010). A plant DNA ligase is an important determinant of seed longevity. *Plant J.* 63, 848–860. doi: 10.1111/j.1365-313X.2010.04 285.x

Wolfe, R. I., and Franckowiak, J. D. (1991). Multiple dominant and recessive genetic marker stocks in spring barley. *Barley Genet. Newsl.* 20, 117–121.

Zeng, Z. B. (1994). Precision mapping of quantitative trait loci. *Genetics* 136, 1457–1468.

Conflict of Interest Statement: The authors declare that the research was conducted in the absence of any commercial or financial relationships that could be construed as a potential conflict of interest.

The reviewer PK and handling Editor declared their shared affiliation, and the handling Editor states that the process nevertheless met the standards of a fair and objective review.

3.8 PAPER 8:

Seed longevity in oilseed rape (*Brassica napus* L.) - genetic variation and QTL mapping

by

Manuela Nagel, Maria Rosenhauer, Evelin Willner, Rod J. Snowdon, Wolfgang Friedt and Andreas Börner

Published in

Plant Genetic Resources: Characterization and Utilization (2011) 9, 260-263

https://doi.org/10.1017/S1479262111000372

ISSN 1479-2621
Plant Genetic Resources: Characterization and Utilization (2011) 9(2); 260–263
doi:10.1017/S1479262111000372

Seed longevity in oilseed rape (*Brassica napus* L.) – genetic variation and QTL mapping

Manuela Nagel[1], Maria Rosenhauer[1], Evelin Willner[2], Rod J. Snowdon[3], Wolfgang Friedt[3] and Andreas Börner[1]*

[1] *Leibniz Institute for Plant Genetics and Crop Plant Research (IPK), Corrensstraße 3, Gatersleben, Germany,* [2] *Leibniz Institute for Plant Genetics and Crop Plant Research, Satellite Collections North, Inselstraße 9, Malchow/Poel, Germany and* [3] *Department of Plant Breeding, Justus Liebig University, Heinrich-Buff-Ring 26–32, Giessen, Germany*

Abstract

Although oilseed rape has become one of the most important oil crops in Europe, little is known regarding the viability of its seed under conditions of long-term storage. We report here an examination of oilseed rape seed longevity performed on a set of 42 accessions housed at the German *ex situ* genebank at IPK, Gatersleben. A comparison of germination between the accessions stored for 26 years showed that viability was in part genetically determined, since it ranged between 42 and 98%. An attempt was made to define the genetic basis of viability by subjecting a mapping population of doubled haploids to three artificial ageing treatments. Quantitative trait loci (QTL) were detected on six chromosomes: N6, N7, N8, N15, N16 and N18. The chromosomal locations of these QTL were compared with their syntenic regions in *Arabidopsis thaliana* in order to explore what genes might underlie genetic variation for longevity.

Keywords: *Brassica napus* L; *ex situ* genebank; quantitative trait loci; seed longevity; synteny

Introduction

Originally developed as a source of oil for energy purposes, oilseed rape (*Brassica napus* L.) has now risen to become one of the most important field crops in Europe, Canada, China and Australia. Its oil is now used in many applications, including biodiesel, culinary oil and livestock feed. Due to its remarkable increase in production, canola has become the focus of much breeding and molecular genetics in recent years (Friedt and Snowdon, 2010).

Until recently, the longevity of *B. napus* seed has been considered to be of little importance. However, as particularly the spring-type canola crop becomes transgenic-based, some concern has been expressed about the growing dominance of transgenic material in the soil seed bank, while at the same time, there is a continuing loss of non-transgenic materials stored in genebanks, as a result of ageing. Seeds from *Brassica* spp. typically lose ~50% of their viability with 7.3 years of storage at 20°C and 50% relative humidity (RH) (Nagel and Börner, 2010), and within 23 years under standard low temperature (−18°C) storage conditions (Walters *et al.*, 2005a). Evidence that seed viability in some crop species is in part genetically determined (Nagel *et al.*, 2010) has prompted the present study of intra-specific variation in *B. napus* seed longevity.

Materials and methods

We tested the seed viability of 42 accessions of *B. napus* ssp. *napus* var. *napus* f. *biennis*, which had been

* Corresponding author. E-mail: boerner@ipk-gatersleben.de

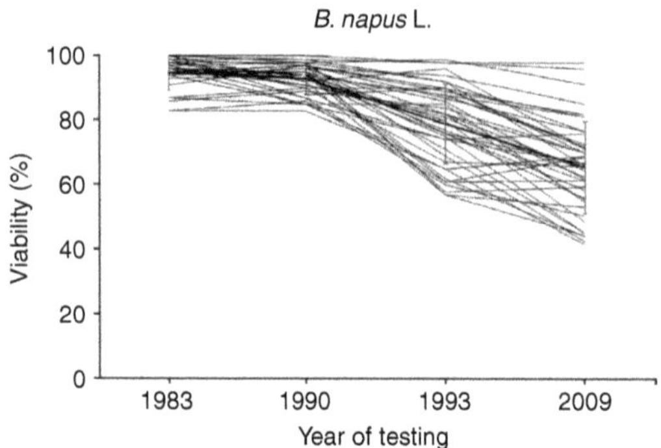

Fig. 1. Mean viability of *B. napus* genebank accessions over different test years. The bold line indicates mean viability, and standard deviations over the years are shown.

multiplied in 1983 and have been maintained in the interim at 7 ± 3°C and 6 ± 2% seed moisture content. Three replicates of 50 seeds/accession were placed on moistened filter paper and germination was monitored after 28 d. Their current viability was compared with historic data collected in 1983, 1990 and 1993 by using arithmetic means and standard deviations.

A set of artificial ageing protocols was applied to a population of 153 doubled haploid lines of the winter oilseed rape YE2-DH mapping population (Badani *et al.*, 2006) to explore the genetic basis of seed longevity. The protocols were: AA1, following Hampton and TeKrony (1995), in which two replicates of 50 seeds each were sealed in glass jars containing 200 ml deionised water to raise the RH above 99%. After holding the jars for 48 h at 42°C, a germination test was conducted as above, with a final count after 7 d. AA2: this was identical to AA1, except that the temperature was 44°C. AA3: following Hay *et al.* (2008), in which two replicates of 100 seeds/line were placed inside a sealable box containing a 8.7 M LiCl solution which ensured that the RH at 20°C stabilized at 47% within 14 d. Artificial ageing was then initiated by changing the solution (7.1 M LiCl), and maintaining the temperature at 45°C (to give an RH of 60%) for 17 d. This was followed by the same germination test as above. A measure of relative viability for each method was calculated by dividing the viability of the aged seeds by their initial viability. The data were subjected to a quantitative trait loci (QTL) analysis, using QGENE software (Nelson, 1997) and a genetic map of, in total, 161 markers comprising amplified fragment length polymorphism and simple-sequence repeat markers (Snowdon *et al.*, unpublished data). A permutation test was used to set the appropriate logarithm of odds (LOD) ratio threshold for each treatment.

Results and discussion

The decay in viability of the genebank accessions over 26 years of storage is shown in Fig. 1. The mean proportion of viable seeds fell from 94.7% in 1983 to 92.9, 79.1 and 65.7% after 7, 10 and 26 years of storage, respectively. The associated standard deviations were 5.4% (1983) 4.6% (1990), 12.5% (1993) and 14.2% (2009). Despite having been grown simultaneously and subjected to the same post-harvest and storage conditions, the accessions nevertheless displayed variation with respect to seed viability, as also occurs in barley, wheat, sorghum, rye and

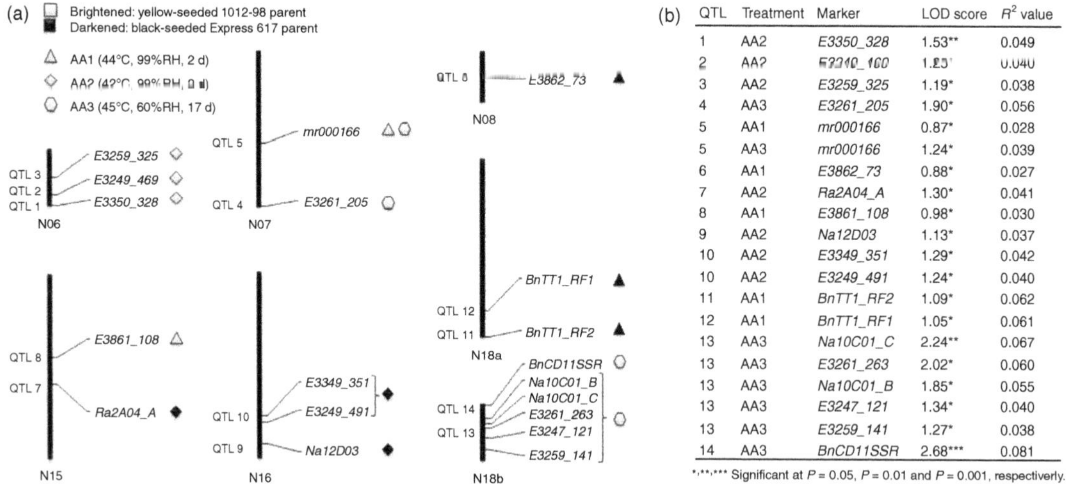

QTL	Treatment	Marker	LOD score	R^2 value
1	AA2	*E3350_328*	1.53**	0.049
2	AA2	[illegible]	[illegible]	0.040
3	AA2	*E3259_325*	1.19*	0.038
4	AA3	*E3261_205*	1.90*	0.056
5	AA1	*mr000166*	0.87*	0.028
5	AA3	*mr000166*	1.24*	0.039
6	AA1	*E3862_73*	0.88*	0.027
7	AA2	*Ra2A04_A*	1.30*	0.041
8	AA1	*E3861_108*	0.98*	0.030
9	AA2	*Na12D03*	1.13*	0.037
10	AA2	*E3349_351*	1.29*	0.042
10	AA2	*E3249_491*	1.24*	0.040
11	AA1	*BnTT1_RF2*	1.09*	0.062
12	AA1	*BnTT1_RF1*	1.05*	0.061
13	AA3	*Na10C01_C*	2.24**	0.067
13	AA3	*E3261_263*	2.02*	0.060
13	AA3	*Na10C01_B*	1.85*	0.055
13	AA3	*E3247_121*	1.34*	0.040
13	AA3	*E3259_141*	1.27*	0.038
14	AA3	*BnCD11SSR*	2.68***	0.081

*,**,*** Significant at $P = 0.05$, $P = 0.01$ and $P = 0.001$, respectiverly.

Fig. 2. QTL interval mapping in the *B. napus* YE2-DH population detected a range of loci with effects on seed longevity after artificial ageing. The choice of artificial ageing protocol (AA1, AA2 and AA3) applied was found to be important. The 14 QTL map to seven different chromosomes (a). Individual QTL effects are tabulated (b).

linseed (Nagel *et al.*, 2009, 2010). We therefore concluded that there is a genotypic component involved in the determination of seed viability.

When the doubled haploid lines from the YE2-DH population were tested using the three artificial ageing methods, 13 significant QTL affecting seed longevity were identified (Fig. 2). Most of the QTL were method-specific. AA1 produced five QTL, mapping to chromosomes N7, N8, N15 and N18a, while AA2 generated six QTL on N6, N15 and N16. By using AA3, four QTL on chromosomes N7 and N18b were found with in particular highest explained phenotypic variation in QTL 14: $R^2 = 0.081$. The only QTL common to AA1 and AA3 was QTL5, on chromosome N7.

The reliability of artificial ageing as a surrogate for long-term storage has been repeatedly discussed in the literature (Delouche and Baskin, 1973; Priestley and Leopold, 1979; McDonald, 1999; Freitas *et al.*, 2006), leading to a number of mutually inconsistent conclusions. Rajjou *et al.* (2008) suggested that a controlled deterioration test protocol, as described by Tesnier *et al.* (2002), mimics many of the molecular and biochemical events experienced during seed ageing. On the other hand, in this study, we showed that even a modest increase in ageing temperature (from 42 to 44°C) can have a major effect on the expression of relevant genes. The QTL identified in AA1 and AA2 mapped to distinct regions from those detected following AA3, which compared with AA1 and AA2 operates on lower seed moisture contents in a dry state, according to Walters *et al.* (2005b).

The *Brassica* consensus map of Parkin *et al.* (2005) was used to estimate potential locations of orthologous loci in the *Arabidopsis thaliana* genome with respect to the *B. napus* QTL. Potential links with *B. napus* QTL were found for the upper part of the *A. thaliana* chromosomes AtC1 and AtC3, and near the lower ends of AtC1 and AtC5. The genetic basis of seed longevity in *A. thaliana* has been ascribed to genes mapping in the upper parts of chromosomes AtC1 and AtC3 (Bentsink *et al.*, 2000; Clerkx *et al.*, 2004b). Other QTL related to germination in the presence of salinity or heat stress, as well as some controlling the rate of germination, are also known in the region of AtC1, which suggests the possibility that tolerance to stress represents an aspect of seed longevity (Clerkx *et al.*, 2004b). Nevertheless, many of the genes involved in the stress response are also participants in the oxidative stress response. An example of this connection has been provided by Thorlby *et al.* (1999), who detected the expression of genes in the region of AtC1 related to tolerance of oxidative stress, as well as of some involved in abscisic acid (ABA) biosynthesis and perception. *A. thaliana* mutants which have lost sensitivity to ABA, or are compromised in its synthesis, tend to show poor seed longevity (Clerkx *et al.*, 2004a). This demonstrates that ABA plays a role in maintaining seed viability during storage. Certain chemical and/or physical properties of the seed-coat also affect germination rate after storage, since the seed of both structural and pigmentation mutants tends to deteriorate faster than that of their wild-type progenitor (Debeaujon *et al.*, 2000; Clerkx *et al.*, 2004a).

References

Badani AG, Snowdon RJ, Baetzel R, Lipsa FD, Wittkop B, Horn R, De Haro A, Font R, Lühs W and Friedt W (2006) Co-localisation of a partially dominant gene for yellow seed colour with a major QTL influencing acid detergent fibre (ADF) content in different crosses of oilseed rape (*Brassica napus*). *Genome* 49: 1499–1509.

Bentsink L, Alonso-Blanco C, Vreugdenhil D, Tesnier K, Groot SPC and Koornneef M (2000) Genetic analysis of seed-soluble oligosaccharides in relation to seed storability of *Arabidopsis*. *Plant Physiology* 124: 1595–1604.

Clerkx EJM, Blankestijn-De Vries H, Ruys GJ, Groot SPC and Koornneef M (2004*a*) Genetic differences in seed longevity of various *Arabidopsis* mutants. *Physiologia Plantarum* 121: 448–461.

Clerkx EJM, El-Lithy ME, Vierling E, Ruys GJ, Blankestijn-De Vries H, Groot SPC, Vreugdenhil D and Koornneef M (2004*b*) Analysis of natural allelic variation of *Arabidopsis* seed germination and seed longevity traits between the accessions Landsberg erecta and Shakdara, using a new recombinant inbred line population. *Plant Physiology* 135: 432–443.

Debeaujon I, Leon-Kloosterziel KM and Koornneef M (2000) Influence of the testa on seed dormancy, germination, and longevity in *Arabidopsis*. *Plant Physiology* 122: 403–413.

Delouche JC and Baskin CC (1973) Accelerated aging techniques for predicting the relative storability of seed lots. *Seed Science and Technology* 1: 427–452.

Freitas RA, Dias DCFS, Oliveira MGA, Dias LAS and Jose IC (2006) Physiological and biochemical changes in naturally and artificially aged cotton seeds. *Seed Science and Technology* 34: 253–264.

Friedt W and Snowdon RJ (2010) Oilseed rape. In: Vollmann J and Rajan J (eds) *Oil crops. Handbook of Plant Breeding*, vol. 4. NY: Springer-Verlag, pp. 91–126.

Hampton JG and TeKrony DM (eds) (1995) *Handbook of Vigour Test Methods*. Zürich: International Seed Testing Association.

Hay FR, Adams J, Manger K and Probert R (2008) The use of non-saturated lithium chloride solutions for experimental control of seed water content. *Seed Science and Technology* 36: 737–746.

McDonald MB (1999) Seed deterioration: physiology, repair and assessment. *Seed Science and Technology* 27: 177–237.

Nagel M and Börner A (2010) The longevity of crop seeds stored under ambient conditions. *Seed Science Research* 20: 1–12.

Nagel M, Vogel H, Landjeva S, Buck-Sorlin G, Lohwasser U, Scholz U and Börner A (2009) Seed conservation in *ex-situ* genebanks – genetic studies on longevity in barley. *Euphytica* 170: 1–10.

Nagel M, Abdur Rehman Arif M, Rosenhauer M and Börner A (2010) Longevity of seeds – intraspecific differences in the Gatersleben genebank collections. *Tagungsband der 60. Tagung der Vereinigung der Pflanzenzüchter und Saatgutkaufleute Österreichs.* Raumberg-Gumpenstein (Austria), pp. 179–181.

Nelson JC (1997) QGene: software for marker-based genomic analysis and breeding. *Molecular Breeding* 3: 239–245.

Parkin IAP, Gulden SM, Sharpe AG, Lukens L, Trick M, Osborn TC and Lydiate DJ (2005) Segmental structure of the *Brassica napus* genome based on comparative analysis with *Arabidopsis thaliana*. *Genetics* 171: 765–781.

Priestley DA and Leopold AC (1979) Absence of lipid oxidation during accelerated aging of soybean seeds. *Plant Physiology* 63: 726–729.

Rajjou L, Lovigny Y, Groot SPC, Belghaz M, Job C and Job D (2008) Proteome-wide characterization of seed aging in *Arabidopsis*: a comparison between artificial and natural aging protocols. *Plant Physiology* 148: 620–641.

Tesnier K, Strookman-Donkers HM, Van Pijlen JG, Van der Geest AHM, Bino RJ and Groot SPC (2002) A controlled deterioration test for *Arabidopsis thaliana* reveals genetic variation in seed quality. *Seed Science and Technology* 30: 149–165.

Thorlby G, Veale E, Butcher K and Warren G (1999) Map positions of SFR genes in relation to other freezing-related genes of *Arabidopsis thaliana*. *Plant Journal* 17: 445–452.

Walters C, Wheeler LM and Grotenhuis JM (2005*a*) Longevity of seeds stored in a genebank: species characteristics. *Seed Science Research* 15: 1–20.

Walters C, Hill LM and Wheeler LJ (2005*b*) Dying while dry: kinetics and mechanisms of deterioration in desiccated organisms. *Integrative and Comparative Biology* 45: 751–758.

3.9 PAPER 9:

Genetic variation for secondary seed dormancy and seed longevity in a set of black-seeded European winter oilseed rape cultivars

by

Jörg Schatzki, Mai Allam, Coretta Klöppel, **Manuela Nagel**, Andreas Börner and Christian Möllers

Published in

Plant Breeding (2013) 132, 174-179

https://doi.org/10.1111/pbr.12023

Plant Breeding
© 2013 Blackwell Verlag GmbH
doi:10.1111/pbr.12023

Genetic variation for secondary seed dormancy and seed longevity in a set of black-seeded European winter oilseed rape cultivars

Jörg Schatzki[1], Mai Allam[2], Coretta Klöppel[1], Manuela Nagel[2], Andreas Börner[2] and Christian Möllers[1,3]

[1]Department of Crop Sciences, Plant Breeding, Georg-August-Universität Göttingen, Von-Siebold-Str. 8, 37075, Göttingen, Germany; [2]Genebank Department, Leibniz Institute of Plant Genetics and Crop Plant Research (IPK Gatersleben), Corrensstr. 3, 06466, Stadt Seeland, Germany; [3]Corresponding author, E-mail: cmoelle2@gwdg.de

With 4 tables

Received July 25, 2012/Accepted October 11, 2012
Communicated by A.-M. Chevre

Abstract

Secondary seed dormancy in oilseed rape is a phenomenon that allows seeds to survive in the soil for many years without germination. Following soil cultivation, dormant seeds may germinate in subsequent years, and they are the main reason for the occurrence of volunteer oilseed rape plants in successive crops. Inheritance of secondary dormancy may be related to seed longevity (SL) in the soil. Genetic reduction in secondary dormancy and SL could provide a mean to reduce the frequency of volunteer plants and especially the dispersal of transgenic oilseed rape. The aim of the present study was to analyse secondary dormancy, germination rate and SL of 28 black-seeded winter oilseed rape cultivars using *in vitro* laboratory tests. The material was tested in field experiments at six different locations in Germany in 2008/2009. Significant effects of the genotype and the location on all traits were found. Heritability was high for secondary dormancy (0.97) and moderate for germination rate (0.70) and SL (0.71). Results indicate that a selection for low secondary dormancy would be effective.

Key words: *Brassica napus* — secondary dormancy — seed germination — seed quality

Before and during harvest of oilseed rape, considerable amounts of seeds may be lost due to pod shattering. In adverse harvest conditions seed losses can reach up to 10 000 seeds per m^2 (Lutman et al. 2005). Shed seeds germinate directly under favourable conditions, but under unfavourable conditions they may become secondary dormant. Those seeds remain viable in the soil for a period of 10 years and longer (Lutman et al. 2003) and lead to an increase in the soil seed bank (Gruber et al. 2010 and references therein). Under favourable conditions, dormant seeds may germinate and occur as a weed or so-called volunteer oilseed rape in succeeding crops for several years (Pekrun et al. 1997a, b, Gruber et al. 2004). Secondary seed dormancy occurs in spring oilseed rape under Canadian growing conditions (Gulden et al. 2003, Fei et al. 2009) and as well in winter oilseed rape under European conditions (Gruber et al. 2010). Volunteer oilseed rape plants may emerge in succeeding crops in such high numbers that herbicide application is required to prevent yield losses. They also will represent a problem if seed quality is changed, for example, from high erucic acid to low erucic acid or from canola ('00')-quality to high oleic acid low linolenic acid quality (HOLL). In this case, quality problems may occur due to the admixture with seeds harvested from those volunteer plants but also due to cross-pollination during flowering period.

The longevity of seeds of oilseed rape in the soil is as well a problem if dispersal of the transgenic crop via seeds and pollen is an issue (Pessel et al. 2001, Colbach et al. 2008). Generally, the half-viability period is estimated to be 7.3 years for different cabbage varieties (*Brassica* spp.) under ambient conditions (20° C at 50% RH; Nagel and Börner 2010), and 25 years for oilseed rape varieties under standard low-temperature genebank conditions (−18°C; Walters et al. 2005). Furthermore, by applying ultra-dry storage with seed moisture contents below 3%, a high viability with over 90% was reported for oilseed rape seeds after 38 years of storage (Pérez-García et al. 2007).

In the literature, several factors are mentioned that influence seed longevity (SL) and trigger secondary seed dormancy at the same time, for example initial seed quality (Ellis and Roberts 1980, Probert et al. 2007), temperature (Pekrun et al. 1997a, Specht and Börner 1998), light (Bazanska and Lewak 1986, López-Granados and Lutman 1998), moisture (Bewley 1997), seed age (Momoh et al. 2002) and anaerobic conditions (Momoh et al. 2002). However, Gulden et al. (2004) reported that the total variation for secondary seed dormancy detected among 16 genotypes was caused predominantly by the genotype, followed by seed size and different harvest regimes. These findings match well with results from viability screenings of the Gatersleben ex situ genebank collection. Different oilseed rape accessions (Nagel et al. 2011) but also barley (Nagel et al. 2009), wheat, rye and flax accessions (Nagel et al. 2010) showed different viability performance after more than 35 years of storage.

A possible link between seed dormancy and SL is discussed controversially (Gräber et al. 2012). Seeds from the non-dormant Arabidopsis *dog1* mutant have a reduced longevity compared to wild-type seeds (Bentsink et al. 2006). In other studies, however, QTL for dormancy (Bentsink et al. 2010) and longevity (Clerkx et al. 2004) did not co-locate.

In the laboratory, secondary seed dormancy can be conveniently induced by incubating the seeds in a polyethylene glycol (PEG) solution for two to four weeks in darkness (Momoh et al. 2002, Gruber et al. 2004, Gulden et al. 2004). Comparing the induction of secondary seed dormancy under *in situ* and under *in vitro* conditions in PEG has led to concurrent results (Gulden et al. 2003, Gruber et al. 2004). Methods that give an idea about long-term survival of seeds within an appropriate time are experimental ageing procedures described by the International Seed Testing Association (Hampton and TeKrony 1995, ISTA 2011).

Incubation of seeds at high-temperature and high-moisture content leads to a rapid decline of seed viability, which is used as indicator for SL under natural conditions (Rajjou et al. 2008).

Although the amount of oilseed rape seeds becoming secondary dormant can be reduced by applying the correct tillage system after harvest (Gruber and Claupein 2006, Gruber et al. 2010), genetically reducing the capacity of oilseed rape cultivars to produce secondary dormant seeds appears attractive. The two efforts, the appropriate tillage system and the use of cultivars with a low capacity to produce secondary dormant seeds, together could provide an instrument to effectively reduce soil seed banks and hence seed and pollen dispersal of transgenic modified and conventionally bred oilseed rape cultivars. In a previous study, large genotypic variation for seed oil and protein content as well as for seed size, seed hull proportion and seed fibre content was detected in a set of black-seeded European winter oilseed rape cultivars tested in field experiments at six different locations in Germany (Dimov et al. 2012). The objective of the present study was to analyse the genetic variation and inheritance for germination rate, secondary seed dormancy and SL in the same set of seed samples from these cultivars and to study correlations to previously recorded seed traits.

Materials and Methods

Plant material and field experiments: The seed material consisted of 28 black-seeded winter oilseed rape cultivars with low erucic acid content in the seed oil and low glucosinolate content in the seeds (Table 3). The material was cultivated at 15 locations in Germany in 2008/2009 (Bundes- und EU-Sortenversuch 1 Winterraps, Gronow et al. 2009). Field experiments were conducted as a randomized complete block design with 4 replicates for each cultivar at each location. Seed samples were taken after combined harvesting of the yield plots. Samples from the 4 replicates of each cultivar at each location were equally mixed and used for Near-Infrared-Reflectance-Spectroscopy (NIRS) analysis. Based on the mean oil content of the seed samples of the locations, seed samples from locations with a low oil content (Langenstein, Ihinger Hof), an intermediate oil content (Hohenschulen, Futterkamp) and a high oil content (Mollenfelde, Sophienhof) were chosen for the analysis of secondary dormancy, germination rate and SL. After harvest, seeds were stored in a cold room at 4°C. Average annual temperature and rainfall in 2009 were 8.9°C and 671 mm at Hohenschulen, 8.9°C and 641 mm at Futterkamp, 9.3°C and 813 mm at Ihinger Hof, 10.3°C and 470 mm at Langenstein, 9.8°C and 634 mm at Sophienhof, 9.3°C and 670 mm at Mollenfelde. Yield plots were harvested between 14.07.2009 (Langenstein) and 01.08.2009 (Sophienhof). For more details about the cultivars and locations, see Gronow et al. (2009).

Seed germination rate: The germination rate (GR) was determined on 2 × 100 seeds per genotype and location. Therefore, filter papers (MN618 with 85 mm radius, 0.32 mm thickness and 140 g/m^2 weight; Macherey-Nagel GmbH, Düren, Germany) were put in plastic Petri dishes (92 × 16 mm; Sarstedt AG & Co., Nürnbrecht, Germany), and 6 ml of deionized water was added. 100 seeds per Petri dish were equally dispersed on the soaked filter paper, and Petri dishes were closed with the corresponding lid. Petri dishes were then stored in cardboard boxes in complete darkness in a climate chamber at 18°C. Germination rate was determined after 1, 5 and 14 days, and number of germinated seeds was summed up to calculate the germination rate in per cent (%).

Then, for testing the viability (data not shown) of the remaining seeds that had not germinated within the 14 days incubation on moist filter paper, Petri dishes were put together in transparent plastic bags and were incubated in a climate chamber for seven days under alternating light and temperature conditions (12 h darkness at 5°C and 12 h light at 25°C). Germinated seeds were counted after 3 and 7 days, and number of completely germinated seeds was summed up as viable seeds.

Induction of secondary seed dormancy: The test for secondary seed dormancy (SD) was performed from December 2009 to February 2010 essentially following the protocol described by Gruber et al. (2004). Before initiating the laboratory test, impurities and broken seeds were removed from the seed lots. For dormancy induction, the same filter papers as indicated above were put in plastic Petri dishes, and 8 ml of a PEG solution with a concentration of 354.37 g/l (AppliChem GmbH, Darmstadt, Germany) was added. The freshly prepared PEG 6000 solution had an osmotic potential of −15 bar (Gruber et al. 2004) equal to the permanent wilting point. 100 seeds per Petri dish were equally dispersed on the soaked filter paper, and Petri dishes were closed with the corresponding lid. All treatments were performed in a climate chamber at a temperature of 18°C and under green light. Green light filters were obtained from the Göttinger Farbfilter GmbH (Bovenden-Lenglern, Germany). Care was taken to prevent any other light entering the growth chamber. Petri dishes were collected and stored for 2 weeks in the same climate chamber in cardboard boxes, carefully wrapped with black plastic foil to protect them from light.

Viability testing of secondary dormancy induced seeds: Viability of the secondary dormancy–induced seeds was determined in three consecutive steps. At first, seeds were rinsed in the Petri dishes with 6 ml distilled water to dilute the PEG solution. Then, the seeds were dispersed in a new Petri dish onto a new filter paper soaked with 6 ml deionized water. After two-day incubation in darkness, germinated seeds with a radicle longer than 2 mm were counted as viable and discarded. The remaining seeds were incubated again, and two days later germinated seeds were counted and discarded again. The remaining seeds were rinsed in 6 ml deionized water and were dispersed in a new Petri dish onto a new filter paper soaked with 6 ml of deionized water. Ten days later, the number of germinated seeds was finally counted. Counting, rinsing and transfer of seeds to new Petri dishes were performed under green light. In between, Petri dishes with seeds were stored in the cardboard boxes in darkness.

Then, for testing the viability of the remaining seeds that had not germinated within the 14 days incubation on moist filter paper, Petri dishes were put together in transparent plastic bags and were incubated in a climate chamber for seven days under alternating light and temperature conditions (12 h darkness at 5°C and 12 h light at 25°C). Germinated seeds were counted after 3 and 7 days, and number of germinated seeds was summed up. Finally, seeds that at this point still had not germinated were incubated in a 0.2% (w/v) solution of 2, 3, 5-triphenyl tetrazolium chloride (Peters 2000), and red stained seeds were counted as viable. The final number of viable seeds consisted of the added number of viable seeds determined in the three steps.

Calculation of the secondary dormancy rate: The viable seeds that did not germinate in deionized water in darkness after the dormancy induction were considered to be secondary dormant, that is, the sum of viable seeds minus the number of seeds germinated in the first step of the viability test. The frequency of dormant seeds (SD) was calculated as SD (%) = (viable seeds - germinated seeds) * 100/viable seeds.

Testing of SL after Accelerated Ageing: The test for SL was performed from February to April 2012 essentially following the 'Accelerated Ageing (AA) Test' protocol described by Hampton and TeKrony (1995). The AA test exposes seeds for short periods to two environmental variables: high temperature and high relative humidity, which cause rapid seed weakening. High vigour seed lots will withstand these extreme stress conditions and deteriorate at a slower rate than low vigour ones (Hampton and TeKrony 1995). The test was carried out using three times 50 seeds per genotype and location. Seeds were placed in air-tight sealed glass jars above 200 ml deionized water causing relative humidity close to 100%. The test was initialized by transferring sealed glass jars into a dark climate chamber at 42 ± 0.5°C for 120 h. Germination tests were performed for control seeds and AA-treated seeds by placing seeds on filter paper soaked with 4 ml deionized water in a climate chamber at

20 ± 0.5°C and 16 h light. After 7 days, incubated, germinated seeds with a radicle longer than 2 mm were counted as viable. SL was regarded as percentage of viable seeds. A high SL value (%) indicates a high germination rate after AA treatment in comparison with the untreated controls. The seed germination rate of the control samples following the storage for more than 2 years at 4°C was still high with values of 94% and above (data not shown).

Statistics: Analysis of variance (ANOVA) and calculation of heritabilities (h^2) were performed using PLABSTAT software (Utz 2011) considering the locations as random. For secondary dormancy and germination rates, ArcSin-transformed data were used due to not being normally distributed. Mean values of the genotypes across the locations were used to calculate Spearman's rank correlation coefficients between traits.

Results

The analysis of variance showed highly significant effects of the locations and the genotypes on secondary seed dormancy and SL of 28 winter oilseed rape cultivars as determined by the *in vitro* tests (Table 1). Highly significant effects of the locations and the genotypes were also found for most of the other traits. Variance components for the genotypes were comparatively large for secondary seed dormancy and thousand-kernel weight, whereas large effects of the locations were observed for SL and oil and protein content. Genotype x location interaction effects were large for germination rate and SL. Heritability was high for all traits investigated, except for germination rate and SL.

The secondary seed dormancy rate and the SL averaged over the genotypes varied considerably between the locations (Table 2). Seed dormancy ranged from 12% for Langenstein up to 38% for Sophienhof. For SL, it ranged from 35.2% for Ihinger Hof to 74.4% for Mollenfelde. A large variation between the locations was also observed for oil content, and there was a positive relation between seed dormancy rate and oil content and a negative relation to protein content (Table 2). In contrast to the other traits, there were only minor differences in the seed germination rate between the locations.

Among the 28 cultivars, there was a large variation for secondary dormancy which ranged from 8% for cultivar 'Iwan' to 56% for cultivar 'DK Secure' (Table 3). A large variation was also observed for SL, ranging from 39% for cultivar 'ES Alienor' to 70.4% for cultivar 'NK Caravel'. Compared to this, the variation in the germination rate with 94.0% to 99.6% was rather narrow. A large variation was also found for thousand-kernel weight, seed oil and protein content and protein content of the defatted meal.

Among the different traits recorded, there was no significant correlation between secondary seed dormancy of the genotypes and their germination rate, SL, thousand-kernel weight, seed oil and protein content as well as protein content of the defatted meal (see Table 4). There was also no correlation to the seed fibre content (NDF, ADF and ADL, seed hull proportion) of those cultivars. For germination rate, the only significant positive correlations were observed with SL, the seed protein content and the protein content of the defatted meal. SL was significantly positive correlated with thousand-kernel weight. Oil and protein content of the seeds was significantly negative correlated.

Table 2: Secondary seed dormancy (SD, %), germination rate (GR, %), seed longevity (SL, %), thousand-kernel weight (TKW, g), oil and protein content of the seed (in% seed dry matter) and protein content of the defatted meal (prot odM, in% seed dry matter) of 28 winter oilseed rape cultivars tested in field experiments in 2008/2009. Depicted are means over six different locations.

Location	SD	GR	SL	TKW[1]	Oil[1]	Protein[1]	Protein odM
Langenstein	12.3	98.5	52.3	4.3	44.5	22.6	40.6
Ihinger Hof	20.8	98.3	35.2	4.4	46.0	22.0	40.7
Hohenschulen	23.4	97.8	66.9	5.0	49.3	18.7	36.8
Futterkamp	30.6	98.2	47.1	4.6	49.7	18.1	35.9
Mollenfelde	32.8	98.8	74.4	4.7	50.6	18.0	36.3
Sophienhof	37.7	97.5	58.0	4.6	50.9	17.1	34.8
LSD5%	4.0	1.0	5.6	0.1	0.41	0.30	0.39

LSD5% least significant difference at P = 5%, odM of defatted meal.
[1]Data taken from Dimov et al. 2012.

Discussion and Conclusions

This study revealed a large and significant variation for secondary seed dormancy among current winter oilseed rape cultivars ranging from <10% to well over 50%. This range confirms results obtained in earlier studies for winter oilseed rape (Pekrun et al. 1997b, Momoh et al. 2002, Gulden et al. 2003, 2004, Gruber et al. 2004). In the study of Gruber et al. (2004), a variation from 3% to 76% was detected in a set of 32 winter oilseed genotypes. Among those genotypes, the secondary dormancy rate of hybrid cultivar 'Elektra' was found to be 26.1%. In the present study, a similar intermediate secondary dormancy rate of 27.8% was determined for this cultivar. Although there was no difference between line cultivars and hybrid cultivars with respect to their secondary dormancy level, it is interesting to note that both semi-dwarf hybrids were among the genotypes with the highest secondary dormancy level. Taken into account that a selection for secondary seed dormancy has never been done in oilseed rape breeding and that genetic variation among modern winter oilseed rape cultivars is considered to be narrow (Bus et al. 2011), the detected large variation is rather surprising. The high heritability of 0.97 indicated that the variation was caused predominantly by the genotypes. A predominant contribution of the genotypes to the overall variation in secondary seed dormancy was also reported by Gulden et al. (2004) for a set of 16 winter/spring oilseed rape genotypes.

Beside the strong genetic component, the analysis of variance (Table 1) showed also a large and significant effect of the location on secondary seed dormancy. The mean values over the locations ranged from 12.3% to 37.7% (Table 2). This was not surprising because the six locations were selected among 15 locations based on large differences in mean seed oil content (Table 2). Seeds harvested from locations with high oil contents and with high yield levels (cf. Dimov et al. 2012) clearly had higher secondary seed dormancy rates. This could indicate that optimal conditions during plant growth and maturation do not only increase seed yield and oil content but also the capacity to form secondary dormant seeds. On the other hand, suboptimal conditions during maturation or a too early harvest may not only affect seed yield and oil content but may lead to a reduced capacity to form secondary dormant seeds. Fei et al. (2007) studied gene expression in maturing seeds of different cultivars in relation to their potential for induction of secondary dormancy and found few differences at the mature stage but a significant number at the transition stage from full-size embryo to mature seed. In this context, it is noteworthy that the seed samples analysed in the present study were derived from yield plots of field experiments and represent a mixture of seeds derived from the main inflorescence as well as from secondary, tertiary, etc.

Table 1: Variance components and heritabilities for secondary seed dormancy (SD, %), germination rate (GR, %), seed longevity (SL, %), thousand-kernel weight (TKW, g), oil and protein content of the seed (in % seed dry matter) and for protein content of the defatted meal (Protein odM, in % seed dry matter) of 28 black-seeded winter oilseed rape cultivars tested in field experiments at six locations.

Source of variance	SD	GR	SL	TKW[1]	Oil[1]	Protein[1]	Protein odM[1]
Location (L)	49.2**	0.43*	192.5**	0.05**	6.9**	5.22**	6.21**
Genotype (G)	125.5**	4.22**	46.5**	0.10**	0.9**	0.36**	1.14**
G × L	25.6	10.83	113.1	0.04	0.6	0.33	0.53
Heritability	0.97	0.70	0.71	0.94	0.90	0.87	0.93

**,*Significant at P = 1%, 10% (F-test, ANOVA), odM of defatted meal.
[1]Data taken from Dimov et al. 2012.

Table 3: Secondary seed dormancy (SD, %), germination rate (GR, %), seed longevity (%), thousand-kernel weight (TKW, g), oil and protein content of the seed (in % seed dry matter) and protein content of the defatted meal (Protein odM, in % seed dry matter) of 28 European winter oilseed rape cultivars tested at six locations in 2008/2009.

Cultivar	Type	SD	GR	SL	TKW[1]	Oil[1]	Protein[1]	Protein odM
'DK Secure'	Hzk	55.5	99.4	60.4	4.2	46.8	20.1	37.6
'NK Caravel'	H	54.8	98.6	70.4	4.5	46.8	19.8	37.2
'PR45D01'	Hzk	54.1	97.5	55.4	4.5	47.8	19.0	36.3
'Zeppelin'	H	51.2	98.8	62.3	4.4	49.6	19.3	38.3
'Arcadia'	L	49.4	98.8	62.0	4.1	48.0	19.4	37.2
'Monolit'	L	48.6	97.1	53.7	5.0	49.5	18.1	35.7
'Cuillin'	H	41.2	97.7	57.3	4.5	48.6	19.9	38.6
'Tassilo'	H	35.3	99.1	59.2	4.5	47.9	20.0	38.4
'Limone'	H	33.2	98.6	63.6	4.7	48.5	19.2	37.1
'PR44W22'	H	32.9	98.6	65.2	4.5	49.1	19.2	37.6
'Lorenz'	L	28.2	97.3	47.1	4.4	50.4	18.4	37.0
'Elektra'	H	27.8	98.8	40.4	4.9	48.0	19.4	38.0
'PR44W18'	H	26.6	98.8	61.7	4.7	48.3	19.3	37.4
'Azur'	L	25.9	98.3	50.4	4.7	49.1	19.1	37.4
'NK Pegaz'	L	17.9	95.6	52.8	4.7	48.3	18.5	35.7
'Bellevue'	L	17.1	99.6	53.2	5.2	48.6	19.9	38.8
'NK Morse'	L	16.5	98.3	58.4	4.6	48.6	18.1	35.1
'Visby'	H	13.9	97.1	42.8	5.0	47.6	19.3	36.8
'NK Aviator'	H	13.5	99.1	59.7	4.0	47.3	19.9	37.7
'Adriana'	L	11.5	99.0	48.9	5.2	49.7	19.2	38.0
'Katabatic'	L	11.4	99.0	59.1	4.5	50.0	19.2	38.3
'Hybrisurf'	H	11.0	99.3	63.9	4.4	48.8	19.9	38.8
'Loveli CS'	L	10.4	94.0	50.7	4.5	49.8	20.2	40.1
'ES Alienor'	L	10.4	97.2	39.0	5.2	47.8	19.5	37.2
'Exotic'	H	10.2	99.6	66.2	5.1	46.6	20.9	39.0
'DK Cabernet'	L	9.9	98.3	47.8	4.2	48.5	18.7	36.2
'Safran'	H	9.2	98.1	59.3	4.4	47.4	19.9	37.7
'Iwan'	L	8.1	97.8	46.8	4.7	48.9	19.7	38.5
Mean	-	26.3	98.2	55.6	4.6	48.5	19.4	37.5
Min	-	8.1	94.0	39.0	4.0	46.8	18.1	35.1
Max	-	55.5	99.6	70.4	5.2	50.4	20.9	40.1
LSD5%	-	8.6	2.1	12.1	0.2	0.9	0.7	0.8

H, Hybrid cultivars; L, Line cultivars; Hzk, semidwarf hybrid cultivars; odM of defatted meal; LSD5% least significant difference at P = 5%.
[1]Data taken from Dimov et al. 2012.

Table 4: Spearman-rank correlation coefficients for secondary seed dormancy (SD), germination rate (GR), seed longevity (SL), thousand-kernel weight (TKW), oil- and protein content of the seed, protein content of the defatted meal (Protein odM), seed hull proportion (SH), neutral detergent fibre (NDF), acid detergent fibre (ADF) and acid detergent lignin (ADL).

	SD	GR	SL	TKW[1]	Oil[1]	Protein[1]	Protein odM	SH[1]	NDF[1]	ADF[1]
GR	0.07									
SL	0.35	0.54**								
TKW[1]	−0.16	−0.09	0.38*							
Oil[1]	−0.10	−0.20	−0.32	0.09						
Protein[1]	−0.17	0.45*	0.28	−0.14	−0.42*					
Protein odM	−0.27	0.50**	0.20	−0.00	0.16	0.77**				
SH[1]	0.25	−0.01	0.09	−0.16	−0.35	−0.24	−0.44*			
NDF[1]	0.19	−0.01	0.16	−0.30	−0.13	−0.33	−0.35	0.77**		
ADF[1]	0.12	0.06	0.19	−0.23	−0.00	−0.26	−0.18	0.66**	0.81**	
ADL[1]	0.00	0.14	0.05	−0.05	−0.07	−0.14	−0.12	0.56**	0.60**	0.84**

**,*Significant at P = 1%, 5%.
[1]Original data were taken from Dimov et al. 2012.

racemes. At harvest, those seeds may be in different maturation stages and hence may show differences in their secondary dormancy capacity.

Although there was a clear positive relation between seed oil content and the secondary seed dormancy rate of the seeds harvested at different locations (Table 2), no correlation between the oil content and the secondary dormancy rate of the genotypes was found (Table 4). There was also no significant correlation to other recorded traits, indicating that an indirect selection for a low secondary seed dormancy capacity will not be feasible. The results are in contrast to those obtained by Gulden et al. (2004), who reported an influence of seed size on the secondary seed dormancy rate. For the seed samples analysed in the present study, Dimov et al. (2012) reported large and significant genotypic differences for seed hull proportion and seed fibre content (NDF, ADF, ADL). Results from this study do not indicate that the seed coat may influence the capacity to form secondary dormant seeds. However, a lack of correlation to those traits may simply be caused by the composition of the plant material used in this study. The analysis of a segregating population derived from a cross between lines with different seed hull and fibre content would provide more meaningful results.

The large differences in the capacity to form secondary dormant seeds raised the question whether these differences might be related to differences in SL found among genotypes stored under ambient (Nagel and Börner 2010) or cold conditions (Nagel et al. 2011). Due to the lack of long-term stored oilseed rape material, this condition was simulated in the present work by short-term incubation of seeds at elevated temperature and humidity. Significant genotypic effects for germination after incubation were found (Table 1). The similar heritability of SL and germination rate (0.71 vs. 0.70) and the significant positive correlation between these two traits suggest a common genetic background. Good seed quality is a general requirement for long-term seed survival in storage (Ellis and Roberts 1980). Although artificially aged seeds of oilseed rape of low quality showed a delay in emergence and days to flowering, plants later on were able to compensate for those effects and no significant differences in 100-grain weight and grain yield per plant were observed (Ghassemi-Golezani et al. 2010). However, seed dormancy does not seem to be affected neither by initial germination rate nor by SL (Table 4), which points to different genetic control mechanisms (Clerkx et al. 2004, Bentsink et al. 2010, Rehman Arif et al. 2012a,b).

Artificial ageing of seeds is generally accepted to mimic extended seed storage because similar changes in the proteome, oxidation of proteins and detoxification of metabolites were observed (Rajjou et al. 2008). However, as soon as ageing conditions are modified, different genomic regions are activated to deal with the stress (Nagel et al. 2011). Therefore, germination behaviour after long-term soil storage might be different to what is observed after artificial ageing. In the present study, analysis of variance showed significant effects of the location on SL indicating that environmental factors as nutrient supply and growth conditions of the mother plant affect longevity of the harvested seeds (Kochanek et al. 2011). Interestingly, there was a significant positive correlation between SL, thousand-kernel weight and germination rate, respectively, but not between thousand-kernel weight and germination rate. Generally, larger seeds are assumed to enhance seed germination, seedling vigour and plant yield, but results from experiments with kale (*Brassica oleracea* L. var acephala DC) did not support the hypothesis that large seeds have superior performance to small seeds or that small seeds have lower vigour (Komba et al. 2007). The observed weak negative relationship between oil content and SL supports previous general findings reported in Walters et al. (2005) and in Nagel and Börner (2010).

In conclusion, applying an *in vitro* test the present study revealed large differences between winter oilseed rape genotypes in the capacity to form secondary dormant seeds and a high heritability for the trait. Hence, selection of genotypes with low seed dormancy should be effective in a breeding programme. However, performing the *in vitro* test for secondary dormancy is laborious and time-consuming, and it is unlikely that this test will be integrated in a practical breeding programme. So far, results do not indicate that an indirect selection for low seed dormancy can be performed applying simpler tests for other traits.

Acknowledgements

The technical assistance of Gunda Asselmeyer is greatly appreciated. Many thanks to Sabine Gruber (University of Hohenheim) for her help in establishing the secondary seed dormancy test. The authors express their gratitude to Sortenförderungsgesellschaft (SFG) in Bonn for providing seed sample material from the Bundes- und EU-Sortenversuch Winterraps 2008/2009 field experiments. We are grateful for the financial support given by the German Federal Ministry of Education and Research (BMBF) FKZ 0315211C.

References

Bazanska, J., and S. Lewak, 1986: Light inhibits germination of rape seeds at unfavourable temperatures. Acta Physiol. Plant. **8**, 145—149.

Bentsink, L., J. Jowett, C. J. Hanhart, and M. Koornneef, 2006: Cloning of *DOG1*, a quantitative trait locus controlling seed dormancy in *Arabidopsis*. Proc. Natl Acad. Sci. USA **103**, 17042—17047.

Bentsink, L., J. Hanson, C. J. Hanhart, H. Blankestijn-de Vries, C. Coltrane, P. Keizer, M. El-Lithy, C. Alonso-Blanco, M. Teresa de Andrése, M. Reymond, F. van Eeuwijk, S. Smeekens, and M. Koornneef, 2010: Natural variation for seed dormancy in Arabidopsis is regulated by additive genetic and molecular pathways. Proc. Natl Acad. Sci. USA **107**, 4264—4269.

Bewley, J. D., 1997: Seed germination and dormancy. Plant Cell **9**, 1055.

Bus, A., N. Körber, R. J. Snowdon, and B. Stich, 2011: Patterns of molecular variation in a species wide germplasm set of *Brassica napus* Theor. Appl. Genet. **123**, 1413—1423.

Clerkx, E. J. M., M. E. El-Lithy, E. Vierling, G. J. Ruys, H. Blankestijn-De Vries, S. P. C. Groot, D. Vreugdenhil, and M. Koornneef, 2004: Analysis of natural allelic variation of Arabidopsis seed germination and seed longevity traits between the accessions Landsberg *erecta* and Shakdara, using a new recombinant inbred line population. Plant Physiol. **135**, 432—443.

Colbach, N., C. Dürr, S. Gruber, and C. Pekrun, 2008: Modelling the seed bank evolution and emergence of oilseed rape volunteers for managing co-existence of GM and non-GM varieties. Eur. J. Agron. **28**, 19—32.

Dimov, Z., E. Suprianto, F. Hermann, and C. Möllers, 2012: Genetic variation for seed hull and fibre content in a collection of European winter oilseed rape material (*Brassica napus* L.) and development of NIRS calibrations. Plant Breeding **131**, 361—368.

Ellis, R. H., and E. H. Roberts, 1980: Improved equations for the prediction of seed longevity. Ann. Bot. **45**, 13—30.

Fei, H., E. Tsang, and A. J. Cutler, 2007: Gene expression profiling during seed maturation in *Brassica napus* in relation to the potential for induction of secondary dormancy. Genomics **89**, 419—428.

Fei, H., Y. Ferhatoglu, E. Tsang, D. Huang, and A. J. Cutler, 2009: Metabolic and hormonal processes associated with the induction of secondary dormancy in *Brassica napus* seeds. Special Issue of the National Research Council of Canada – Plant Biotechnology Institute. Botany **87**, 585—596.

Ghassemi-Golezani, K., S. Khomari, B. Dalil, A. Hosseinzadeh-Mahootchy, and A. Chadordooz-Jeddi, 2010: Effects of seed aging on field performance of winter oilseed rape. J. Food Agric. Environ. **8**, 175—178.

Gräber, K., K. Nakabayashi, E. Miatton, G. Leubner-Metzger, and W. J. J. Soppe, 2012: Molecular mechanisms of seed dormancy. Plant, Cell Environ., **35**, 1769—1786.

Gronow, J., W. Sauermann, and G. Barthelmes, 2009: Sortenversuche 2009 – Mit Winterraps, Futtererbsen und Sonnenblumen. UFOP Schriften, Heft 37/Agrar (E-Book). Available at: http://www.ufop.de/medien/downloads/agrar-info/ufop-schriften/ (last accessed on July 25, 2012).

Gruber, S., and W. Claupein, 2006: Effect of soil tillage intensity on seedbank dynamics of oilseed rape compared with plastic pellets as reference material. J. Plant Dis. Prot., Special Issue **20**, 273—280.

Gruber, S., C. Pekrun, and W. Claupein, 2004: Seed persistence of oilseed rape (*Brassica napus*): variation in transgenic and conventionally bred cultivars. J. Agric. Sci. **142**, 29—40.

Gruber, S., A. Bühler, J. Möhring, and W. Claupein, 2010: Sleepers in the soil – Vertical distribution by tillage and long-term survival of oilseed rape seeds compared with plastic pellets. Eur. J. Agron. **33**, 81—88.

Gulden, R. H., S. J. Shirtliffe, and A. G. Thomas, 2003: Secondary seed dormancy prolongs persistence of volunteer canola in western Canada. Weed Sci. **51**, 904—913.

Gulden, R. H., A. G. Thomas, and S. J. Shirtliffe, 2004: Relative contribution of genotype, seed size and environment to secondary seed dormancy potential in Canadian spring oilseed rape (*Brassica napus*). Weed Res. **44**, 97—106.

Hampton, J. G., and D. M. TeKrony, 1995: Handbook of vigor test methods, 3rd edn. The International Seed Testing Assoc., Zurich.

ISTA, 2011: International Rules for Seed Testing. International Seed Testing Association, Bassersdorf, Switzerland.

Kochanek, J., K. J. Steadman, R. J. Probert, and S. W. Adkins, 2011: Parental effects modulate seed longevity: exploring parental and offspring phenotypes to elucidate pre-zygotic environmental influences. New Phytol. **191**, 223—233.

Komba, C. G., B. J. Brunton, and J. G. Hampton, 2007: Effect of seed size within seed lots on seed quality in kale. Seed Sci. Technol. **35**, 244—248.

López-Granados, F., and P. J. W. Lutman, 1998: Effect of environmental conditions on the dormancy and germination of volunteer oilseed rape seed (*Brassica napus*). Weed Sci. **46**, 419—423.

Lutman, P. J. W., S. E. Freeman, and C. Pekrun, 2003: The long-term persistence of seeds of oilseed rape (*Brassica napus*) in arable fields. J. Agric. Sci. **141**, 231—240.

Lutman, P. J. W., K. Berry, R. W. Payne, E. Simpson, J. B. Sweet, G. T. Champion, M. J. May, P. Wightman, K. Walker, and M. Lainsbury, 2005. Persistence of seeds from crops of conventional and herbicide tolerant oilseed rape (*Brassica napus*). Proc. R. Soc. B **272**, 1909 1915.

Momoh, E. J. J., W. J. Zhou, and B. Kristiansson, 2002: Variation in the development of secondary dormancy in oilseed rape genotypes under conditions of stress. Weed Res. **42**, 446—455.

Nagel, M., and A. Börner, 2010: The longevity of crop seeds stored under ambient conditions. Seed Sci. Res. **20**, 1—12.

Nagel, M., H. Vogel, S. Landjeva, G. Buck-Sorlin, U. Lohwasser, U. Scholz, and A. Börner, 2009: Seed conservation in ex-situ genebanks – genetic studies on longevity in barley. Euphytica **170**, 1—10.

Nagel, M., M. A. Rehman Arif, M. Rosenhauer, and A. Börner, 2010: Longevity of seeds - intraspecific differences in the Gatersleben genebank collections, Proceedings of the 60th conference of plant breeders and seed traders association in Austria 2009, Raumberg-Gumpenstein. pp. 179—181.

Nagel, M., M. Rosenhauer, E. Willner, R. J. Snowdon, W. Friedt, and A. Börner, 2011: Seed longevity in oilseed rape (*Brassica napus* L.) - genetic variation and QTL mapping. Plant Gen. Resour. **9**, 260—263.

Pekrun, C., P. J. W. Lutman, and K. Baeumer, 1997a: Germination behaviour of dormant oilseed rape seeds in relation to temperature. Weed Res. **37**, 419—431.

Pekrun, C., P. J. W. Lutman, and K. Baeumer, 1997b: Induction of secondary dormancy in rape seeds (*Brassica napus* L.) by prolonged imbibition under conditions of water stress or oxygen deficiency in darkness. Eur. J. Agron. **6**, 245—255.

Pérez-García, F., M. E. Gonzalez-Benito, and C. Gomez-Campo, 2007: High viability recorded in ultra-dry seeds of 37 species of Brassicaceae after almost 40 years of storage. Seed Sci. Technol. **35**, 143—153.

Pessel, D., J. Lecomte, V. Emeriau, M. Krouti, A. Messean, and P. H. Gouyon, 2001: Persistence of oilseed rape (*Brassica napus* L.) outside of cultivated fields. Theor. Appl. Genet. **102**, 841—846.

Peters, J., 2000: Tetrazolium testing handbook. Association of Official Seed Analysts: Handbook on Seed Testing, Contribution 29. Available at: http://www.aosaseed.com

Probert, R., J. Adams, J. Coneybeer, A. Crawford, and F. Hay, 2007: Seed quality for conservation is critically affected by pre-storage factors. Aust. J. Bot. **55**, 326—335.

Rajjou, L., Y. Lovigny, S. P. C. Groot, M. Belghaz, C. Job, and D. Job, 2008: Proteome-wide characterization of seed aging in *Arabidopsis*: a comparison between artificial and natural aging protocols. Plant Physiol. **148**, 620—641.

Rehman Arif, M. A., M. Nagel, K. Neumann, B. Kobiljski, U. Lohwasser, and A. Börner, 2012a: Genetic studies of seed longevity in hexaploid wheat using segregation and association mapping approaches. Euphytica **186**, 1—13.

Rehman Arif, M. A., K. Neumann, M. Nagel, B. Kobiljski, U. Lohwasser, and A. Börner, 2012b: An association mapping analysis of dormancy and pre-harvest sprouting in wheat. Euphytica, DOI: 10.1007/s10681-012-0705-1.

Specht, C. E., and A. Börner, 1998: An interim report on a long term storage experiment on rye (*Secale cereale* L) under a range of temperatures and atmospheres. Genet. Resour. Crop Evol. **45**, 483—488.

Utz, H. F. (2011) PLABSTAT (Version 3A), A computer program for statistical analysis of plant breeding experiments. Institute of Plant Breeding, Seed Science and Population Genetics, University of Hohenheim, Stuttgart, Germany. Available at: https://plant breeding.uni-hohenheim.de/software.html.

Walters, C., L. M. Wheeler, and J. M. Grotenhuis, 2005: Longevity of seeds stored in a genebank: species characteristics. Seed Sci. Res. **15**, 1—20.

3.10 PAPER 10:

QTL analysis of falling number and seed longevity in wheat (*Triticum aestivum* L.)

by

Andreas Börner, **Manuela Nagel**, Monika Agacka-Mołdoch, Peter Ulrich Gierke, Michael Oberforster, Theresa Albrecht and Volker Mohler

Published in

Journal of Applied Genetics (2018) 59, 35-42

https://doi.org/10.1007/s13353-017-0422-5

To view supplementary material for this article, please visit

https://link.springer.com/article/10.1007%2Fs13353-017-0422-5

Journal of Applied Genetics (2018) 59:35–42
https://doi.org/10.1007/s13353-017-0422-5

PLANT GENETICS • ORIGINAL PAPER

QTL analysis of falling number and seed longevity in wheat (*Triticum aestivum* L.)

Andreas Börner[1] • Manuela Nagel[1] • Monika Agacka-Mołdoch[1,2] • Peter Ulrich Gierke[3] • Michael Oberforster[4] • Theresa Albrecht[5] • Volker Mohler[5]

Received: 26 July 2017 / Revised: 28 November 2017 / Accepted: 6 December 2017 / Published online: 14 December 2017

Abstract
Pre-harvest sprouting (PHS) and seed longevity (SL) are complex biological processes of major importance for agricultural production. In the present study, a recombinant inbred line (RIL) population derived from a cross between the German winter wheat (*Triticum aestivum* L.) cultivars History and Rubens was used to identify genetic factors controlling these two physiological seed traits. A falling number (FN) test was employed to evaluate PHS, while SL was measured using a germination test (and the speed of germination) after controlled deterioration. FN of the population was assessed in four environments; SL traits were measured in one environment. Four major quantitative trait loci (QTL) for FN were detected on chromosomes 4D, 5A, 5D, and 7B, whereas for SL traits, a major QTL was found on chromosome 1A. The FN QTL on chromosome 4D that coincided with the position of the dwarfing gene *Rht-D1b* only had effects in environments that were free of PHS. The remaining three QTL for FN were mostly pronounced under conditions conducive to PHS. The QTL on the long arm of chromosome 7B corresponded to the major gene locus controlling late maturity α-amylase (LMA) in wheat. The severity of the LMA phenotype became truly apparent under sprouting conditions. The position on the long arm of chromosome 1A of the QTL for SL points to a new QTL for this important regenerative seed trait.

Keywords Controlled deterioration · Dormancy · Germination · Pre-harvest sprouting · QTL mapping

Communicated by: Barbara Naganowska

Electronic supplementary material The online version of this article (https://doi.org/10.1007/s13353-017-0422-5) contains supplementary material, which is available to authorized users.

✉ Andreas Börner
boerner@ipk-gatersleben.de

✉ Volker Mohler
volker.mohler@lfl.bayern.de

[1] Leibniz Institute of Plant Genetics and Crop Plant Research, Corrensstr. 3, OT Gatersleben, 06466 Stadt Seeland, Germany

[2] Institute of Soil Science and Plant Cultivation, State Research Institute, ul. Czartoryskich 8, 24-100 Puławy, Poland

[3] Landesanstalt für Landwirtschaft und Gartenbau, Dezernat 52, Prüf- und Anerkennungsstelle für Saat- und Pflanzgut, Schiepziger Str. 29, 06120 Halle (Saale), Germany

[4] Austrian Agency for Health and Food Safety (AGES), Institute for Sustainable Plant Production, Spargelfeldstr. 191, 1220 Vienna, Austria

[5] Bavarian State Research Center for Agriculture (LfL), Institute for Crop Science and Plant Breeding, Am Gereuth 6, 85354 Freising, Germany

Introduction

Pre-harvest sprouting (PHS) in cereals such as wheat (*Triticum aestivum* L.) is a precocious germination of grains while still in the spikes of the mother plant. It appears in certain years when humid conditions due to rainfall occur before harvest and can result in tremendous decline in grain quality and yield. The biological reason for PHS is the early breakage of seed dormancy. The trait is of quantitative nature and there are many studies of mapping quantitative trait loci (QTL) for PHS resistance in wheat (e.g., Kulwal et al. 2004, 2012; Mares et al. 2005; Rasul et al. 2009; Rehman Arif et al. 2012b, 2017; Lohwasser et al. 2013; Albrecht et al. 2015).

Seed longevity (SL), the time span a seed remains viable, is important for the seed industry but also for ex situ germplasm conservation. Compared to other crops, SL of wheat can be classified as moderate: the half-viability period is at room temperature 7.6 years (Nagel and Börner 2010) and at cold storage 54 years (Walters et al. 2005).

Again, the trait is inherited quantitatively. Genetic studies, however, are rare. First results for wheat have been published by Landjeva et al. (2010) and Rehman Arif et al. (2012a, 2017).

Whether there is a causal relationship between seed dormancy and SL is under debate. In the literature, controversial results have been published for different crops. First attempts on mapping QTL for dormancy along with SL were made in rice (*Oryza sativa* L.) by Miura et al. (2002). QTL for both traits were located on different chromosomes, evidencing independent genetic control. The same conclusion was drawn from other studies in rice (Guo et al. 2004; Zeng et al. 2006) and Arabidopsis (Clerkx et al. 2004).

In wheat, comparable studies were performed using different mapping population types. Landjeva et al. (2010) investigated a set of 85 'Chinese Spring'/*Aegilops tauschii* introgression lines where D genome segments of bread wheat were substituted for their homoeologues from *Ae. tauschii* (Pestsova et al. 2006). The authors identified two sections in the centromere regions of chromosomes 1DS and 5DL determining SL. When investigating the same set for dormancy, however, a locus was found on chromosome 6DL (Lohwasser et al. 2013). Recombinant inbred lines (RILs) of the International Triticeae Mapping Initiative (ITMI) mapping population Opata 85 × W7984 grown in two different seasons were investigated by Rehman Arif et al. (2012a). QTL for SL were determined on chromosomes 1A, 1D, 2A, 2D, 3B, 3D, 6B, and 7A, whereas a locus for dormancy was detected on chromosome 4A. An association mapping panel consisting of 96 wheat cultivars and genotyped with 874 DArT markers was phenotyped for PHS, dormancy, and SL by Rehman Arif et al. (2012a, b). Totals of 56 and 37 markers were associated with PHS (and/or seed dormancy) and SL, respectively. Interestingly, 11 markers were found to be connected to both traits. Recently, Rehman Arif et al. (2017) investigated another panel of 183 wheat accessions genotyped with 2134 DArT markers for PHS, dormancy, and SL. A huge number of marker–trait associations were found. Nineteen and 11 of the marker–trait associations for PHS and dormancy, respectively, overlapped those for SL.

In this study, a bi-parental population was investigated for possible overlaps in the genetic control of PHS and SL using genetic linkage-based QTL analysis. Since the population was segregating for the major gibberellic acid-insensitive dwarfing gene *Rht-D1* on chromosome 4D and the grain texture-controlling gene *Pinb-D1* on chromosome 5D, we could also address the question whether the considered alleles of these important wheat breeding signatures have significant effects on the target traits.

Materials and methods

Plant material

Previous work established a mapping population of 103 RILs derived from a cross between winter wheat genotypes History and Rubens (Holzapfel et al. 2008). In the present study, 82 RILs were used. Rubens carries the dwarfing allele *Rht-D1b*, whereas History possesses the wild-type allele *Rht-D1a*. Both cultivars have mutant alleles at *Pinb-D1* (Rubens: *Pinb-D1b*; History: *Pinb-D1d*). History is a late maturity α-amylase (LMA)-prone genotype (Mohler et al. 2014), where LMA is the untimely synthesis of α-amylase during the middle stages of grain development that is retained through to harvest, causing a reduction in falling number (FN) (Mares and Mrva 2014). In addition, *Rht-D1b* is known to have a masking effect on the expression of LMA.

Field trials

The population along with the parents was grown at three locations in Germany (Pulling, Oberhummel, and Rosenthal) and Grabenegg in Austria during the 2010/2011, 2012/2013, and 2013/2014 cropping seasons. The field trials in Germany followed a randomized complete block design with two blocks, whereas in Austria, an unreplicated augmented design with standard cultivars (History, Rubens, and JB Asano) as replicates was used. Plot sizes were 1.25 m^2. Spikes for trait measurements were hand-harvested at physiological maturity (German locations) or after plants had experienced an artificial rain treatment lasting several days with breaks using SuperNet™ (Netafim, Fresno, CA, USA) sprinklers (Grabenegg location).

Falling number determination

FN of wheat samples from each plot were determined on a Falling Number instrument FN 1700 (Perten Instruments, Hägersten, Sweden), according to the standard method no. 107/1 of the International Association for Cereal Science and Technology.

Germination test

Seed germination for control treatment (C) and after controlled deterioration (CD) was performed for each entry from Pulling 2011 in four replications (independent experiments) of 50 seeds per line. The seeds were placed in pleated filter paper strips (Munktell & Filtrak GmbH, Barenstein, Germany; 50 pleats, 2000 × 110 mm/20 mm depth). Pleated paper strips were inserted in plastic boxes with a flat filter paper (Munktell & Filtrak GmbH, Barenstein, Germany; C 160, 105 × 155 mm) on the ground for even water distribution below pleated paper strips. The boxes were covered with clear plastic lids against evaporation. The water saturation was 85% of the maximum capacity from the pleated paper strip and the flat ground paper. Germination tests were performed in a climate chamber (Dresdner Kühlanlagenbau GmbH, Dresden, Germany). The temperature was constant at 20 ± 2 °C and the phase of illumination was 8 h per day. In order to determine the germination speed, germinated seeds (visible radical emergence) were counted after 2, 4, and 8 days. Based on these counts, the area under the curve (AUC; the integration of the fitted curve between $t = 0$ and a defined endpoint $t = x$) after 192 h of germination was determined using the germination software GERMINATOR (Joosen et al. 2010). On day 8, total germination (TG) and the number of seedlings showing a normal appearance [normal germination (NG)] according to the International Seed Testing Association (ISTA, 2015) were recorded.

Controlled deterioration

For CD treatment, four replicates of 50 seeds per line were equilibrated to 47% relative humidity for 10 days, above 8.7 M (i.e., non-saturated) LiCl (Hay et al. 2008). CD was initiated by increasing the temperature to 45 °C for 19 days, using a 7.1 M LiCl solution (60% relative humidity) to maintain a stable seed moisture content (MC). The CD period was estimated on the basis of the viability equation $v = K_i - p/10^{K_E - C_W \log m - C_H t - C_Q t^2}$ using wheat specific constants $K_E = 9.42$, $C_W = 5.859$, $C_H = 0.0329$, $C_Q = 0.000478$ (Ellis et al. 1990), seed equilibrium MC $m = 10.2\%$, $t = 45$ °C, initial viability of $K_I = 95\%$, and final viability of $v = 50\%$.

Genetic mapping

A genetic map for History/Rubens (Fig. S1) was re-calculated with JoinMap® software version 4.0 (Kyazma B.V., Wageningen, The Netherlands). Linkage of loci was claimed at a logarithm of the odds ratio (LOD) score ≥ 3.0, with a maximum recombination fraction of 0.4. The Haldane mapping function was used to convert the recombination fraction into map distance. Map construction used the data set of 491 molecular markers from the study of Miedaner et al. (2012), which included microsatellite, amplified fragment length polymorphism (AFLP), and Diversity Arrays Technology (DArT) markers. The protocols for generating and scoring microsatellite and AFLP markers were described by Schmolke et al. (2005). The procedure for establishing DArT markers can be found in Akbari et al. (2006). The marker data set was complemented with ten single nucleotide polymorphism (SNP) markers for *Pinb-D1b*, *Pinb-D1d*, IWA1952 (Chr. 1A), IWB13256 (Chr. 3A), IWB10587 (Chr. 4A), IWA2363 (Chr. 5A), IWA7005 (Chr. 7A), IWB6028 (Chr. 2B), IWA7318 (Chr. 5B), and IWA2219 (Chr. 6B). DNA from primary leaves was extracted as described by Huang et al. (2000). All SNP assays were developed by Fluidigm Corporation (South San Francisco, CA, USA). Segregation of SNP markers was followed on an EP1 genotyping platform using 192.24 Dynamic Array Integrated Fluidic Circuits. All SNP genotyping analysis protocols can be found in the user guide published by the manufacturer (https://www.fluidigm.com).

Phenotypic data analysis

Analysis of variance (ANOVA), Wilcoxon rank-sum test, and least squares means (LSmeans) for each environment were computed using R (R Core Team 2015).

QTL analysis

The MultiQTL software package version 2.6 (MultiQTL Ltd., Haifa, Israel) was used for QTL analysis. FN values of RILs of each single environment adjusted for block effects were analyzed using the multiple-environment option. For seed longevity traits, arithmetic means of the four technical replicates were used. Relative values of RILs for the germination-related traits were determined by dividing the means after CD by the means of the corresponding controls. The testing for a QTL at each location across the genome involved a likelihood ratio test, rescaled as an LOD score, with or without adjusting for multiple QTL simultaneously; the first procedure is known as multiple interval mapping (MIM; Kao et al. 1999), the latter as simple interval mapping (SIM; Lander and Botstein 1989). For declaring a QTL significant, statistical significance thresholds (*P*) for each QTL were empirically determined by permutations of the data set (Churchill and Doerge 1994; Doerge and Churchill 1996). LOD support intervals (95%) of QTL were estimated by bootstrapping with 10,000 samples (Lebreton and Visscher 1998). QTL were first detected by SIM using a single QTL model (one QTL per linkage group). Significance levels for detected QTL were determined by 1000 permutations. The MIM procedure included all chromosomes from SIM that harbored a QTL with an LOD score surpassing the 95% ($P < 0.05$) critical threshold. QTL obtained with MIM were tested for significance ($P < 0.001$) with a permutation test ($N = 10{,}000$). Finally, a two-linked QTL model in SIM on each chromosome was applied, first testing if two QTL were significant and then testing if two QTL were more significant than one QTL. The allelic substitution effect (here the additive effect and the difference between the two homozygote genotypic values) and the percentage of phenotypic variance explained (PVE) by the QTL were calculated by MIM. QTL that showed PVE > 10% in at least one environment were regarded as major QTL. Genetic linkage maps including QTL support intervals were drawn with MapChart 2.1 software (Voorrips 2002).

Results

Analysis of phenotypic data

Falling number

The density distribution of FN was continuous for all environments (Fig. 1). Trait values were generally lower in Pulling 2011 and Grabenegg 2014 than in Oberhummel 2013 and Rosenthal 2013, suggesting the presence of environmental factors that triggered starch degradation. Under optimum harvest conditions such as which occurred in Oberhummel 2013, Rubens (422.5 ± 13.4 [standard error] s) was found to display higher FN than History (361.3 ± 13.4 s). The ANOVA for each environment showed a significant effect of the genotype on FN (Tables S1–S4).

Seed longevity

The distributions of the RILs together with the positions of the parents for the germination-related traits are given in Fig. 2. Under control conditions, both parents and most of the RILs reached high germination percentages, ranging between 90 and 100% for both TG and NG. After CD treatment, the germination percentages declined (TG and NG). The germination speed (AUC) of control seeds was similar for both parents and most of the progeny. After the deterioration treatment, the speed of both parents and the RILs was reduced. A Wilcoxon rank-sum test showed that the distributions of the RILs including the parental lines before and after treatment were significantly different ($P < 0.001$) for TG, NG, and AUC.

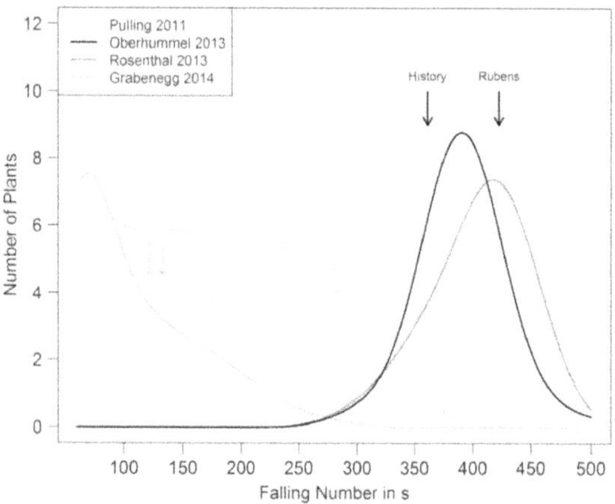

Fig. 1 Distributions for the trait falling number in four environments for recombinant inbred lines derived from the History/Rubens cross. Ratings of the parents are indicated by arrows, except for Rosenthal 2011, for which no parental data were available

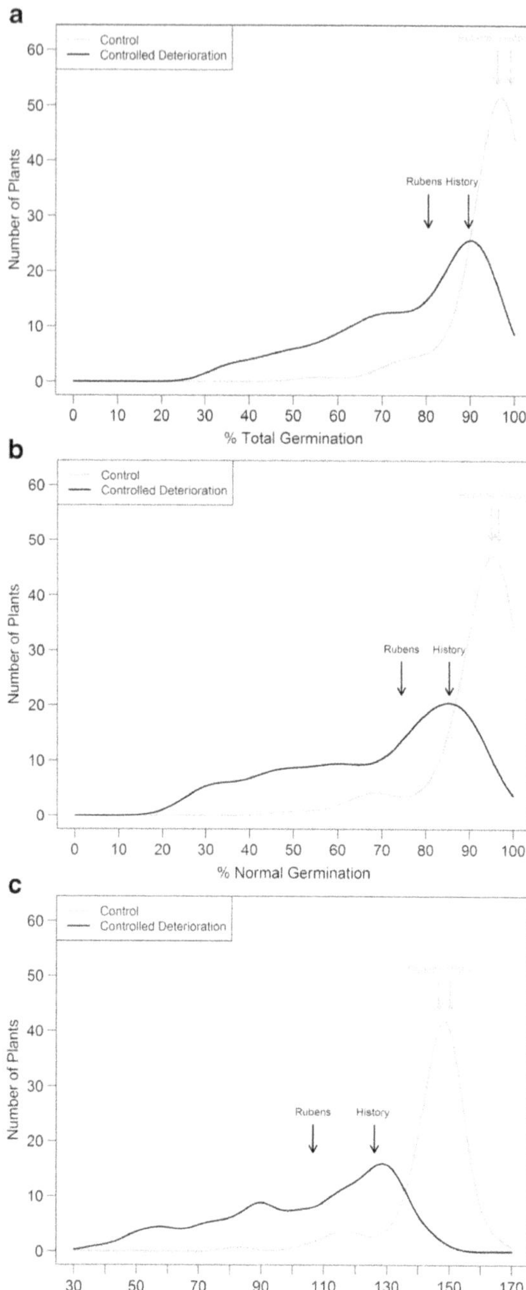

Fig. 2 Distributions for the traits % total germination (**a**), % normal germination (**b**), and area under the curve (AUC) after 192 h of germination (**c**) of untreated seeds and seeds after controlled deterioration for recombinant inbred lines derived from the History/Rubens cross. Ratings of the parents are indicated by arrows

Table 1 Quantitative trait loci (QTL) for falling number in the History/Rubens recombinant inbred line population using the multiple-environment option in multiple interval mapping. Markers closest to the QTL are underlined

Chr	Marker interval (QTL peak)	LOD[a]	Pulling 2011		Oberhummel 2013		Rosenthal 2013		Grabenegg 2014		QTL allele donor
			PVE[b]	Eff[c]	PVE[b]	Eff[c]	PVE[b]	Eff[c]	PVE[b]	Eff[c]	
4D	wPt-0710–*Rht-D1*	5.1	0	0	18.1	30	9.2	25	0	0	Rubens
5A	gwm186–P7560-439	11.6	16.0	56	4.2	14	7.6	22	12.2	37	Rubens
5D	P7551-267–wmc574	7.4	18.3	60	1.1	7	3.2	15	15.0	41	Rubens
7B	P7455-236–P7553-711	9.1	14.3	53	6.8	18	3.2	14	20.4	47	Rubens

[a] $\log_{10}$ likelihood ratio. Since the multiple-environment option considered all data sets simultaneously, a QTL logarithm of the odds ratio (LOD) value common to all environments was determined

[b] Percentage of the phenotypic variation explained by the QTL at the peak of the LOD profile

[c] Allelic substitution effect (difference between homozygous genotypic classes) in seconds

QTL mapping

Four major QTL for FN were identified on chromosomes 4D, 5A, 5D, and 7B employing a single QTL model in MIM (Table 1; Fig. 3). The use of a two-linked QTL model in SIM and testing of the critical sub-models did not reveal the presence of linked QTL on the four target chromosomes. The proportion of phenotypic variance explained by each QTL was greater or almost 10% in two of the four environments. The most significant QTL with an LOD score of 11.6 was assigned to the marker

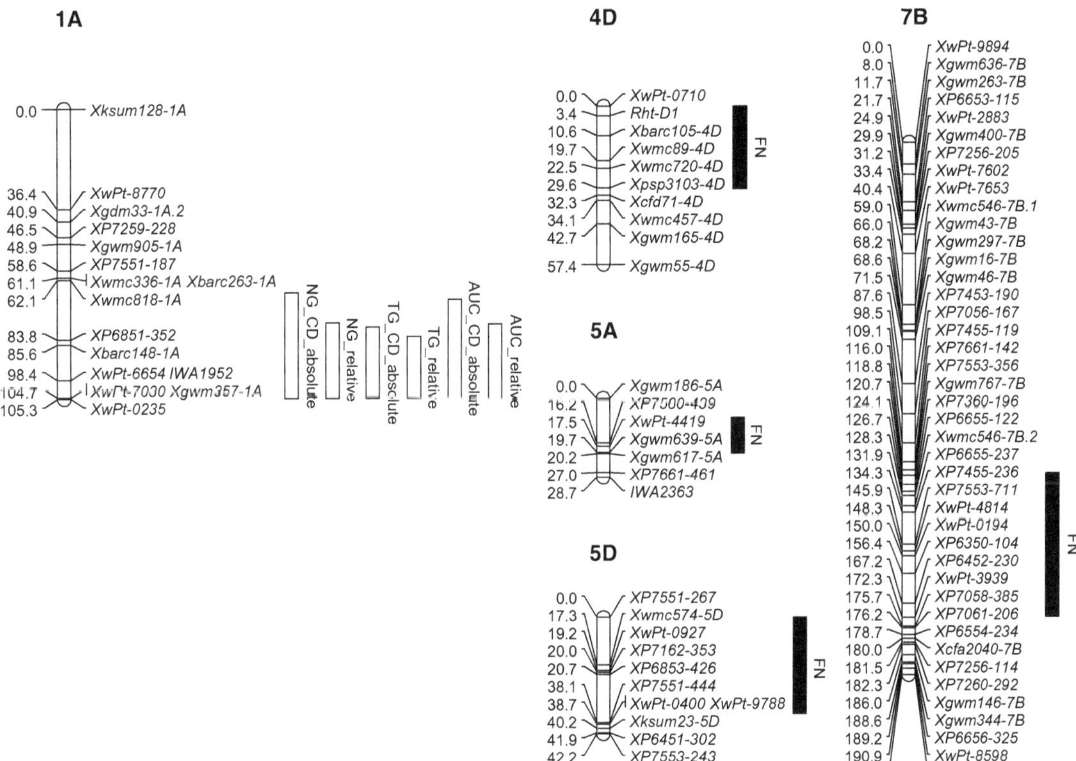

Fig. 3 Genetic chromosome maps of the History/Rubens recombinant inbred line population showing locations of quantitative trait loci (QTL) for falling number (FN), % total germination (TG) after controlled deterioration (CD), % normal germination (NG) after CD, and germination speed (AUC) after CD. QTL were diagrammed as bars representing confidence intervals (95%) of QTL map location on the left side of chromosomes. Absolute map positions in centimorgans (cM) and marker names are shown on the left and right sides, respectively, of each genetic map

Table 2 Quantitative trait loci (QTL) for % total germination (TG) and % normal germination (NG) of seeds after controlled deterioration (CD) in the History/Rubens recombinant inbred line population. Markers closest to the QTL are underlined

Chr	Marker interval (QTL peak)	TG_CD			Relative TG			NG_CD			Relative NG			QTL allele donor
		LOD[a]	PVE[b]	Eff[c]	LOD[a]	PVE[b]	Eff[c]	LOD[a]	PVE[b]	Eff[c]	LOD[a]	PVE[b]	Eff[c]	
1A	wPt-6654–wPt-7030	4.2	22.1	16.0	4.3	21.4	0.13	3.9	21.0	18.2	4.5	24.0	0.17	History

[a] $\log_{10}$ likelihood ratio

[b] Percentage of the phenotypic variation explained by the QTL at the peak of the logarithm of the odds ratio (LOD) profile

[c] Allelic substitution effect (difference between homozygous genotypic classes) in seconds

interval gwm186–P7560-439 on chromosome 5A and accounted for 4.2 to 16.0% of the variation in the four environments. The QTL on chromosomes 5D and 7B were detected with LOD scores of 7.4 and 9.4 and exhibited their highest additive effects, 60 and 53 s, respectively, in Pulling 2011. With respect to the environment, the QTL on chromosome 4D close to the *Rht-D1* locus was exclusively expressed in Oberhummel 2013 and Rosenthal 2013, whereas the highest phenotype contributions of the QTL on chromosomes 5A, 5D, and 7B were observed in Pulling 2011 and Grabenegg 2014. At all loci, higher FN was conferred by the Rubens allele. For the SL traits TG, NG, and AUC, a single but similar main-effect QTL was located on the long arm of chromosome 1A using a single QTL model in SIM (Tables 2 and 3; Fig. 3). The QTL peaks were located close to co-segregating markers wPt-6654 and IWA1952. For each trait, the QTL explained more than 20% of the phenotypic variation. QTL support intervals obtained with data sets for relative trait values were always smaller than those obtained with data sets for absolute trait values. The highest LOD score for the QTL was obtained with the relative AUC data set. The QTL allele that increased germination and germination speed was inherited from History. The LOD curves for chromosomes carrying significant QTL are shown in Fig. S2.

Discussion

It is well known that traits determining seed physiology are under polygenic control. This was confirmed in the present study and significant QTL for all traits under study were detected. The validity of the detected QTL is, in particular, evidenced for the FN QTL on chromosome 7B by recent findings of Mohler et al. (2014). Using a high-pI α-amylase-specific enzyme-linked immunosorbent assay (Mares and Mrva 2008), the authors revealed the presence of LMA in History. Therefore, it is very likely that the FN QTL on chromosome 7B corresponds to the major LMA QTL at the distal end of chromosome 7B (Mrva and Mares 2001; Emebiri et al. 2010). The FN QTL on chromosomes 5A and 5D had pronounced effects only in the two PHS environments Pulling 2011 and Grabenegg 2014; thus, they can be clearly designated as QTL for PHS. Based on similar chromosome locations, both QTL may have been described before. The QTL on chromosome 5A was located in the centromere region just as the QTL reported by Groos et al. (2002) and Nakamura et al. (2007). Jaiswal et al. (2012) identified a putative marker–trait association between PHS and microsatellite marker cfd102 located on chromosome 5D. It appears that we could confirm this QTL since cfd102 and the QTL confidence interval marker wmc574 from our study were located within a 3-cM interval on a reference genetic map (Maccaferri et al. 2015). Only one putative major QTL on 1AL, which needs further investigation, was identified for the SL-related traits. On the same chromosome arm, QTL for SL were reported (Rehman Arif et al. 2012a, 2017). Whether some of them might correspond to the locus detected in the History/Rubens mapping population can only be speculated.

One aim of the study was to find out whether two major wheat breeding signatures, namely *Rht-D1* and *Pinb-D1*, show pleiotropic effects on the seed traits investigated. The

Table 3 Quantitative trait loci (QTL) for the area under the curve (AUC) after 192 h of germination of seeds after controlled deterioration (CD) in the History/Rubens recombinant inbred line population. Markers closest to the QTL are underlined

Chr	Marker interval (QTL peak)	AUC_CD			Relative AUC			QTL allele donor
		LOD[a]	PVE[b]	Eff[c]	LOD[a]	PVE[b]	Eff[c]	
1A	wPt-6654–wPt-7030	3.9	21.3	25.0	5.8	30.1	0.19	History

[a] $\log_{10}$ likelihood ratio

[b] Percentage of the phenotypic variation explained by the QTL at the peak of the logarithm of the odds ratio (LOD) profile

[c] Allelic substitution effect (difference between homozygous genotypic classes) in seconds

FN locus detected on chromosome 4D was found to be related to the *Rht-D1b* dwarfing allele and increased FN. The QTL was detected in environments Oberhummel 2013 and Rosenthal 2013 where PHS was absent, but not in environments Pulling 2011 and Grabenegg 2014, where PHS occurred (Table 1). Concurrently, the effect of the major LMA locus on chromosome 7BL was highest in the latter two locations. A general effect of *Rht-D1b* on the expression of LMA is well known, and the increased expression of LMA might be due to a cold shock that occurred after flowering, which made the effect of *Rht-D1b* redundant (Mares and Mrva 2008). On the other hand, the large effect of the 7BL QTL observed here might also be related to the occurrence of PHS. *Rht-D1b* does not show exceptionally strong dormancy (Flintham and Gale 1982; Flintham et al. 1997, both cited in Van De Velde et al. 2017), suggesting that the modifying effect of *Rht-D1b* on LMA expression is also no longer working once processes promoting germination have been initiated. This raises the question whether *Rht-B1c* with its pronounced effect on dormancy or new useful *Rht-B1* mutations with impact on dormancy such as those reported by Van De Velde et al. (2017) would still exert a modifying effect on the LMA phenotype under conditions conducive to PHS. Grain hardness depends on the degree of adhesion between the starch granules and the protein matrix. The lower adhesion observed in genotypes with soft endosperm causes enhanced capillary forces, resulting in faster water uptake than into hard grains (Manley et al. 2011). This behavior may affect germination. However, no QTL for both FN and SL at *Pinb-D1* were detected, indicating the equivalence of effects of *Pinb-D1b* and *Pinb-D1d* on kernel hardness. However, this result does not rule out that the wild-type allele *Pinb-D1a* may have an effect on the target traits.

Another aim of the study was to find out whether the studied grain traits were under common genetic control. Considering only the results of the History/Rubens mapping population, the traits PHS and SL seem to be inherited independently. However, by including other studies in the comparison, overlaps appear. DArT marker wPt-9788 closely located to the PHS QTL on chromosome 5D in this study was found to be associated with SL by Rehman Arif et al. (2017). Also, the major dwarfing gene *Rht-D1*, being related to regular FN here, was found to have an effect on germination speed after CD, especially when being in combination with *Rht-B1* (Nagel et al. 2013). Moreover, DArT marker wPt-7108 located in the QTL interval of the LMA region on chromosome 7BL (Mohler et al. 2014) is associated with a QTL region for SL on chromosome 7BL (Rehman Arif et al. 2017). Respecting the present major QTL for SL on chromosome 1AL, the associated marker wPt-6654 was found to be related to PHS in an association mapping study of Albrecht et al. (2015). Therefore, it can be concluded that there are indications for a partial overlap in the genetic control of SL and PHS. However, this study also showed that diversity panels or multi-parental populations due to greater allele diversity may be more suited for investigating traits assumed to be under common genetic control.

Acknowledgements We thank Susanne Wüllner and Drs. Bianca Büttner and Günther Schweizer from LfL for running the Fluidigm marker assays. Prof. Thomas Miedaner (University of Hohenheim) kindly provided DArT marker data. We thank Dr. Johannes Schacht (Limagrain GmbH) for growing the History/Rubens RIL population.

Author contributions AB and VM conceived the research. MN, MA-M, PUG, TA, and MO performed and analyzed experiments. VM performed QTL analysis. VM, TA, and AB wrote the manuscript. All authors commented on the manuscript.

Funding Parts of the work presented here were supported by the German Federation of Industrial Research Associations (AiF), the German Federal Ministry for Economic Affairs and Technology, the Gemeinschaft zur Förderung der privaten deutschen Pflanzenzüchtung e.V. (GFP), the Austrian Research Promotion Agency (FFG), and the Austrian Seed Association (CORNET project 834766; IGF no. 75 EN/1).

Compliance with ethical standards

Conflict of interest The authors declare that they have no conflicts of interest.

Ethical approval This article does not contain any studies with human participants or animals performed by any of the authors.

References

Akbari M, Wenzl P, Caig V, Carling J, Xia L, Yang S, Uszynski G, Mohler V, Lehmensiek A, Kuchel H, Hayden MJ, Howes N, Sharp P, Vaughan P, Rathmell B, Huttner E, Kilian A (2006) Diversity arrays technology (DArT) for high-throughput profiling of the hexaploid wheat genome. Theor Appl Genet 113:1409–1420

Albrecht T, Oberforster M, Kempf H, Ramgraber L, Schacht J, Kazman E, Zechner E, Neumayer A, Hartl L, Mohler V (2015) Genome-wide association mapping of preharvest sprouting resistance in a diversity panel of European winter wheats. J Appl Genet 56:277–285

Churchill GA, Doerge RW (1994) Empirical threshold values for quantitative trait mapping. Genetics 138:963–971

Clerkx EJM, El-Lithy ME, Vierling E, Ruys GJ, Blankestijn-De Vries H, Groot SPC, Vreugdenhil D, Koornneef M (2004) Analysis of natural allelic variation of Arabidopsis seed germination and seed longevity traits between the accessions Landsberg *erecta* and Shakdara, using a new recombinant inbred line population. Plant Physiol 135:432–443

Doerge RW, Churchill GA (1996) Permutation tests for multiple loci affecting a quantitative character. Genetics 142:285–294

Ellis RH, Hong TD, Roberts EH, Tao KL (1990) Low moisture content limits to relations between seed longevity and moisture. Ann Bot 65: 493–504

Emebiri LC, Oliver JR, Mrva K, Mares D (2010) Association mapping of late maturity α-amylase (LMA) activity in a collection of synthetic hexaploid wheat. Mol Breed 26:29–39

Flintham JE, Gale MD (1982) The Tom Thumb dwarfing gene, *Rht3* in wheat, I. Reduced pre-harvest damage to breadmaking quality. Theor Appl Genet 62:121–126

Flintham JE, Börner A, Worland AJ, Gale MD (1997) Optimizing wheat grain yield: effects of *Rht* (gibberellin-insensitive) dwarfing genes. J Agric Sci 128:11–25

Groos C, Gay G, Perretant MR, Gervais L, Bernard M, Dedryver F, Charmet G (2002) Study of the relationship between pre-harvest sprouting and grain color by quantitative trait loci analysis in a white × red grain bread-wheat cross. Theor Appl Genet 104:39–47

Guo LB, Zhu LH, Xu YB, Zeng DL, Wu P, Qian Q (2004) QTL analysis of seed dormancy in rice (*Oryza sativa* L.). Euphytica 140:155–162

Hay FR, Adams J, Manger K, Probert R (2008) The use of non-saturated lithium chloride solutions for experimental control of seed water content. Seed Sci Technol 36:737–746

Holzapfel J, Voss H-H, Miedaner T, Korzun V, Häberle J, Schweizer G, Mohler V, Zimmermann G, Hartl L (2008) Inheritance of resistance to Fusarium head blight in three European winter wheat populations. Theor Appl Genet 117:1119–1128

Huang X, Zeller FJ, Hsam SLK, Wenzel G, Mohler V (2000) Chromosomal location of AFLP markers in common wheat utilizing nulli-tetrasomic stocks. Genome 43:298–305

International Seed Testing Association (ISTA) (2015) International rules for seed testing. ISTA, Bassersdorf, Switzerland

Jaiswal V, Mir RR, Mohan A, Balyan HS, Gupta PK (2012) Association mapping for pre-harvest sprouting tolerance in common wheat (*Triticum aestivum* L.) Euphytica 188:89–102

Joosen RVL, Kodde J, Willems LAJ, Ligterink W, van der Plas LHW, Hilhorst HWM (2010) GERMINATOR: a software package for high-throughput scoring and curve fitting of Arabidopsis seed germination. Plant J 62:148–159

Kao CH, Zeng ZB, Teasdale RD (1999) Multiple interval mapping for quantitative trait loci. Genetics 152:1203–1216

Kulwal PL, Singh R, Balyan HS, Gupta PK (2004) Genetic basis of pre-harvest sprouting tolerance using single-locus and two-locus QTL analyses in bread wheat. Funct Integr Genomics 4:94–101

Kulwal P, Ishikawa G, Benscher D, Feng Z, Yu LX, Jadhav A, Mehetre S, Sorrells ME (2012) Association mapping for pre-harvest sprouting resistance in white winter wheat. Theor Appl Genet 125:793–805

Lander ES, Botstein D (1989) Mapping mendelian factors underlying quantitative traits using RFLP linkage maps. Genetics 121:185–199

Landjeva S, Lohwasser U, Börner A (2010) Genetic mapping within the wheat D genome reveals QTL for germination, seed vigour and longevity, and early seedling growth. Euphytica 171:129–143

Lebreton CM, Visscher PM (1998) Empirical nonparametric bootstrap strategies in quantitative trait loci mapping: conditioning on the genetic model. Genetics 148:525–535

Lohwasser U, Rehman Arif MA, Börner A (2013) Discovery of loci determining pre-harvest sprouting and dormancy in wheat and barley applying segregation and association mapping. Biol Plant 57:663–674

Maccaferri M, Zhang J, Bulli P, Abate Z, Chao S, Cantu D, Bossolini E, Chen X, Pumphrey M, Dubcovsky J (2015) A genome-wide association study of resistance to stripe rust (*Puccinia striiformis* f. sp. *tritici*) in a worldwide collection of hexaploid spring wheat (*Triticum aestivum* L.) G3 (Bethesda) 5:449–465

Manley M, du Toit G, Geladi P (2011) Tracking diffusion of conditioning water in single wheat kernels of different hardnesses by near infrared hyperspectral imaging. Anal Chim Acta 686:64–75

Mares D, Mrva K (2008) Late-maturity α-amylase: low falling number in wheat in the absence of preharvest sprouting. J Cereal Sci 47:6–17

Mares DJ, Mrva K (2014) Wheat grain preharvest sprouting and late maturity alpha-amylase. Planta 240:1167–1178

Mares D, Mrva K, Cheong J, Williams K, Watson B, Storlie E, Sutherland M, Zou Y (2005) A QTL located on chromosome 4A associated with dormancy in white- and red-grained wheats of diverse origin. Theor Appl Genet 111:1357–1364

Miedaner T, Risser P, Paillard S, Schnurbusch T, Keller B, Hartl L, Holzapfel J, Korzun V, Ebmeyer E, Utz HF (2012) Broad-spectrum resistance loci for three quantitatively inherited diseases in two winter wheat populations. Mol Breed 29:731–742

Miura K, Lin S, Yano M, Nagamine T (2002) Mapping quantitative trait loci controlling seed longevity in rice (*Oryza sativa* L.). Theor Appl Genet 104:981–986

Mohler V, Albrecht T, Mrva K, Schweizer G, Hartl L (2014) Genetic analysis of falling number in three bi-parental common winter wheat populations. Plant Breed 133:448–453

Mrva K, Mares DJ (2001) Quantitative trait locus analysis of late maturity α-amylase in wheat using the doubled haploid population cranbrook x halberd. Aust J Agric Res 52:1267–1273

Nagel M, Börner A (2010) The longevity of crop seeds stored under ambient conditions. Seed Sci Res 20:1–12

Nagel M, Behrens AK, Börner A (2013) Effects of *Rht* dwarfing alleles on wheat seed vigour after controlled deterioration. Crop Pasture Sci 64:857–864

Nakamura S, Komatsuda T, Miura H (2007) Mapping diploid wheat homologues of *Arabidopsis* seed ABA signaling genes and QTLs for seed dormancy. Theor Appl Genet 114:1129–1139

Pestsova EG, Börner A, Röder MS (2006) Development and QTL assessment of *Triticum aestivum–Aegilops tauschii* introgression lines. Theor Appl Genet 112:634–647

R Core Team (2015) R: a language and environment for statistical computing. R Foundation for Statistical Computing, Vienna, Austria

Rasul G, Humphreys DG, Brûlé-Babel A, McCartney CA, Knox RE, DePauw RM, Somers DJ (2009) Mapping QTLs for pre-harvest sprouting traits in the spring wheat cross 'RL4452/AC Domain'. Euphytica 168:363–378

Rehman Arif MA, Nagel M, Neumann K, Kobiljski B, Lohwasser U, Börner A (2012a) Genetic studies of seed longevity in hexaploid wheat using segregation and association mapping approaches. Euphytica 186:1–13

Rehman Arif MA, Neumann K, Nagel M, Kobiljski B, Lohwasser U, Börner A (2012b) An association mapping analysis of dormancy and pre-harvest sprouting in wheat. Euphytica 188:409–417

Rehman Arif MA, Nagel M, Lohwasser U, Börner A (2017) Genetic architecture of seed longevity in bread wheat (*Triticum aestivum* L.). J Biosci 42:81–89

Schmolke M, Zimmermann G, Buerstmayr H, Schweizer G, Miedaner T, Korzun V, Ebmeyer E, Hartl L (2005) Molecular mapping of Fusarium head blight resistance in the winter wheat population Dream/Lynx. Theor Appl Genet 111:747–756

Van De Velde K, Chandler PM, Van Der Straeten D, Rohde A (2017) Differential coupling of gibberellin responses by *Rht-B1c* suppressor alleles and *Rht-B1b* in wheat highlights a unique role for the DELLA N-terminus in dormancy. J Exp Bot 68:443–455

Voorrips RE (2002) MapChart: software for the graphical presentation of linkage maps and QTLs. J Heredity 93:77–78

Walters C, Wheeler LM, Grotenhuis JM (2005) Longevity of seeds stored in a genebank: species characteristics. Seed Sci Res 15:1–20

Zeng DL, Guo LB, Xu YB, Yasukumi K, Zhu LH, Qian Q (2006) QTL analysis of seed storability in rice. Plant Breed 125:57–60

3.11 PAPER 11:

Effects of *Rht* dwarfing alleles on wheat seed vigour after controlled deterioration

by

Manuela Nagel, Anne-Kathrin Behrens and Andreas Börner

Published in

Crop and Pasture Science (2013) 64, 857-864

https://doi.org/10.1071/CP13041

CSIRO PUBLISHING
Crop & Pasture Science, 2013, **64**, 857–864
http://dx.doi.org/10.1071/CP13041

Effects of *Rht* dwarfing alleles on wheat seed vigour after controlled deterioration

Manuela Nagel[A,B], *Anne-Kathrin Behrens*[A], *and Andreas Börner*[A]

[A]Leibniz Institute of Plant Genetics and Crop Plant Research (IPK Gatersleben), Corrensstraße 3, 06466 Stadt Seeland, Germany.
[B]Corresponding author. Email: Nagel@ipk-gatersleben.de

Abstract. Reduced height (*Rht*) alleles, commonly known as the 'Green Revolution' genes, have facilitated wheat breeding programs and achieved globally a more than 10% wheat yield increase. However, studies in barley indicate that shorter plant habits are associated with reduced seed vigour and longevity. Therefore, wheat seeds of six near-isogenic lines (NIL) carrying the dwarfing alleles *Rht-B1b*, *Rht-D1b*, *Rht-B1c*, *Rht-B1b+-D1b*, *Rht-B1c+-D1b* and the wild-type allele *Rht-B1a+-D1a*, each in four background cultivars, were stressed by controlled deterioration. Seed vigour expressed as root and shoot lengths, time to 50% (T50) and time between 16 and 84% (T16-84) germination showed significant changes after treatment. However, after controlled deterioration only a combination of *Rht* alleles highly affected T16-84 and T50, which followed the general pattern *Rht-B1c+-D1b* followed by > *Rht-B1c* > *Rht-B1b+-D1b* > *Rht-B1b* > *Rht D1b* = *Rht-B1a+-D1a* (wild type). Interestingly, only under control conditions seed vigour correlated positively with thousand-kernel weight, which decreased with severity of *Rht* type. Further, the seed length was not affected by the different NIL. In conclusion, NIL carrying combinations of *Rht* alleles tend to influence seed vigour, which could influence seed longevity. Therefore, plant breeders but especially genebank managers should consider that the genetic background of genotypes may affect seed deterioration processes, which could be an economically important aspect in future.

Additional keywords: *Rht1*, *Rht2*, *Rht3*, seed storage, seed viability, *Triticum aestivum* L.

Received 29 January 2013, accepted 12 October 2013, published online 26 November 2013

Introduction

High seed viability and vigour are important requirements for efficient agricultural production due to effects on uniform field and stand establishment, weed repression and, consequently, yield. However, as soon as seeds achieve maturity, even pre-harvest, deterioration and viability losses are initiated. Several biotic and abiotic factors can modify the deterioration rate; among those: seed moisture content and temperature (Ellis and Roberts 1980), storage atmosphere (Barzali *et al.* 2005; Groot *et al.* 2012), pathogens (McGee 2000), the plant species (Walters *et al.* 2005; Nagel and Börner 2010) and the genotype (Revilla *et al.* 2006; Nagel *et al.* 2010).

To examine the genotypic effect, genetic architecture of seed vigour and longevity was investigated in *Arabidopsis* (Bentsink *et al.* 2000; Clerkx *et al.* 2004), rice (Miura *et al.* 2002; Zeng *et al.* 2006; Xue *et al.* 2008), soybean (Singh *et al.* 2008), maize (Revilla *et al.* 2009), lettuce (Schwember and Bradford 2010), barley (Nagel *et al.* 2009) and wheat (Rehman Arif *et al.* 2012). In common, the ability to resist or to succumb to deterioration processes was inherited by few major loci. In barley, Nagel *et al.* (2009) found one major locus on chromosome 2HL, which carries the *Zeo1* gene determining a short plant habit including compact spikes and reduced fertility. The dwarf parent contributed alleles to a reduced viability, which leads to the assumption that mechanisms influencing plant height also affect seed viability and vigour.

Dwarfing alleles have been detected among others in *Arabidopsis* and maize (Peng *et al.* 1999), oilseed rape (Zeng *et al.* 2011), rye (Börner *et al.* 1993), rice (Sasaki *et al.* 2002) and barley (Börner *et al.* 1999) but the most famous 'Green Revolution' alleles are carried by wheat (Gale and Youssefian 1985). These so-called reduced height (*Rht*) genes increased grain yield of at least 10% (Chapman *et al.* 2007) and can be found in almost 100% of modern US wheat cultivars (Guedira *et al.* 2010) and over 70% of the worldwide grown cultivars (Evans 1998).

Dwarfing alleles affect the elongation of internode cells in response to endogenous gibberellins (GA) (Appleford and Lenton 1991), which lead to reduced stem growth rates. In association with an enhanced photosynthetic capacity, due to higher concentrations of soluble protein, chlorophyll and RuBisco (Morgan *et al.* 1990), the assimilate partitioning to developing ears is increased and enhances a greater floret survival at anthesis and a higher grain number per ear (Youssefian *et al.* 1992*a*, 1992*b*). The most common *Rht-B1b* and *Rht-D1b* alleles (Guedira *et al.* 2010), which are carried by 'semidwarf' genotypes, have in addition to shorter culms, a higher lodging resistance, an improved harvest index and, finally, a higher yield per plant (McClung *et al.* 1986). Not commercially used are 'Tom Thumb' alleles, *Rht-B1c*, which cause an extreme reduction in plant height, though under certain backgrounds they reduce incidence to pre-harvest sprouting (Flintham and Gale 1982).

On the basis of the detected results in barley, the aims of this paper were to investigate how the genetic background in conjunction with the most common *Rht-B1b*, *Rht-D1b* and the *Rht-B1c* alleles (previously known as *Rht1*, *Rht2* and *Rht3*, respectively), influence wheat seed vigour and, consequently, seed longevity. We applied controlled deterioration (CD), a vigour test which is generally used to predict field establishment. Seeds of four different cultivars each carrying six combinations of *Rht* alleles were exposed to this stress treatment using high temperature and moisture content. Significant reductions of seed vigour were discussed in relationship to seed longevity, the different severities of *Rht* alleles and the effect of barley dwarfing alleles.

Materials and methods

Six near-isogenic lines (NIL) in four background cultivars ('April Bearded', 'Bersée', 'Maris Huntsman' and 'Maris Widgeon') were developed by Flintham *et al.* (1997) and used for experiments in 2011. Each set of NIL comprises lines carrying one of *Rht-B1b*, *Rht-D1b*, *Rht-B1c*, *Rht-B1b+-D1b*, *Rht-B1c+-D1b* and *Rht-B1a+-D1a* (wild-type) alleles. The material was multiplied in experimental plots at the Leibniz Institute of Plant Genetics and Crop Plant Research (IPK Gatersleben) in 2007 and has been stored at −18°C after processing steps as harvest, cleaning and drying to 6 ± 2%.

Seed vigour was investigated after the CD procedure described by Hampton and TeKrony (1995). Four independent reps of 25 seeds were adjusted to 18% seed moisture content (smc, fresh weight basis) according to the formula (1):

$$m_{H_2O} = \left(\left(\frac{(100 - smc_{initial})}{(100 - smc_{final})} \right) - 1 \right) * m_{seeds} \quad (1)$$

where m_{H_2O} is the amount of deionised water to be added, $smc_{initial}$ is the initial smc tested according to ISTA (2010) protocols, smc_{final} is the smc to be achieved and m_{seeds} the initial seed weight. Wetted seeds were shaken and equilibrated at 7°C for 24 h and sealed in aluminium bags at 42°C for another 24 h (CD1) and 48 h (CD2). Experimentally aged seeds (CD1 and CD2) and seeds of untreated control were germinated on moistened filter at 20 ± 1°C (14 h light) for 8 days. Vigorous, healthy seedlings without any aberration according to ISTA (2010) were counted of all NIL and analysed for mean root and shoot lengths (in cm) and percentage of normal seedlings. A second set of CD2 seeds were placed in an incubator (day: 25 ± 1°C, 14 h; night: 20 ± 1°C, 10 h) for another 8 days and germinated seeds counted daily when radicle lengths were >2 mm. Germination speed assessed by time to 50% germination (T50 in h) and time between 16 and 84% germination (T16-84 in h), which shows uniformity of germination by excluding the effect of fast and slow germinating seeds were calculated using the Germinator package of Joosen *et al.* (2010).

Seed parameters, here thousand-kernel weight (TKW, g), seed area (cm^2), seed length (cm) and seed width (cm), were measured by digital image analysis of MARVIN seed analyser (GTA Sensorik GmbH, Germany) using four reps of 100 seeds.

Mean and standard deviation of four reps were calculated by Microsoft Excel 2010 and analysed by ANOVA using SigmaPlot 12.0. Differences were declared as significant (*) at $P < 1\%$ and highly significant ($^+$) at $P < 0.1\%$. The pairwise multiple comparison procedures of the Holm–Sidak method gave additional information about significant differences between cultivars, NIL and treatments at overall significance level $P = 5\%$. Spearman correlation on ranks elucidated significant relationship between different traits at $P < 5\%$.

Results

Seedling length but not root length is affected by Rht *genes*

Rht alleles are commonly known to reduce plant height, which are further dependent on allele composition and genetic background. The present study investigates the effects of *Rht* alleles on seedling length after 8 days of germination and development.

Significant differences in shoot length appeared between cultivars, NIL and treatments (Table 1). Across treatments and cultivars, length reduced from *Rht-B1a+-D1a* > *Rht-D1b* > *Rht-B1b* > *Rht-B1b+-D1b* > *Rht-B1c* > *Rht-B1c+-D1b* (Table 2) and was mostly consistent with results obtained at the adult stage by Flintham *et al.* (1997), except for the longer shoots of *Rht-B1b* > *Rht-D1b*. The cultivars 'April Bearded' and 'Maris Huntsman' produced the longest shoots followed by 'Bersée' and 'Maris Widgeon' (Table 3), which was also characterised as the shortest plant at the adult stage (Flintham *et al.* 1997). In relation to the control the strongest shoot reductions after seed CD treatment were measured for the wild type (*Rht-B1a+-D1a*) showing shortening by 13.5% after CD1 and by 15.9% after CD2. The *Rht-B1c+-D1b* double mutant (reduced by 5.5%) and *Rht-D1b* (reduced by 10.2%) showed the slightest effects after CD1 and CD2, respectively, which does not indicate a relationship between *Rht* genes and sensitivity to seed deterioration. Varietal background and associated significant *Rht* × cultivar interaction influenced treatment power.

Table 1. Variance components for seed parameter [thousand-kernel weight (TKW, g), seed area (cm^2), seed length (cm), seed width (cm)] and seed vigour data [shoot length (shoot, cm), root length (root, cm), normal seedlings (%NS), time to 50% germination (T50, h), time between 16 and 84% germination (T16-84, h)] of four background cultivars, six near-isogenic lines (NIL) and treatments (Control, CD1, CD2)

*$P < 0.1$; **$P < 0.01$ (F-test, ANOVA)

Source of variance	TKW	Area	Length	Width	T50	T16-84	Shoot	Root	%NS
Cultivar	21.8*	103.9*	116.8*	69.9*	54.4*	18.9*	72.8*	31.8*	6.7**
NIL	13.4*	7.3**	0.8	17.6*	3.0	2.0	219.4*	1.7	6.7*
Treatment	–	–	–	–	53.8*	33.7*	48.2*	49.4*	33.5*

Table 2. Averages for six near-isogenic lines (NIL) in four background cultivars are shown. Data include seed parameters [thousand-kernel weight (TKW, g), seed area (cm²), seed length (cm), seed width (cm)] and vigour data [shoot length (shoot, cm), root length (root, cm), normal seedlings (%NS)] of an untreated control and after controlled deterioration applying 18% smc, 42°C for 24 h (CD1) and 48 h (CD2)

Germination speed [time to 50% germination (T50, h), time between 16 and 84% germination (T16-84, h)] was additionally analysed in CD2. Least significant difference (l.s.d. 5%) at $P = 0.05$

NIL	Seed parameter				Control					CD1			CD2				
	TKW	Area	Length	Width	T50	T16-84	Shoot	Root	%NS	Shoot	Root	%NS	T50	T16-84	Shoot	Root	%NS
Rht-B1a+-D1a	40.45	17.52	6.12	3.13	46.03	13.64	9.39	13.60	87.11	8.12	12.63	79.00	48.54	18.69	7.90	12.81	80.50
Rht-B1b	36.99	16.94	6.09	3.01	46.44	15.83	7.65	13.65	88.18	6.84	12.92	84.22	50.94	16.20	6.52	12.32	81.75
Rht D1b	34.25	16.66	6.08	2.94	46.35	18.36	7.60	13.68	89.00	7.00	13.44	83.00	48.50	18.93	6.83	12.92	82.00
Rht-B1c	30.04	16.53	6.10	2.87	44.15	14.43	5.42	13.73	88.25	4.85	12.94	79.75	50.75	23.15	4.66	12.27	85.23
Rht-B1b+-D1b	28.69	15.90	6.02	2.77	45.80	15.69	6.60	13.83	95.50	5.95	13.11	86.75	51.43	23.71	5.80	12.77	80.75
Rht-B1c+-D1b	27.13	16.23	6.11	2.79	44.52	12.66	4.90	13.88	86.75	4.63	13.53	76.86	59.17	29.42	4.17	12.13	73.00
l.s.d. 5%	1.42	0.21	0.04	0.03	0.98	1.62	0.22	0.26	1.89	0.18	0.22	2.07	1.89	1.85	0.20	0.29	2.30

Table 3. Averages for four background cultivars each with six near-isogenic lines are shown. Data include seed parameters [thousand-kernel weight (TKW, g), seed area (cm²), seed length (cm), seed width (cm)] and vigour data [shoot length (shoot, cm), root length (root, cm), normal seedlings (%NS)] of an untreated control and after controlled deterioration applying 18% smc, 42°C for 24 h (CD1) and 48 h (CD2)

Germination speed [time to 50% germination (T50, h), time between 16 and 84% germination (T16-84, h)] was additionally analysed in control and CD2. Least significant difference (l.s.d. 5%) at $P = 0.05$

Cultivar	Seed parameter				Control					CD1			CD2				
	TKW	Area	Length	Width	T50	T16-84	Shoot	Root	%NS	Shoot	Root	%NS	T50	T16-84	Shoot	Root	%NS
April Bearded	25.60	14.18	5.66	2.59	42.69	11.07	7.10	13.31	87.33	7.01	12.79	83.31	44.35	16.41	6.64	12.09	84.17
Bersée	32.44	16.83	6.00	3.02	48.73	17.99	6.33	13.62	90.62	5.35	12.87	81.17	65.95	31.33	4.75	11.81	73.00
Maris Huntsman	38.14	18.28	6.54	3.05	46.49	18.53	7.47	14.34	91.41	6.53	13.58	83.00	46.89	21.18	6.74	13.83	85.33
Maris Widgeon	35.53	17.22	6.14	3.02	44.29	12.80	6.78	13.65	87.12	6.04	13.14	78.90	49.04	17.80	5.79	12.43	79.65
l.s.d. 5%	1.16	0.17	0.03	0.03	0.80	1.33	0.18	0.21	1.53	0.15	0.18	1.69	1.54	1.51	0.16	0.24	1.88

Additive effects of shoot length in the control and after CD1 and CD2 treatments were observed when *Rht-D1b* was combined with *Rht-B1b* and *Rht-B1c* (Fig. 1*a*). In contrast root lengths were only influenced by treatments and cultivars but did not change between NIL of the same cultivar (Fig. 1*b*), which has been confirmed by Kubo *et al.* (2005).

Seed vigour of NIL is reduced after controlled deterioration

Seed vigour is considered as 'the sum of those properties which determine the level of activity and performance during germination and seedling emergence' (Powell 1988) and has been defined here by T50, T16-84, % normal seedlings and root and shoot length, which were discussed above.

T50, T16-84 and % normal seedlings were significantly influenced by the cultivar and the CD treatment (Table 4). In relation to control % normal seedlings of the double mutants *Rht-B1b+-D1b* decreased by 11.4% after CD1 (wild type: 9.3%) and *Rht-B1b+-D1b* and *Rht-B1c+-D1b* and by 15.4 and 15.9% after CD2, respectively (wild type: 7.6%). T50 increased for the mutants carrying *Rht-B1c*, *Rht-B1b+-D1b*, *Rht-B1c+-D1b* alleles by 14.9, 12.3 and 32.9%, respectively, for T50 (wild type: 5.5%) and T16-84 by 60.4, 51.1 and 132.4%, respectively (wild type 37.0%) after CD2 (Fig. 1*c*, *d*). In the control, different NIL of the same cultivars did not contribute to any change of T50; but after stress treatment the highest significant increase for T50 and T16-84 was visible for 'Bersée' NIL (Table 3).

The weight and the shape are influenced by Rht *alleles*

Rht alleles are widely known to have pleiotropic effects and impact, next to plant height, on ear components as grain yield per ear and TKW (Flintham *et al.* 1997). In agreement with these results TKW decreases significantly in dependence of severity of dwarfism and cultivar (Table 1). The weight depletion follows a similar pattern as shoot length and reduces from *Rht-B1a+-D1a* > *Rht-B1b* > *Rht-D1b* > *Rht-B1b+-D1b* > *Rht-B1c* > *Rht-B1c+-D1b* (Table 2). Highly significant correlations ($P < 0.01$) were found between shoot length ($r = 0.58$), T16-84 ($r = 0.53$) and TKW for control; and root length ($r = 0.52$) and TKW for CD2. Interestingly, the seed length hardly changed between the NIL (Fig. 1*e*). Weak significant depletions were only observed for seed area and width when *Rht* alleles are enhanced (Figs 1*d*, 2).

Discussion

Dwarfism of plants has been already discovered in old Greece and Rom (Pelton 1964) but was not efficiently used up to 1935 when the first wheat dwarf cultivar 'Norin 10' was patented in Japan (Gale and Youssefian 1985). Semidwarfing genes lead to the higher yield potential (Flintham *et al.* 1997) and are present in the majority of modern cultivars (Evans 1998; Knopf *et al.* 2008; Guedira *et al.* 2010). Their plant height is reduced by shortened internodes, which is controlled by few genes as *Rht-B1b* and *RhtD1b* (Börner and Mettin 1989). The more extreme *Rht-B1c*

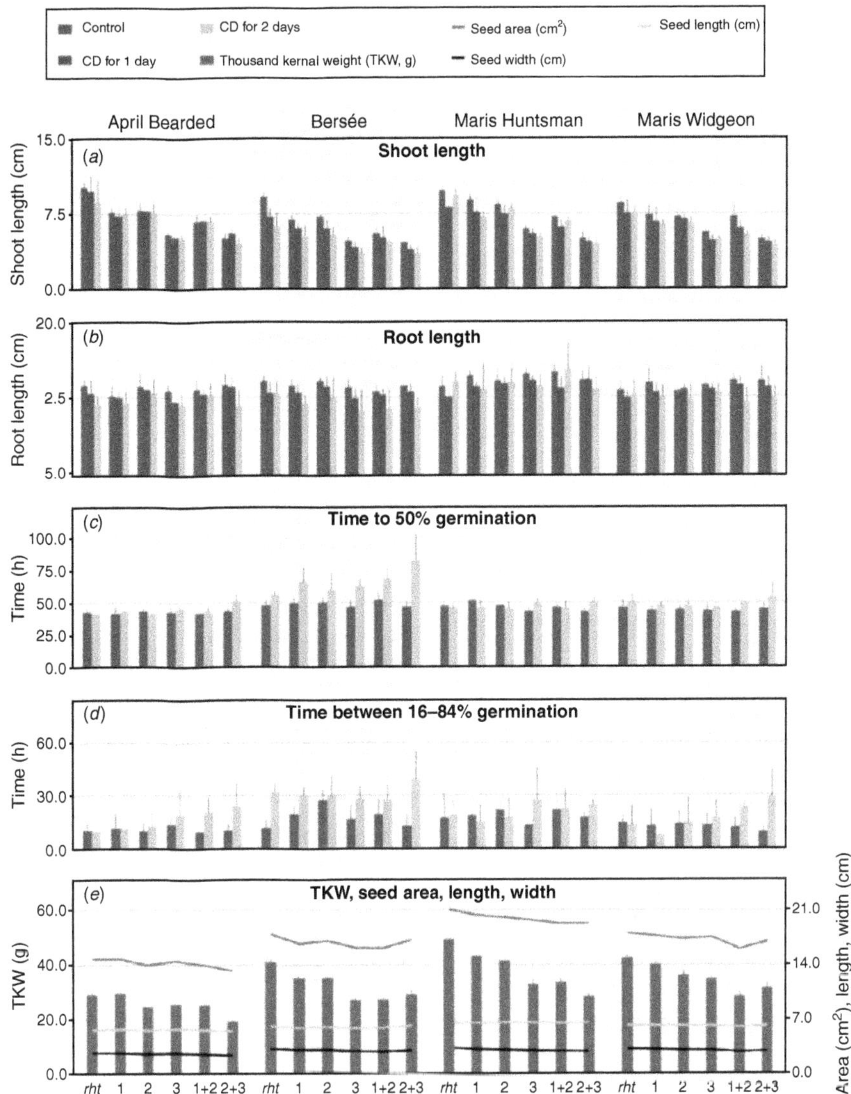

Fig. 1. Changes in seed vigour of four sets of near-isogenic *Rht* lines after controlled deterioration (CD). Seed vigour was assessed as mean ± s.d. of four reps of (*a*) shoot length, (*b*) root length, (*c*) time to 50% germination and (*d*) time between 16 and 84% germination. (*e*) Thousand-kernel weight (TKW), seed area, length and width provides additional information about seed shape. *rht* - *Rht-B1a+-D1a*, 1 - *Rht-B1b*, 2 - *Rht-D1b*, 3 – *Rht-B1c*, 1+2 – *Rht B1b+ D1b*, 2+3 – *Rht-B1c+-D1b*.

dwarfs show additional diminution in the number of cells per organ (Paolillo *et al.* 1991).

Different severities of Rht *alleles*

The present study confirms an overall reduction of seedling lengths in dependency of the *Rht* type. Principally, *Rht-B1* and *Rht-D1* are orthologues of the *Arabidopsis Gibberellin Insensitive* gene (Peng *et al.* 1999) and encode DELLA proteins, which react on the presence of endogenous bioactive GA (Pearce *et al.* 2011). These are important growth regulators, which promote α-amylase production for endosperm starch hydrolysis during seed germination and facilitate cell elongation in different organs and tissues throughout plant growth and development (Karssen *et al.* 1989). The mutated *Rht-B1b* and *Rht-D1b* alleles contain amino acid substitutions (Peng *et al.* 1999) or in the case of *Rht-B1c* genomic insertions in the DELLA/TVHYNP motifs, which reduce GA sensitivity and

Table 4. Average of four background cultivars and six near-isogenic lines are shown for vigour data [shoot length (shoot, cm), root length (root, cm), normal seedlings (%NS), time to 50% germination (T50, h) and time between 16 and 84% germination (T16-84, h)] of an untreated control and after controlled deterioration applying 18% smc, 42°C for 24 h (CD1) and 48 h (CD2)

Least significant difference (l.s.d. 5%) at $P = 0.05$

Treatment	Shoot	Root	%NS	T50	T16-84
Control	6.92	13.73	89.13	45.55	15.10
CD 1	6.23	13.10	81.60	–	–
CD2	5.98	12.54	80.54	51.56	21.68
l.s.d. 5%	0.07	0.08	0.81	0.58	0.80

enhance repression of growth (Pearce *et al.* 2011; Wu *et al.* 2011) by for example reducing the time available for cell elongation (Tonkinson *et al.* 1995).

The different severities of *Rht* genes and their additive effects lead to distinct levels of inhibition of GA actions (Flintham *et al.* 1997) and different degrees of inhibition of elongation and finally seedling lengths.

Reduction of seedling and root length as indicators of viability loss

The stress treatments CD1 and CD2 contribute generally to a reduced appearance of normal seedlings in all NIL, which go along with significant reductions in shoot and root length. The CD procedure exposes seed to high temperature and humidity levels and accelerates biochemical processes, which are accompanied by H_2O_2 accumulation and viability loss in wheat and sunflower. An additional decrease of antioxidative enzymes as superoxide dismutase, catalase and glutathione reductase lead to an increase of lipid peroxidation (Kibinza *et al.* 2006; Lehner *et al.* 2008), which is seen as one of the most detrimental reactions during seed ageing (McDonald 1999). It causes loss of membrane integrity, and further accumulation of reactive oxygen species results in an impairment of RNA and protein syntheses (Hendry 1993; Bailly 2004; Bailly and Kranner 2011) and DNA degradation (Kranner *et al.* 2011). The loss of functions and the diversity of deleterious effects facilitate growth restriction including anomalous development combined with reduced % normal seedlings, shoot and root length.

However, over all cultivars the CD treatments showed on shoot and root length either stronger effects on the wild type or no effect across the NIL, respectively, which is assumed to be related to different impacts of GA on root and shoot growth. The GA are required for both but remarkable effects on root growth could be only elucidated when the endogenous level of GA was decreased by inhibitors (Tanimoto 2005, 2012). Changes in GA concentration after stress or storage treatments have not been investigated so far and might be interesting for future studies.

Expression of Rht *genes influences germination speed under stress*

The vigour parameters T50 and T16-84 indicate that cultivars carrying *Rht-B1c* or *Rht-B1b+-D1b* and *Rht-B1c+-D1b* alleles show a deceleration of germination speed after CD. In particular, a T50 extension appeared simultaneously with increased severity

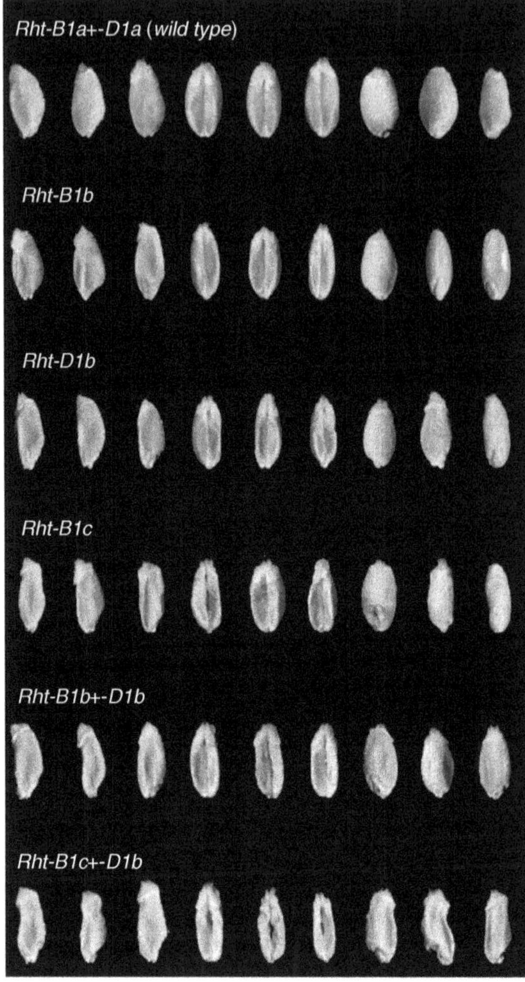

Fig. 2. Change in seed shape of six near-isogenic lines in the background cultivar 'Maris Huntsman'. Significant changes were detected for thousand-kernel weight, seed width and area but not for seed length.

of *Rht* alleles of the 'Bersée' NIL and confirms the significant influence of the genetic background (Pereira *et al.* 2002; Addisu *et al.* 2009; Nagel *et al.* 2009). However, molecular events as translational capacity, mobilisation of seed storage reserves and detoxification accompany the CD treatment and show similarities to natural seed ageing (Rajjou *et al.* 2008). Consequently some *Rht* alleles are assumed to have reduced seed longevity. At the same time CD acts also as a vigour test to predict field establishment (Hampton and TeKrony 1995), the relationship between reduced seed vigour and the dwarfing alleles were confirmed by Addisu *et al.* (2009). Especially under stressed environments, semidwarf cultivars carrying *Rht-B1b* or *Rht-D1b* alleles showed reduced stand establishment and lower yield, which has been partly attributed to low seedling vigour (Allan

1980; Botwright *et al.* 2001; Rebetzke *et al.* 2001). The reduced seed vigour indicates that some *Rht* alleles directly affect germination and seed deterioration processes or are genetically related to genes which do.

This assumption is supported by Nagel *et al.* (2009) who found a quantitative trait loci (QTL) for seed longevity/seed vigour in barley, which is associated with the *Zeo1* gene. Plants which carry the dominant alleles are characterised by a reduced plant height (Lundqvist *et al.* 1997) and decreased seed vigour (Nagel *et al.* 2009). On the basis of the current results we speculate that the *Zeo1* gene might be involved in similar mechanism as *Rht* genes. A common feature might be the regulation of GA, which are major signalling factors during seed germination (Bewley 1997) and are part of the dwarfing mechanism (Pearce *et al.* 2011). However, the results do not indicate if the vigour depletion is triggered either by a reduction of GA sensitivity of DELLA proteins, by a decreased GA concentration or other mechanisms affecting signalling and germination. Further, the GA-insensitive dwarfing genes in wheat are mapped on the short arms of chromosomes 4B (*Rht-B1*) and 4D (*Rht-D1*) whereas the barley *Zeo1* gene is located on the long arm of chromosome 2H and not at the same position as the *Rht-H1* in the centromeric region of 2HS (Börner *et al.* 1998).

Seed length is not affected by kernel weight

Rht alleles affect the fertility of distal florets within spikelet and promote, due to increased fertility, a reduction of TKW (Flintham and Gale 1983; Zhang *et al.* 2013). Interestingly, the current data indicate that the reduction of seed weight does not appear simultaneously with a decrease in seed length. We speculate that the basic kernel length is retained over all severities of *Rht* alleles and lateral packages are built up as soon as storage components are available.

Additionally, the reduced TKW, which goes along with the shrivelled phenotypes especially of the double mutants, might play a role for the reduced germination speed. Several reports indicate a relationship between high TKW and enhanced seed vigour and stand establishment based on the availability of higher amounts of seed reserves (Akinci *et al.* 2008). However, a clear agreement has not been found (Schutte *et al.* 2008) and results do not show correlations between TKW and germination speed after CD. Allan (1980) showed that semidwarf cultivars have in addition to the reduced weight also a lower seed protein content, which might affect indirectly seed vigour.

Conclusions and Outlook

The present study showed that, dependent on the genetic background, only combinations of *Rht* alleles affect seed vigour and could have final impacts on seed longevity. We assume that some modern cultivars could be reduced in those properties, which are economically important features for the seed industry. However, the severity of the vigour depletion after CD was not as strong as it had been indicated in the barley study but could be explained by the different species we used for the experiments and different loci which responded in the respective study. For future experiments, it would be interesting to investigate barley dwarfs of different background cultivars. Further, the analyses of GA-sensitive wheat lines carrying *Rht8* alleles, which are assumed to have higher vigour due to longer coleoptile (Rebetzke and Richards 2000; Jain and Yadav 2009), would give an additional aspect to the analyses.

Acknowledgements

We kindly acknowledge Sibylle Pistrick and Anita Winger for their great experimental support; Annett Marlow and Stefanie Thumm for the careful handling and multiplication of the seed material and Gudrun Schütze for the preparation of the seed images.

References

Addisu M, Snape JW, Simmonds JR, Gooding MJ (2009) Reduced height (*Rht*) and photoperiod insensitivity (*Ppd*) allele associations with establishment and early growth of wheat in contrasting production systems. *Euphytica* **166**, 249–267. doi:10.1007/s10681-008-9838-7

Akinci C, Yildirim M, Bahar B (2008) The effects of seed size on emergence and yield of durum wheat. *Journal of Food Agriculture and Environment* **6**, 234–237.

Allan RE (1980) Influence of semi-dwarfism and genetic background on stand establishment of wheat. *Crop Science* **20**, 634–638. doi:10.2135/cropsci1980.0011183X002000050022x

Appleford NEJ, Lenton JR (1991) Gibberellins and leaf expansion in near-isogenic wheat lines containing *Rht1* and *Rht3* dwarfing alleles. *Planta* **183**, 229–236. doi:10.1007/BF00197793

Bailly C (2004) Active oxygen species and antioxidants in seed biology. *Seed Science Research* **14**, 93–107. doi:10.1079/SSR2004159

Bailly C, Kranner I (2011) Analyses of reactive oxygen species and antioxidants in relation to seed longevity and germination. In 'Seed dormancy: methods and protocols. Methods in molecular biology'. Vol. 773. (Ed. AR Kermode) pp. 343–67. (Humana Press: New York)

Barzali M, Lohwasser U, Niedzielski M, Börner A (2005) Effects of different temperatures and atmospheres on seed and seedling traits in a long-term storage experiment on rye (*Secale cereale* L.). *Seed Science and Technology* **33**, 713–721.

Bentsink L, Alonso-Blanco C, Vreugdenhil D, Tesnier K, Groot SPC, Koornneef M (2000) Genetic analysis of seed-soluble oligosaccharides in relation to seed storability of *Arabidopsis*. *Plant Physiology* **124**, 1595–1604. doi:10.1104/pp.124.4.1595

Bewley JD (1997) Seed germination and dormancy. *The Plant Cell* **9**, 1055–1066. doi:10.1105/tpc.9.7.1055

Börner A, Mettin D (1989) Genetische Grundlagen der Halmverkurzung (Dwarfismus) beim Weizen und Möglichkeiten der züchterischen Nutzung. *Die Kulturpflanze* **37**, 29–55. doi:10.1007/BF01984580

Börner A, Gale MD, Appleford NEJ, Lenton JR (1993) Gibberellin status and responsiveness in shoots of tall and dwarf genotypes of diploid rye (*Secale cereale*). *Physiologia Plantarum* **89**, 309–314. doi:10.1111/j.1399-3054.1993.tb00159.x

Börner A, Korzun V, Worland AJ (1998) Comparative genetic mapping of loci affecting plant height and development in cereals. *Euphytica* **100**, 245–248. doi:10.1023/A:1018364425150

Börner A, Korzun V, Malyshev S, Ivandic V (1999) Molecular mapping of two dwarfing genes differing in their GA response on chromosome 2H of barley. *Theoretical and Applied Genetics* **99**, 670–675. doi:10.1007/s001220051283

Botwright TL, Rebetzke GJ, Condon AG, Richards RA (2001) Influence of variety, seed position and seed source on screening for coleoptile length in bread wheat (*Triticum aestivum* L.). *Euphytica* **119**, 349–356. doi:10.1023/A:1017527911084

Chapman SC, Mathews KL, Trethowan RM, Singh RP (2007) Relationships between height and yield in near-isogenic spring wheats that contrast for major reduced height genes. *Euphytica* **157**, 391–397. doi:10.1007/s10681-006-9304-3

Clerkx EJM, El-Lithy ME, Vierling E, Ruys GJ, Blankestijin-De Vries H, Groot SPC, Vreugdenhil D, Koornneef M (2004) Analysis of natural allelic variation of *Arabidopsis* seed germination and seed longevity traits between the accessions Landsberg erecta and Shakdara, using a new recombinant inbred line population. *Plant Physiology* **135**, 432–443. doi:10.1104/pp.103.036814

Ellis RH, Roberts EH (1980) Improved equations for the prediction of seed longevity. *Annals of Botany* **45**, 13–30.

Evans LT (1998) 'Feeding the ten billion – plants and population growth.' (Cambridge University Press: Cambridge, UK)

Flintham JE, Gale MD (1982) The Tom Thumb dwarfing gene, *Rht3* in wheat: 1. Reduced preharvest damage to breadmaking quality. *Theoretical and Applied Genetics* **62**, 121–126. doi:10.1007/BF00293343

Flintham JE, Gale MD (1983) The Tom Thumb dwarfing gene *Rht3* in wheat. *Theoretical and Applied Genetics* **66**, 249–256.

Flintham JE, Borner A, Worland AJ, Gale MD (1997) Optimizing wheat grain yield: effects of *Rht* (gibberellin-insensitive) dwarfing genes. *The Journal of Agricultural Science* **128**, 11–25. doi:10.1017/S0021859696003942

Gale MD, Youssefian S (1985) Dwarfing genes in wheat. In 'Progress in plant breeding, 1'. (Ed. GE Russell) pp. 1–35. (Butterworths: London)

Groot SPC, Surki AA, de Vos RCH, Kodde J (2012) Seed storage at elevated partial pressure of oxygen, a fast method for analysing seed ageing under dry conditions. *Annals of Botany* **110**, 1149–1159. doi:10.1093/aob/mcs198

Guedira M, Brown-Guedira G, Van Sanford D, Sneller C, Souza E, Marshall D (2010) Distribution of *Rht* genes in modern and historic winter wheat cultivars from the Eastern and Central USA. *Crop Science* **50**, 1811–1822. doi:10.2135/cropsci2009.10.0626

Hampton JG, TeKrony DM (1995) 'Handbook of vigour test methods.' (International Seed Testing Association: Zürich, Switzerland)

Hendry GAF (1993) Oxygen, free radical processes and seed longevity. *Seed Science Research* **3**, 141–153. doi:10.1017/S0960258500001720

ISTA (2010) 'International rules for seed testing.' (International Seed Testing Association: Bassersdorf, Switzerland)

Jain N, Yadav R (2009) Distribution of *Rht* genes and their effects on coleoptile length and height in wheat (*Triticum aestivum*). *Indian Journal of Agricultural Sciences* **79**, 391–393.

Joosen RVL, Kodde J, Willems LAJ, Ligterink W, van der Plas LHW, Hilhorst HWM (2010) GERMINATOR: a software package for high-throughput scoring and curve fitting of Arabidopsis seed germination. *The Plant Journal* **62**, 148–159. doi:10.1111/j.1365-313X.2009.04116.x

Karssen CM, Zagorski S, Kepczynski J, Groot SPC (1989) Key role for endogenous gibberellins in the control of seed germination. *Annals of Botany* **63**, 71–80.

Kibinza S, Vinel D, Come D, Bailly C, Corbineau F (2006) Sunflower seed deterioration as related to moisture content during ageing, energy metabolism and active oxygen species scavenging. *Physiologia Plantarum* **128**, 496–506. doi:10.1111/j.1399-3054.2006.00771.x

Knopf C, Becker H, Ebmeyer E, Korzun V (2008) Occurrence of three dwarfing *Rht* genes in German winter wheat varieties. *Cereal Research Communications* **36**, 553–560. doi:10.1556/CRC.36.2008.4.4

Kranner I, Chen HY, Pritchard HW, Pearce SR, Birtic S (2011) Inter-nucleosomal DNA fragmentation and loss of RNA integrity during seed ageing. *Plant Growth Regulation* **63**, 63–72. doi:10.1007/s10725-010-9512-7

Kubo K, Jitsuyama Y, Iwama K, Watanabe N, Yanagisawa A, Elouafi I, Nachit MM (2005) The reduced height genes do not affect the root penetration ability in wheat. *Euphytica* **141**, 105–111. doi:10.1007/s10681-005-6161-4

Lehner A, Mamadou N, Poels P, Come D, Bailly C, Corbineau F (2008) Changes in soluble carbohydrates, lipid peroxidation and antioxidant enzyme activities in the embryo during ageing in wheat grains. *Journal of Cereal Science* **47**, 555–565. doi:10.1016/j.jcs.2007.06.017

Lundqvist U, Franckowiak D, Konishi T (1997) New and revised descriptions of barley genes. *Barley Genetics Newsletter* **26**, 22–516.

McClung AM, Cantrell RG, Quick JS, Gregory RS (1986) Influence of the *Rht1* semidwarf gene on yield, yield components, and grain protein in durum wheat. *Crop Science* **26**, 1095–1099. doi:10.2135/cropsci1986.0011183X002600060001x

McDonald MB (1999) Seed deterioration: physiology, repair and assessment. *Seed Science and Technology* **27**, 177–237.

McGee DC (2000) 'Pathology of seed deterioration.' (Crop Science Society of America: Madison, WI)

Miura K, Lin SY, Yano M, Nagamine T (2002) Mapping quantitative trait loci controlling seed longevity in rice (*Oryza sativa* L.). *Theoretical and Applied Genetics* **104**, 981–986. doi:10.1007/s00122-002-0872-x

Morgan JA, LeCain DR, Wells R (1990) Semidwarfing genes concentrate photosynthetic machinery and affect leaf gas exchange of wheat. *Crop Science* **30**, 602–608.

Nagel M, Börner A (2010) The longevity of crop seeds stored under ambient conditions. *Seed Science Research* **20**, 1–12. doi:10.1017/S0960258509990213

Nagel M, Vogel H, Landjeva S, Buck-Sorlin G, Lohwasser U, Scholz U, Börner A (2009) Seed conservation in *ex situ* genebanks—genetic studies on longevity in barley. *Euphytica* **170**, 5–14. doi:10.1007/s10681-009-9975-7

Nagel M, Rehman Arif MA, Rosenhauer M, Börner A (2010) Longevity of seeds – intraspecific differences in the Gatersleben genebank collections. In 'Proceedings of the 60th Conference of Plant Breeders and Seed Traders Association in Austria 2009'. Raumberg-Gumpenstein. (Eds P Ruckenbauer, A Brandstetter, M Geppner, H Grausgruber, K Buchgraber) pp. 179–181. (Vereinigung der Pflanzenzüchter und Saatgutkaufleute Österreichs: St. Pölten, Austria)

Paolillo DJ, Sorrels ME, Keyes GJ (1991) Gibberellic acid sensitivity determines the length of the extension zone in wheat leaves. *Annals of Botany* **67**, 479–485.

Pearce S, Saville R, Vaughan SP, Chandler PM, Wilhelm EP, Sparks CA, Al-Kaff N, Korolev A, Boulton MI, Phillips AL, Hedden P, Nicholson P, Thomas SG (2011) Molecular characterization of *Rht-1* dwarfing genes in hexaploid wheat. *Plant Physiology* **157**, 1820–1831. doi:10.1104/pp.111.183657

Pelton J (1964) Genetic and morphogenetic studies of angiosperm single-gene dwarfs. *Botanical Review* **30**, 479–512. doi:10.1007/BF02858541

Peng J, Richards DE, Hartley NM, Murphy GP, Devos KM, Flintham JE, Beales J, Fish LJ, Worland AJ, Pelica F, Sudhakar D, Christou P, Snape JW, Gale MD, Harberd NP (1999) 'Green revolution' genes encode mutant gibberellin response modulators. *Nature* **400**, 256–261. doi:10.1038/22307

Pereira MJ, Pfahler PL, Barnett RD, Blount AR, Wofford DS, Littell RC (2002) Coleoptile length of dwarf wheat isolines: Gibberellic acid, temperature, and cultivar interactions. *Crop Science* **42**, 1483–1487. doi:10.2135/cropsci2002.1483

Powell AA (1988) Seed vigour and field establishment. *Advances in Research and Technology of Seeds* **2**, 29–61.

Rajjou L, Lovigny Y, Groot SPC, Belghaz M, Job C, Job D (2008) Proteome-wide characterization of seed aging in *Arabidopsis*: a comparison between artificial and natural aging protocols. *Plant Physiology* **148**, 620–641. doi:10.1104/pp.108.123141

Rebetzke GJ, Richards RA (2000) Gibberellic acid-sensitive dwarfing genes reduce plant height to increase kernel number and grain yield of wheat. *Australian Journal of Agricultural Research* **51**, 235–245. doi:10.1071/AR99043

Rebetzke GJ, Appels R, Morrison AD, Richards RA, McDonald G, Ellis MH, Spielmeyer W, Bonnett DG (2001) Quantitative trait loci on chromosome 4B for coleoptile length and early vigour in wheat (*Triticum aestivum* L.). *Australian Journal of Agricultural Research* **52**, 1221–1234. doi:10.1071/AR01042

Rehman Arif MA, Nagel M, Neumann K, Kobiljski B, Lohwasser U, Börner A (2012) Genetic studies of seed longevity in hexaploid wheat using segregation and association mapping approaches. *Euphytica* **186**, 1–13. doi:10.1007/s10681-011-0471-5

Revilla P, Velasco P, Malvar RA, Cartea ME, Ordas A (2006) Variability among maize (*Zea mays* L.) inbred lines for seed longevity. *Genetic Resources and Crop Evolution* **53**, 771–777. doi:10.1007/s10722-004-5542-1

Revilla P, Butron A, Rodriguez VM, Malvar RA, Ordas A (2009) Identification of genes related to germination in aged maize seed by screening natural variability. *Journal of Experimental Botany* **60**, 4151–4157. doi:10.1093/jxb/erp249

Sasaki A, Ashikari M, Ueguchi-Tanaka M, Itoh H, Nishimura A, Swapan D, Ishiyama K, Saito T, Kobayashi M, Khush GS, Kitano H, Matsuoka M (2002) Green revolution: a mutant gibberellin-synthesis gene in rice – new insight into the rice variant that helped to avert famine over thirty years ago. *Nature* **416**, 701–702. doi:10.1038/416701a

Schutte BJ, Regnier EE, Harrison SK (2008) The association between seed size and seed longevity among maternal families in *Ambrosia trifida* L. populations. *Seed Science Research* **18**, 201–211. doi:10.1017/S0960258508082974

Schwember AR, Bradford KJ (2010) Quantitative trait loci associated with longevity of lettuce seeds under conventional and controlled deterioration storage conditions. *Journal of Experimental Botany* **61**, 4423–4436. doi:10.1093/jxb/erq248

Singh RK, Raipuria RK, Bhatia VS, Rani A, Pushpendra , Husain SM, Chauhan D, Chauhan GS, Mohapatra T (2008) SSR markers associated with seed longevity in soybean. *Seed Science and Technology* **36**, 162–167.

Tanimoto E (2005) Regulation of root growth by plant hormones – roles for auxin and gibberellin. *Critical Reviews in Plant Sciences* **24**, 249–265. doi:10.1080/07352680500196108

Tanimoto E (2012) Tall or short? Slender or thick? A plant strategy for regulating elongation growth of roots by low concentrations of gibberellin. *Annals of Botany* **110**, 373–381. doi:10.1093/aob/mcs049

Tonkinson CL, Lyndon RF, Arnold GM, Lenton JR (1995) Effect of the *Rht3* dwarfing gene on dynamics of cell extension in wheat leaves, and its modification by gibberellic acid and paclobutrazol. *Journal of Experimental Botany* **46**, 1085–1092. doi:10.1093/jxb/46.9.1085

Walters C, Wheeler LM, Grotenhuis JM (2005) Longevity of seeds stored in a genebank: species characteristics. *Seed Science Research* **15**, 1–20. doi:10.1079/SSR2004195

Wu J, Kong X, Wan J, Liu X, Zhang X, Guo X, Zhou R, Zhao G, Jing R, Fu X, Jia J (2011) Dominant and pleiotropic effects of a *GAI* gene in wheat results from a lack of interaction between DELLA and GID1. *Plant Physiology* **157**, 2120–2130. doi:10.1104/pp.111.185272

Xue Y, Zhang SQ, Yao QH, Peng RH, Xiong AS, Li X, Zhu WM, Zhu YY, Zha DS (2008) Identification of quantitative trait loci for seed storability in rice (*Oryza sativa* L.). *Euphytica* **164**, 739–744. doi:10.1007/s10681-008-9696-3

Youssefian S, Kirby EJM, Gale MD (1992*a*) Pleiotropic effects of the GA-insensitive *Rht* dwarfing genes in wheat: 2. Effects on leaf, stem, ear and floret growth. *Field Crops Research* **28**, 191–210. doi:10.1016/0378-4290(92)90040-G

Youssefian S, Kirby EJM, Gale MD (1992*b*) Pleiotropic effects of the GA-insensitive *Rht* dwarfing genes in wheat: 1. Effects on development of the ear, stem and leaves. *Field Crops Research* **28**, 179–190. doi:10.1016/0378-4290(92)90039-C

Zeng DL, Guo LB, Xu YB, Yasukumi K, Zhu LH, Qian Q (2006) QTL analysis of seed storability in rice. *Plant Breeding* **125**, 57–60. doi:10.1111/j.1439-0523.2006.01169.x

Zeng X, Zhu L, Chen Y, Qi L, Pu Y, Wen J, Yi B, Shen J, Ma C, Tu J, Fu T (2011) Identification, fine mapping and characterisation of a dwarf mutant (bnaC.dwf) in *Brassica napus*. *Theoretical and Applied Genetics* **122**, 421–428. doi:10.1007/s00122-010-1457-8

Zhang F, Jiang Y-Z, Yu S-B, Ali J, Paterson AH, Khush GS, Xu J-L, Gao Y-M, Fu B-Y, Lafitte R, Li Z-K (2013) Three genetic systems controlling growth, development and productivity of rice (*Oryza sativa* L.): a reevaluation of the 'Green Revolution'. *Theoretical and Applied Genetics* **126**, 1011–1024. doi:10.1007/s00122-012-2033-1

3.12 PAPER 12:

Genetic architecture of seed longevity in bread wheat (*Triticum aestivum* L.)

by

Mian A. R. Arif, **Manuela Nagel**, Ulrike Lohwasser and Andreas Börner

Published in

Journal of Biosciences (2017) 42, 81-89

https://doi.org/10.1007/s12038-016-9661-6

To view supplementary material for this article, please visit
https://link.springer.com/article/10.1007%2Fs12038-016-9661-6

Genetic architecture of seed longevity in bread wheat (*Triticum aestivum* L.)

MIAN ABDUR REHMAN ARIF[1,2], MANUELA NAGEL[1], ULRIKE LOHWASSER[1] and ANDREAS BÖRNER[1,*]

[1]*Leibniz Institute of Plant Genetics and Crop Plant Research, Gatersleben, Germany*

[2]*Nuclear Institute for Agriculture and Biology, Faisalabad, Pakistan*

**Corresponding author (Email, boerner@ipk-gatersleben.de)*

The deterioration in the quality of *ex situ* conserved seed over time reflects a combination of both physical and chemical changes. Intraspecific variation for longevity is, at least in part, under genetic control. Here, the grain of 183 bread wheat accessions maintained under low-temperature storage at the IPK-Gatersleben genebank over some decades have been tested for their viability, along with that of fresh grain subjected to two standard artificial ageing procedures. A phenotype–genotype association analysis, conducted to reveal the genetic basis of the observed variation between accessions, implicated many regions of the genome, underling the genetic complexity of the trait. Some, but not all, of these regions were associated with variation for both natural and experimental ageing, implying some non-congruency obtains between these two forms of testing for longevity. The genes underlying longevity appear to be independent of known genes determining dormancy and pre-harvest sprouting.

[Arif MAR, Nagel M, Lohwasser U and Börner A 2017 Genetic architecture of seed longevity in bread wheat (*Triticum aestivum* L.). *J. Biosci.* **42** 81–89]

1. Introduction

Some 1,750 *ex situ* genebanks have been established over the last decades to combat the continuing erosion of genetic variation experienced by crop plants. They are estimated to currently curate >7.4 million accessions (FAO 2010), of which ~45% are cereal species (Börner *et al.* 2014). The loss in quality of stored seed can be slowed, but not stopped, by controlling the storage environment (temperature, relative humidity and gaseous composition). The deterioration in viability experienced during long-term low-temperature storage is due to damage to the membranes, to the DNA and to the action of a variety of enzymes and other proteins (Coolbear 1995; McDonald 1999). Species producing very long-lived seed have evolved a number of structural and chemical features to limit the rate of this decay (Bartosz 1981). Among the agents responsible for seed ageing identified to date, lipid peroxidation is the most well documented (Davies 2005), but oxidative damage to DNA and proteins has also been identified as being causal (Rao *et al.* 1987; Bailly *et al.* 2008).

Standard experimental ageing procedures have been elaborated by the International Seed Testing Association (ISTA) to speed the assessment of viability loss. Both the accelerated ageing (AA) and the controlled deterioration (CD) tests have been used to reveal the genetic basis of longevity in barley (Nagel *et al.* 2009, 2015), oilseed rape (Nagel *et al.* 2011) and wheat (Rehman Arif *et al.* 2012a). Although a proteomics-based analysis has shown that in aged *Arabidopsis thaliana* seed, the two tests highlight a number of common molecules (Rajjou *et al.* 2008), there is no consensus as yet as to whether the two protocols generate comparable outcomes (Walters 1998; McDonald 1999; Black *et al.* 2006).

The genetic basis of most of the important traits in both cereal and non-cereal crops is complex (Börner *et al.* 2002). Seed longevity belongs to this class of trait (Dickson 1980; Clerkx *et al.* 2004b); furthermore, it is

Keywords. Accelerated ageing; association mapping; controlled deterioration; conventional storage; *Triticum aestivum*

Supplementary materials pertaining to this article are available on the Journal of Biosciences Website.

DOI: 10.1007/s12038-016-9661-6

Published online: 20 January 2017

strongly affected by the environment experienced by the mother plant during seed development and by the seed during both the immediate post-harvest period and the period of *ex situ* storage (Contreras *et al.* 2008, 2009). Two approaches are typically taken to determine the number and location of genes underlying variation in a quantitatively inherited trait: these are biparental linkage mapping and phenotype–genotype association analysis (Zhu *et al.* 2008). Both these approaches have been taken to address the inheritance of longevity in a number of crop species. Examples include *A. thaliana* (Bentsink *et al.* 2000; Clerkx *et al.* 2004a, b), rice (Miura *et al.* 2002; Zeng *et al.* 2006; Xue *et al.* 2008; Li *et al.* 2014), soybean (Singh *et al.* 2008), barley (Nagel *et al.* 2009, 2015), *Aegilops tauschii* (Landjeva *et al.* 2010), maize (Revilla *et al.* 2009), lettuce (Schwember and Bradford 2010), oilseed rape (Nagel *et al.* 2011), wheat (Rehman Arif *et al.* 2012a) and tobacco (Agacka *et al.* 2015). Dormancy – the inability of a mature cereal grain to germinate until it has aged sufficiently (Simpson 1990), and pre-harvest sprouting (PHS) – its tendency to germinate prematurely, have both been intensively studied at both the physiological and the genetic level (Kulwal *et al.* 2005; Lohwasser *et al.* 2005, 2013; Mares *et al.* 2009; Rehman Arif *et al.* 2012b). The relationship between seed longevity and either dormancy or PHS, however, has not been fully explored. In rice, Miura *et al.* (2002) concluded that dormancy and longevity are independent of each other, as was also the case for wheat (Rehman Arif *et al.* 2012a). Here, the objectives were to characterize variation for longevity in wheat grain samples stored *ex situ* over a long period, and to identify phenotype–genotype associations in grain which had been either stored long term or exposed to artificial ageing.

2. Materials and methods

2.1 *Plant material*

A set of 183 hexaploid wheat (129 spring type, 54 winter type) accessions (supplementary table 1) was selected from the collection maintained at the IPK genebank and last multiplied in 1974. These were the oldest seed lots available in the storage. The grain has been stored at 0 ±1°C and the grains' moisture content (GMC) maintained at 8±2%. Historical germination data are available from 1978 and 1998, and a new set of data was generated in 2008. The majority of the collection (177 accessions) was grown out to produce fresh material in 2010. Each accession was represented by 30 plants arranged as five plants per pot; the plants were manually harvested and hand-threshed.

2.2 *Standard germination test*

The conventional germination test procedure followed the standard protocol (ISTA 2008), in which the grains are laid between two layers of moist filter paper, which is then formed into a roll and held in a Jacobsen apparatus. The temperature of the water bath of the equipment was 25 ±1°C during the day and 23±1°C during the night. Germination success was recorded after eight days. The test was applied to a sample of 100 grains from the 1974 harvest, and to three replicates of 100 grains per treatment from the materials regenerated in 2010.

2.3 *AA and CD tests*

For the AA test, three replicates of 100 grains per accession were laid on a rack within a sealed glass jar, which contained 200 mL deionized water to ensure the maintenance of near 100% relative humidity. The jars were exposed to 43±0.5°C for 72 h, after which the grains' ability to germinate was tested by an standard germination test (SGT) (ISTA 2008). For the CD test, initial GMC was first determined using an ISTA approved protocol, then increased to 18% by the addition of an amount of deionized water based on the expression, $mH_2O=[100-GMC_I\ (\%)/100-GMC_T\ (\%)] \times W_I$, where mH_2O is the quantity of water added; GMC_I is the initial GMC (%); GMC_T is the target GMC (18%) and W_I the initial weight of the 100 grain sample (g). The grain was left to equilibrate for 2 h at room temperature after which they were chilled to 7°C for 22 h. Three single replicates of 100 grains per accession were then sealed in an aluminium foil pouch and held at 43±0.5°C for 72 h. Finally, the grains' ability to germinate was tested by an SGT. Longevity was expressed by dividing the rate recorded following the treatment (AA or CD) by that of the non-treated control to give a relative AA (RAA) and a relative CD (RCD).

2.4 *Dormancy and PHS tests*

Spikes were harvested at Zadoks stage 92 over a period of 10 days. To assay for dormancy, a sample of 60 fresh grains per accession was laid on moist filter paper in a plastic box held under a 12 h photoperiod for either 7 days at 20°C (D20) or 14 days at 10°C (D10). A dormancy index (DI) (Strand 1965) was derived from the expression (2×D10+D20)/3, where D10 and D20 represented the proportion of successfully germinated grains at each temperature. For the PHS test, five freshly harvested spikes per accession were laid over wet sand and held for 14 days in two replicates which followed each other. For the interpretation of the data, a rating of seven score points was used where one score meant no sprouting and seven score meant complete sprouting. Each accession's PHS score was derived from the mean of these two replicates.

2.5 *Genotyping*

DNA was extracted from three grains per accession of the 1974 material. The grains were crushed and the DNA extracted using a commercial kit (QIAGEN, Germany). The DNA was transferred to Triticarte Pty. Ltd. (now Diversity Arrays Technology Pty Ltd.) Canberra, Australia. for DArT (Diversity Array Technology) genotyping. All information about sequences and chromosomal location of the markers are given by *http://www.diversityarrays.com*.

2.6 *Population structure and phenotype/genotype association analysis*

A sub-set of 161 DArT markers, chosen to define a set of 5 cM intervals (*http://www.diversityarrays.com*) was used to characterize the population structure of the germplasm panel, based on STRUCTURE software (Pritchard *et al.* 2000). The admixture model with a burn-in period of 10,000 iterations and 10,000 MCMC (Markov Chain Monte Carlo) approach duration to test for *K* over the range of 1–15 was used. PAUP software (Swofford 2002) was used to portray genetic relationships between accessions. Associations between individual markers and each trait were calculated using TASSEL v2.01 software (Bradbury *et al.* 2007) assuming either the general linear model (GLM) or the mixed linear model (MLM) (Yu *et al.* 2006). The EMMA method (efficient mixed-model association) (Kang *et al.* 2008) was applied and default settings were chosen for the MLM parameters. Markers assigned a significant ($p \leq 0.05$) or highly significant ($p \leq 0.01$) association with both GLM and MLM models were designated as credible for all traits except for RAA and RCD where an association was declared as significant or highly significant when it appeared in 2 out of 3 replicates and the mean.

2.7 *Statistical analysis*

All statistical analyses of the data were performed using SPSS v17.0 software (SPSS Inc. 1999). Candidate genes mapping within a genomic region associated with a trait were identified from the deletion bin maps prepared in wheat cv. Chinese Spring *(http://wheat.pw.usda.gov/pubs/2004/Genetics/Bioinfo/)*.

3. Results

3.1 *Genomic distribution of DArT markers*

The 2,134 polymorphic DArT markers used for genotyping cover a genetic distance of 2,875 cM (supplementary figure 1); they were non-uniformly distributed between the wheat subgenomes: 931 were B genome loci, 824 A genome loci and 379 D genome loci. A total of 166 markers defined more than one locus (156 defined two loci, seven three loci and three four loci). The most well represented homoeologous group was group 6 (382 markers) and the least well represented was group 4 (150 loci).

3.2 *Population structure*

The STRUCTURE analysis recognized six groups (Q1–Q6) (figure 1). Q1 clustered 30 accessions of mainly Asian descent (25 spring and five winter types), and were clustered in the PAUP (Phylogenetic Analysis using Parsimony)-based phylogenetic tree (supplementary figure 2). Q2 harboured 48 accessions (14 spring, 34 winter) of diverse provenance outside of Asia. The PAUP tree assigned most of these to the 'winter wheat worldwide' section, although eight belonged to the 'winter wheat temperate region' section and four to the 'spring wheat temperate region' section. The provenance of most of the 19 accessions in Q3 (15 spring, four winter) was from the Eastern Mediterranean; in the PAUP tree, these accessions clustered within the 'spring wheat South Europe' section. The accessions within Q4 (36 spring, one winter) were bred in Italy and other temperate countries; most clustered within the 'spring wheat South Europe' section, but seven fell into the 'winter wheat worldwide' and two into the 'winter wheat temperate region' sections. Accessions in Q5 (18 spring, one winter) had diverse provenance; the PAUP analysis placed 11 in the 'spring wheat temperate region' section, six in the 'spring wheat Middle/East Asia' section and one each in the 'spring wheat temperate region' and 'winter wheat temperate region' sections. Finally, the small Q6 (seven spring, one winter) harboured exclusively material bred in Greece, all of which were included in 'spring wheat South Europe' section. The remaining 23 accessions (17 spring, six winter) formed a 'mixed group' and were scattered throughout the PAUP tree.

3.3 *Variation for the ability to germinate*

The mean germination rate of all accessions as tested by the SGT carried out in 2008 was 56.4±1.72% (range: 0–94%). A comparison with the outcomes of the 1978 and 1998 SGTs is shown in figure 2 and supplementary table 2. In 1978, the germination rate was substantially higher (87.1±0.67, range:

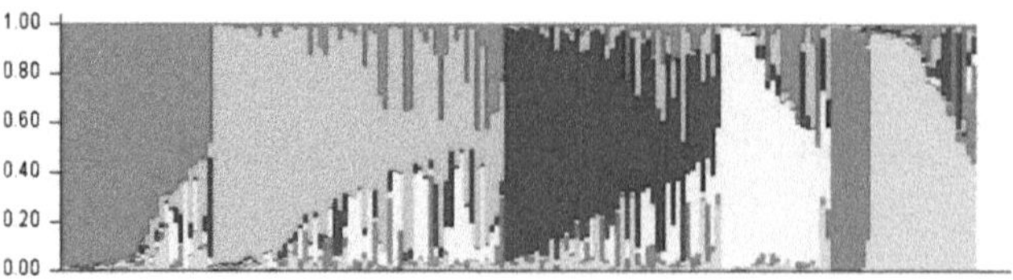

Figure 1. STRUCTURE-based analysis of the germplasm set.

57–99%), and the rate remained high in 1998 (91.1±0.91%, range: 22–100%). Variation for germination capacity as seen from the perspective of the accessions' provenance (Asia, South Europe, Rest of Europe, and Americas), botanical variety (*aestivum*, *lutescens*, *ferrugineum* and 'others'), and vernalization requirement (spring/winter) is illustrated in supplementary figure 3. While there was no influence of either of the former two categorizations on performance, it appeared that the accession's vernalization requirement does exert an influence. For grain harvested in 2010, the overall mean germination rate was 85.7±0.88% (range: 26–99%) (supplementary table 3).

The AA treatment reduced the global mean germination rate to 66.3±1.47 (range: 3–95%), and the Q group means also differed significantly from one another. The mean RAA was 76.8±1.46 (range: 4–100%), with significant differences between both Q group and geographical provenance group means. Similarly, the CD treatment reduced the global germination to 58.6±1.70 (range: 5–96%), once again with significant differences between both Q group and geographical provenance group means. The mean RCD was 67.6 ±1.69 (range: 5–100%) but here there was a significant difference between geographical provenance group means, but not between Q group means.

As summarized in supplementary table 4 and figure 3, the mean percentage dormancy rates (D10 and D20) were, respectively, 11.9±1.28 (range: 0–95%) and 75.6±2.18 (range: 0–100%). The DI ranged from 0 to 96 with mean of 33.1 ±1.30. For PHS, the mean score for was 4.0±0.11 (range: 1.2–6.9).

3.4 *Association analysis*

Altogether, the TASSEL analysis identified 101 marker-trait associations (MTAs) in the STG test carried out in 1978 (supplementary figure 4), of which 25 were highly significant ($p<0.01$). The homoeologous group 2 chromosomes harboured the highest number of MTAs (29), while the group 4 chromosomes only harbored four MTAs. Analysis of the 2008 STG data revealed 103 MTAs (29 highly significant), distributed throughout the genome. In this case the homoeologous group harbouring the highest number of MTAs was group 6 (24), and the

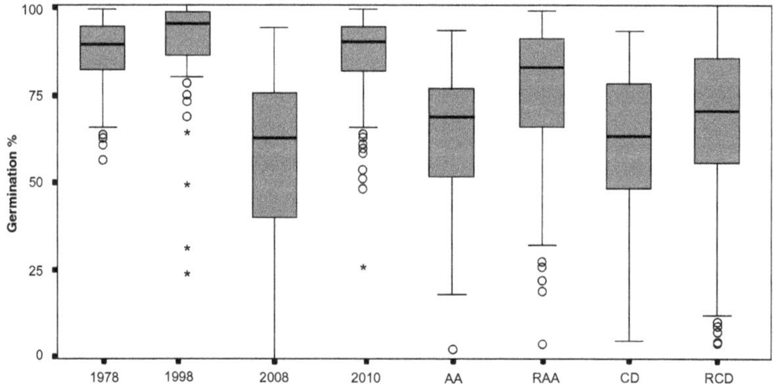

Figure 2. Box plots illustrating the variation recorded for germination % in SGTs conducted in 1978, 1998 and 2008 of materials harvested in 1974 and of fresh grains (2010), along with the germination performance following after artificial ageing.

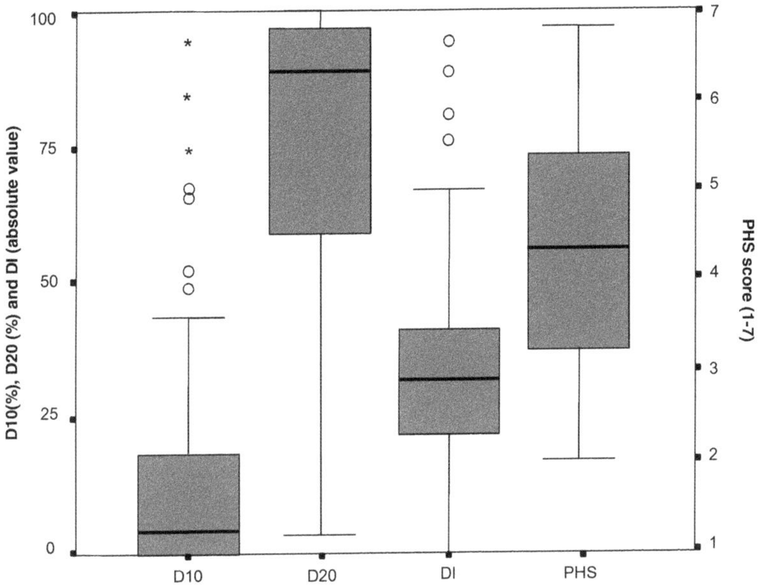

Figure 3. Box plots illustrating the variation recorded for dormancy (D10, D20, DI) and PHS.

least well populated one was again group 4 (three). The same analysis applied to the STG carried out with grain harvested in 2010 revealed 95 significant MTAs (11 highly significant); here, the most well populated homoeologous group was group 6 (26) and the least well populated was group 5 (one).

Since there was very high correlation between AA and RAA as well as CD and RCD (data not shown), only

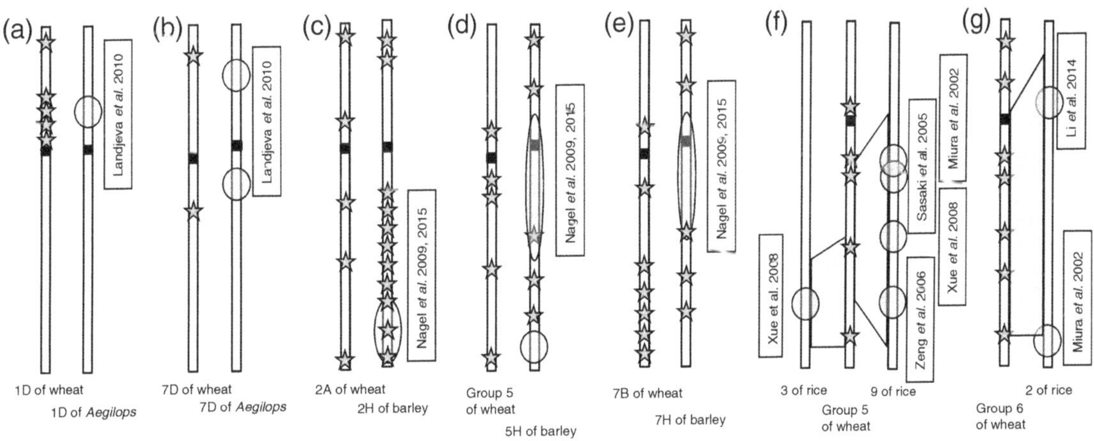

Figure 4. Synteny of longevity loci between (**a**) chromosomes 1D of wheat and 1D of *Aegilops*, (**b**) 7D of wheat and 7D of *Aegilops*, (**c**) chromosomes 2A of wheat and 2H of barley, (**d**) homoeologous group 5 of wheat and chromosome 5H of barley, (**e**) chromosomes 7B of wheat and 7H of barley, (**f**) homoeologous group 5 of wheat and chromosomes 3 and 9 of rice and (**g**) homoeologous group 6 of wheat and chromosome 2 of rice. Small squares indicate centromeres on wheat and barley. Stars on wheat chromosomes represent areas in association with longevity detected in this study and on barley detected in Nagel *et al.* (2015). Shaded circles indicate seed longevity QTLs identified in previous studies either in barley or rice. Stein *et al.* (2007) is used to compare wheat with rice.

associations with RAA and RCD are presented. Analysis of the RAA germination data identified 74 significant MTAs (ten highly significant), with the most well represented homoeologous group being group 3 (23) and the least well represented one group 5 (four). For RCD, there were 97 significant MTAs (11 highly significant), with 38 mapping to the group 3 chromosomes and just two to the group 4 chromosomes. With respect to DI, 78 significant MTAs (19 highly significant) were uncovered, of which 20 mapped to the group 6 chromosomes and only four to each to groups 1 and 4 chromosomes. Finally, for PHS, 110 significant MTAs (30 highly significant) were identified, of which 36 mapped to the group 2 chromosomes and only seven to the group 7 chromosomes.

4. Discussion

A core responsibility of genebanks is to conserve crop germplasm, which is increasingly threatened by genetic erosion. This implies a regular assessment of germination capacity (Nagel *et al.* 2009). The rate of deterioration of a given seed lot's quality depends on many factors, both environmental and genetic. An example of how variable this can be can be seen in the behaviour of the materials multiplied in 1974 and tested 34 years later: some accessions failed to germinate, while others were almost 100% viable (supplementary table 2). Since the materials had all been handled identically, it is assumed that most of this variation reflects genetic variation for seed longevity. The association analysis revealed that many regions of the genome harbour genes influencing longevity, whether this was assayed by either natural or experimental ageing.

The combination of genetic and physical mapping in wheat has generated a series of deletion bins, each associated with a set of DArT markers (Francki *et al.* 2009; Semagn *et al.* 2006; Mantovani *et al.* 2008; Hai-Chun *et al.* 2009). Altogether, 15 bins were identified as likely containing genes influencing longevity (supplementary table 5). Bins 1AS-0.86-1.00 and C-1DS3-0.48 that carried two associated DArT markers of homoeologous group 1 chromosomes contain the genes *Glu-A3* and its orthologues and a number of enzymes (various dehydrogenases, chalcone synthase, lipoxygenase, cellulose synthase, protein kinases, NADH-oxidoreductase, reverse transcriptase and some ethylene forming enzymes). Of particular interest with respect to longevity could be chalcone synthase, lipoxygenase and cellulose synthase. Chalcone synthase (*Chs*) along with another defense responsive (DR) gene, flavanol 7-*O*-methyl transferase, has been mapped to the short arms of chromosomes 1B and 1D by Li *et al.* (1999).

Two associated markers belonging to homoeologous group 2 chromosomes were assigned to 2AS5-0.78-1.00 and 2AL1-0.85-1.00 that carried *elicitor-responsive gene, Vrga1, Enod* (early nodulation gene) and genes influencing the production and level of enzymes like NADH dehydrogenase, glutamate dehydrogenase, pyruvate decarboxylase, peroxidases, superoxide dismutase (SOD) and chaperonins. Li *et al.* (1999) reported *per2* (peroxide 2), *sod* (superoxide dismutase), *wip* (wound induced protein) and other DR genes to be located on all the three homoeologous group 2 chromosomes. In fact, genes *cbp1* and *cbp2* (chitinase binding proteins) were mapped to the long arms of 2A and 2B that might correspond to the MTAs on long arms of 2A and 2B in this investigation. Nagel *et al.* (2009 and 2015) also found QTLs and associations for longevity on chromosome 2H of barley where they reported a dehydration responsive element binding (*DREB*) protein probably to be a candidate gene for longevity.

Nine associated markers of group 3 chromosomes were assigned to bins 3AS2-0.23, 3AL5-0.78-1.00, 3BS1-0.33-0.57, 3BL7-0.63-1.00 and 3DS6-0.55-1.00, which contain genes such as a biostress resistance related protein, a putative plasma membrane protein and others responsible for the production of enzymes involved in amino acid synthesis. Biparental mapping has also identified loci affecting longevity on chromosome 3B near centromere and 3DL (Rehman Arif *et al.* 2012a). The chromosome 4B bin 4BS4-0.37 harbours genes encoding a protein kinase like protein, a dehydrin/LEA group 2 like protein and putative a phosphoesterase. While no bin identification was possible for the group 5 chromosomes, defence genes have been mapped on each of the homoeologous group 5 chromosomes (Li *et al.* 1999). The MTA on 5BS in this investigation might correspond to the *Tha3* gene whereas one of the MTAs on 5BL might correspond to *Lpx*. Two markers of homoeologous group 6 chromosomes associated with longevity were placed in 6BS-Sat and C-6BL3-0.36. The distal end of 6BL carries two DR genes (*Tha1* and *Cbp1*) (Li *et al.* 1999). Polyphenol oxidase (*Ppo*) was mapped to both 6A and 6D chromosomes and hydroxyl-proline rich protein (*Hrp*) and ribosome inactivating protein (*Rip*) were mapped to chromosomes 6A and 6D, respectively. *Tha1* and *Cbp1* seem to be in comparable location to the MTA on the distal end of 6B but it could not be assigned to any deletion bin.

Finally, 5 DArT markers of group 7 were placed in 7AS1-0.89.1.00, 7BS1-27-1.00, 7BL10-0.78-1.00 and 7DS4-0.61-1.0 bins. A cluster of DR genes have been reported to occur on the long arm of group 7 chromosomes, especially *Tha1, Tha2, Cht1b*, and *Cat* reside within a relatively small segment on the distal end of 7BL (Li *et al.* 1999). Faris *et al.*

(1999) found a strong QTL for tan spot infection on 7BL corresponding to the same cluster of DR genes. Longevity MTAs on 7BL and genes reported by Li *et al.* (1999) are in comparable locations. Similarly, two MTAs on 7AL might correspond to *Tha2* gene (Li *et al.* 1999).

Li *et al.* (1999) found that most of the DR genes in the genome are present in clusters along the distal regions of the chromosomes. The same is observed in this study i.e., some seed longevity MTAs are present in clusters of two or more MTAs such as on 1AS, 2AL, 2BS, 2DL, 3AL, 3D, 4BS, 5AL, 6AL, 6D, 7AS, 7BL and 7DS whereas the others were distributed throughout the entire chromosomes (supplementary figure 4). Rehman Arif *et al.* (2012a) also assume enzymes like peroxidases (Gulen and Eris 2004), glutamate dehydrogenase (Skopelitis *et al.* 2006), alcohol dehydrogenase (Kato-Noguchi 2001) and aldehyde dehydrogenase (Sunkar *et al.* 2003) to be responsible for seed longevity which along with other functions protecting plants against various kinds of stress (Houde *et al.* 2006). Consequently, seeds expressing these enzymes more effectively may be better equipped to live longer and maintain their genetic and membrane integrity. It can be concluded that seed longevity is in part under the control of the morphological features of the seeds such as cell wall and cell membrane.

The associated markers with the germination rate following long term low temperature storage and the artificial ageing treatments were not similar. The implication is that different deterioration mechanisms are involved in these processes (Walters 1998; McDonald 1999). *A. thaliana* seed responds quite predictably to a range of ageing regimes (Clerkx *et al.* 2004a, 2004b), while proteomic comparisons have revealed that the extent of protein carbonylation is strongly increased whatever the storage conditions are, and the carbonylation targets mostly overlap (Rajjou *et al.* 2008). In contrast, linkage mapping in lettuce has failed to show any overlap between the loci responsible for deterioration resulting from natural and artificial ageing (Schwember and Bradford 2010) Here, ten MTAs were shared between long term storage and experimental ageing (six with AA and four with CD) (supplementary figure 4). The conclusion is that in wheat, deterioration during long term low temperature storage is dissimilar from that induced by artificial ageing, although a few genes are likely important for both. In all, 23 genomic locations were identified where the 2008 SGT- and the AA/CD- MTAs lay within 5 cM of one another.

The genetics of cereal grain longevity has actively been researched in rice (Miura *et al.* 2002; Sasaki *et al.* 2005, 2015; Zeng *et al.* 2006; Xue *et al.* 2008), barley (Nagel *et al.* 2009, 2015) and in a set of wheat/*Aegilops tauschii* introgression lines (Landjeva *et al.* 2010). The latter study identified loci which match in location to the MTAs identified here on the proximal ends of the 1DS (figure 4a) and 7DL (figure 4b) chromosome arms. The barley data include similar locations to the MTAs identified on chromosomes 2A (figure 4c), 5A and 5B (figure 4d) and 7B (figure 4e). In rice, longevity loci have been identified on eight of its 12 chromosomes (Miura *et al.* 2002; Sasaki *et al.* 2005; Zeng *et al.* 2006; Xue *et al.* 2008), while Li *et al.* (2014) have identified MTAs with respect to both conventional and high temperature ageing. The syntenic relationship between wheat chromosome 5A and 5B and rice chromosomes 3 and 9, and between chromosome 6B and rice chromosome 2 (figures 4f and g) highlights some related genomic regions which harbour genes underlying longevity.

The nexus between longevity and dormancy/PHS remains controversial. In rice, Siddique *et al.* (1988) argued that the connection is real, while Miura *et al.* (2002) were able to show by linkage mapping that the relevant genes are non-identical. In *A. thaliana* mutants affecting testa pigmentation and/or structure show both a reduced level of dormancy and a more rapid rate of deterioration (Debeaujon *et al.* 2000). Here, only 11 of the dormancy and 19 of the PHS MTAs overlapped those for longevity, indicating only a limited degree of similarity with respect to the genetic determination of these traits.

Genome-wide association mapping can provide a powerful means of revealing the genetics of complex, quantitatively inherited traits. The present investigation of the genetic basis for the longevity of the wheat grain has shown that the deterioration in viability does have a genetic component. The recognition that distinct sets of genomic regions harbouring the genes active during the process of natural and induced ageing indicates that the mechanisms involved in the artificial ageing protocols do not fully mirror those operating during long term low temperature storage, although the sharing of some genomic regions does imply that experimental ageing can to, an extent, be predictive. In wheat, there appears to be little evidence for either dormancy or PHS to be associated with longevity.

References

Agacka M, Nagel M, Doroszewska T, Lewis RS and Börner A 2015 Mapping quantitative trait loci determining seed longevity in tobacco (*Nicotiana tabacum* L.). *Euphytica* **202** 479–486

Bailly C, El-Maarouf-Bouteau H and Corbineau F 2008 From intracellular signaling networks to cell death: the dual role of reactive oxygen species in seed physiology. *C. R. Biol.* **331** 806–814

Bartosz G 1981 Non-specific reactions: molecular basis of ageing. *J. Theor. Biol.* **91** 233–235

Bentsink L, Alonso-Blanco C, Vreugdenhil D, Tesnier K, Groot SPC and Koornneef M 2000 Genetic analysis of seed-soluble oligosaccharides in relation to seed storability of *Arabidopsis*. *Plant Physiol.* **124** 1595–1604

Black M, Bewley JD and Halmer 2006 *The encyclopedia of seeds: science, technology and uses* (Wallingford, UK: CAB International)

Börner A, Schumann E, Fürste A, Cöster H, Leithold B, Röder MS and Weber WE 2002 Mapping of quantitative trait loci determining agronomic important characters in hexaploid wheat (*Triticum aestivum* L.). *Theor. Appl. Genet.* **105** 921–936

Börner A, Khlestkina EK, Chebotar S, Nagel M, Rehman Arif MA, Kobiljski B, Lohwasser and Röder MS 2014 Molecular markers in management of *ex situ* PGR - a case study. *J. Biosci.* **37** 871–877

Bradbury PJ, Zhang Z, Kroon DE, Casstevens TM, Ramdoss Y and Buckler ES 2007 TASSEL: software for association mapping of complex traits in diverse samples. *Bioinformatics* **23** 2633–2635

Clerkx EJM, Blankestijn-De VH, Ruys GJ, Groot SPC and Koornneef M 2004a Genetic differences in seed longevity of various *Arabidopsis* mutants. *Physiol. Plant.* **121** 448–461

Clerkx EJM, El-Lithy ME, Vierling E, Ruys GJ, Blankestijn-De VH, Groot SPC, Vreugdenhil D and Koornneef M 2004b Analysis of natural allelic variation of *Arabidopsis* seed germination and seed longevity traits between the accessions Landsberg *erecta* and Shakdara, using a new recombinant inbred line population. *Plant Physiol.* **135** 432–443

Contreras S, Bennett MA, Tay D and Metzger JD 2008 Maternal light environment during seed development affects lettuce seed weight, germinability, and storability. *Hortic. Sci.* **43** 845–852

Contreras S, Bennet MA, Metzger JD, Tay D and Nerson H 2009 Red to far-red ratio during seed development affects lettuce seed germinability and longevity. *Hortic. Sci.* **44** 130–134

Coolbear P 1995 Mechanisms of seed deterioration; in *Seed Quality: Basic mechanisms and agricultural implications* (eds) AS Basra (Food Product Press, New York) pp 223–277

Davies MJ 2005 The oxidative environment and protein damage. *Biochim. Biophys. Acta* **1703** 93–109

Debeaujon I, Léon-Kloosterzie KM and Koornneef M 2000 Influence of the testa on seed dormancy, germination and longevity in *Arabidopsis. Plant Physiol.* **122** 403–413

Dickson MH 1980 Genetic aspects of seed quality. *Hortic. Sci.* **15** 771–774

FAO 2010 The second report on the state of the world's plant genetic resources for food and agriculture (Rome: Commission on Genetic Resources for Food and Agriculture, Food and Agriculture Organization of the United Nations)

Faris JD, Li WL, Liu DJ, Chen PD and Gill BS 1999 Candidate gene analysis of quantitative disease resistance in wheat. *Theor. Appl. Genet* **9** 219–225

Francki GM, Walker E, Crawford AC, Broughton S, Ohm HW, Barclay I, Wilson RE and McLean R 2009 Comparison of genetic and cytogenetic maps of hexaploid wheat (*Triticum aestivum* L.) using SSR and DArT markers. *Mol. Gen. Genomics* **281** 181–191

Gulen H and Eris A 2004 Effect of heat stress on peroxidase activity and total protein content in strawberry plants. *Plant Sci.* **166** 739–744

Hai-Chun J, Bayon C, Kostya K, Berry S, Wenzl P, Huttner E, Kilian A and Hammond-Kosack KE 2009 DArT markers: diversity analyses, genomes comparison, mapping and integration with SSR markers in *Triticum monococcum. BMC Genomics* **10** 458

Houde M, Belcaid M, Ouellet F, Danyluk J, Monroy AF, Dryanova A, Gulick P, Bergeron A, *et al.* 2006 Wheat EST resources for functional genomics for abiotic stress. *BMC Genomics* **7** 149

ISTA 2008 *International rules for seed testing* (Bassersdorf: International Seed Testing Association)

Kang HM, Zaitlen NA, Wade CM, Kirby A, Heckermann D, Daly MJ and Eskin E 2008 Efficient control of population structure in model organism association mapping. *Genetics* **178** 1709–1723

Kato-Noguchi H 2001 Wounding stress induces alcohol dehydrogenase in maize and lettuce seedlings. *Plant Growth Regul.* **35** 285–288

Kulwal PL, Kumar N, Gaur A, Khurana P, Khurana JP, Tyagi AK, Balyan HS and Gupta PK 2005 Mapping of a major QTL for pre-harvest sprouting tolerance on chromosome 3A in bread wheat. *Theor. Appl. Genet.* **111** 1052–1059

Landjeva S, Lohwasser U and Börner A 2010 Genetic mapping within the wheat D genome reveals QTLs for germination, seed vigour and longevity, and early seedling growth. *Euphytica* **171** 129–143

Li WL, Fairs JD, Chitoor JM, Leach JE, Hullbert SH, Liu DJ, Chen PD and Gill BS 1999 Genomic mapping of defense response genes in wheat. *Theor. Appl. Genet.* **98** 226–233

Li G, Na YW, Kwon SW and Park YJ 2014 Association analysis of seed longevity in rice under conventional and high-temperature germination conditions. *Plant Syst. Evol.* **300** 389–402

Lohwasser U, Röder MS and Börner A 2005 QTL mapping for the domestication traits pre-harvest sprouting and dormancy in wheat (*Triticum aestivum* L.). *Euphytica* **143** 247–249

Lohwasser U, Rehman Arif MA and Börner A 2013 Discovery of loci determining pre-harvest sprouting and dormancy in wheat and barley applying segregation and association mapping. *Biol. Plant.* **57** 663–674

Mantovani P, Maccaferri M, Sanguineti MC, Tuberosa R, Catizone I, Wenzl P, Thomson B, Carling J, *et al.* 2008 An integrated DArT-SSR linkage map of durum wheat. *Mol. Breed.* **22** 629–648

Mares DJ, Rathjen J, Mrva K and Cheong J 2009 Genetic and environmental control of dormancy in white-grained wheat (*Triticum aestivum* L.). *Euphytica* **168** 311–318

McDonald MB 1999 Seed deterioration: physiology, repair and assessment. *Seed Sci. Technol.* **27** 177–237

Miura K, Lyn SY, Yano M and Nagamine T 2002 Mapping quantitative trait loci controlling seed longevity in rice (*Oryza sativa* L.). *Theor. Appl. Genet.* **104** 981–986

Nagel M, Vogel H, Landjeva S, Buck-Sorlin G, Lohwasser U, Scholz U and Börner A 2009 Seed conservation in *ex situ* genebanks - genetic studies on longevity in barley. *Euphytica* **170** 5–14

Nagel M, Rosenhauer M, Willner E, Snowdon RJ, Friedt W and Börner A 2011 Seed longevity in oilseed rape (*Brassica napus* L.) - genetic variation and QTL mapping. *Plant Genet. Res.* **9** 260–263

Nagel M, Kranner I, Neumann K, Rolletschek H, Seal CE, Colville L and Fernández-Marín BA 2015 Genome-wide association mapping and biochemical markers reveal that seed ageing and longevity are intricately affected by genetic background and developmental and environmental conditions in barley. *Plant Cell Environ.* **38** 1011–1022

Pritchard JK, Stephens M and Donnelly P 2000 Inference of population structure using multilocus genotypic data. *Genetics* **155** 945–959

Rajjou L, Lovigny Y, Groot SPC, Belghazi M, Job C and Job D 2008 Proteome-wide characterization of seed aging in *Arabidopsis*: a comparison between artificial and natural aging protocols. *Plant Physiol.* **148** 620–641

Rao NK, Roberts EH and Ellis RH 1987 Loss of viability in lettuce seeds and the accumulation of chromosome damage under different storage conditions. *Ann. Bot.* **60** 85–96

Rehman Arif MA, Nagel M, Neumann K, Kobiljski B, Lohwasser U and Börner A 2012a Genetic studies of seed longevity in hexaploid wheat using segregation and association mapping approaches. *Euphytica* **186** 1–13

Rehman Arif MA, Neumann K, Nagel M, Kobiljski B, Lohwasser U and Börner A 2012b An association mapping analysis of dormancy and pre-harvest sprouting in wheat. *Euphytica.* **188** 409–417

Revilla P, Butrón A, Rodríguez VM, Malvar RA and Ordás A 2009 Identification of genes related to germination in aged maize seed by screening natural variability. *J. Exp. Bot.* **60** 4151–4157

Sasaki K, Fukuta Y and Sato T 2005 Mapping of quantitative trait loci controlling seed longevity of rice (*Oryza sativa* L.) after various periods of seed storage. *Plant Breed.* **124** 361–366

Sasaki K, Takeuchi Y, Miura K, Yamaguchi T, Ando T, Ebitani T, Higashitani A, Yamaya T, *et al.* 2015 Fine mapping of major quantitative trait locus, *qLG-9*, that controls seed longevity in rice. (*Oryza sativa* L.). *Theor. Appl. Genet.* **128** 769–778

Schwember AR and Bradford KJ 2010 Quantitative trait loci associated with longevity of lettuce seeds under conventional and controlled deterioration storage conditions. *J. Exp. Bot.* **61** 4423–4436

Semagn K, Bjørnstad Å, Skinnes H, Marøy AG, Tarkegne Y and William M 2006 Distribution of DArT, AFLP, and SSR markers in a genetic linkage map of a doubled-haploid hexaploid wheat population. *Genome.* **49** 545–555

Siddique SB, Seshu DV and Pardee WD 1988 Rice cultivar variability in tolerance for accelerated ageing of seed. *IRRI Research Paper Series.* **131** 2–7

Simpson GM 1990 *Seed dormancy in grasses* (New York: Cambridge University Press)

Singh RK, Raipuria RK, Bhatia VS, Rani A, Pushpendra HSM, Chauhan D, Chauhan GS and Mohopatra T 2008 SSR markers associated with seed longevity in soybean. *Seed Sci. Technol.* **36** 162–167

Skopelitis DS, Paranychianakis NV, Paschalidis KA, Pliakonis ED, Delis ID, Yakoumakis ID, Kouvarakis A, Papadakis AK, Stephanou EG and Roubelakis-Angelakis KA 2006 Abiotic stress generates ROS that signal expression of anionic glutamate dehydrogenases to form glutamate for proline synthesis in tobacco and grapevine. *Plant Cell* **18** 2767–2781

SPSS Inc. 1999 SPSS Base 10.0 for Windows User's Guide. SPSS Inc., Chicago, Illinois

Stein N, Prasad M, Scholy U, Thiel T, Zhang H, Wolf M, Kota R, Varshney R, *et al.* 2007 A 1,000-loci transcript map of the barley genome: new anchoring points for integrative grass genomics. *Theor. Appl. Genet.* **114** 823–839

Strand E 1965 Studies on seed dormancy in barley. *Meldinger fra Norges Landbrukshogskole Hoegskole.* **44** 1–23

Sunkar R, Bartels D and Kirch H 2003 Overexpression of a stress-inducible aldehyde dehydrogenase gene from Arabidopsis thaliana in transgenic plants improves stress tolerance. *Plant J.* **35** 452–464

Swofford D 2002 *Paup*: Phylogenetic analysis using parsimony (*and others methods), version 4* (Sunderland: Sinuauer Associates)

Walters C 1998 Understanding the mechanisms and kinetics of seed ageing. *Seed Sci. Res.* **8** 223–244

Xue Y, Zhang SQ, Yao QH, Peng RH, Xiong AS, Li X, Zhu WM, Zhu YY, *et al.* 2008 Identification of quantitative trait loci for seed storability in rice (*Oryza sativa* L.). *Euphytica* **164** 739–744

Yu J, Pressoir G, Briggs WH, Bi IV, Yamasaki M, Doebley JF, McMullen MD, Gaut BS, *et al.* 2006 A unified mixed-model for association mapping that accounts for multiple levels of relatedness. *Nat. Genet.* **38** 203–208

Zeng DL, Guo LB, Xu YB, Yasukumi K, Zhu LH and Qian Q 2006 QTL analysis of seed storability in rice. *Plant Breed.* **125** 57–60

Zhu C, Gore M, Buckler ES and Yu J 2008 Status and prospects of association mapping in plants. *Plant Genome.* **1** 5–20

MS received 24 January 2016; accepted 02 November 2016

Corresponding editor: Utpal Nath

3.13 PAPER 13:

Novel loci and a role for nitric oxide for seed dormancy and pre-harvest sprouting in barley

by

Manuela Nagel, Ahmad M. Alqudah, Marlene Bailly, Loic Rajjou, Sibylle Pistrick, Gabriele Matzig, Andreas Börner and Ilse Kranner

Published in

Plant, Cell & Environment (2019) 42, 1318-1327

https://doi.org/10.1111/pce.13483

To view supplementary material for this article, please visit

https://onlinelibrary.wiley.com/doi/10.1111/pce.13483

Received: 7 April 2018 | Accepted: 16 November 2018
DOI: 10.1111/pce.13483

ORIGINAL ARTICLE

WILEY

Novel loci and a role for nitric oxide for seed dormancy and preharvest sprouting in barley

Manuela Nagel[1] | Ahmad M. Alqudah[1] | Marlène Bailly[2] | Loïc Rajjou[2] | Sibylle Pistrick[1] | Gabriele Matzig[1] | Andreas Börner[1] | Ilse Kranner[3]

[1] Genebank Department, Leibniz Institute of Plant Genetics and Crop Plant Research (IPK Gatersleben), Seeland, Germany

[2] Institut Jean-Pierre Bourgin, INRA, AgroParisTech, CNRS, Université Paris-Saclay, 78000 Versailles Cedex, France

[3] Department of Botany and Center for Molecular Biosciences (CMBI), University of Innsbruck, Innsbruck, Austria

Correspondence
M. Nagel, Genebank Department, Leibniz Institute of Plant Genetics and Crop Plant Research (IPK Gatersleben), Corrensstraße 3, OT Gatersleben, D-06466 Seeland, Germany.
Email: nagel@ipk-gatersleben.de

Funding information
FP7 Environment, Grant/Award Number: (311840 "EcoSeed"); LabEx, Grant/Award Number: ANR-10-LABX-0040-SPS

Abstract

Barley is used for food and feed, and brewing. Nondormant seeds are required for malting, but the lack of dormancy can lead to preharvest sprouting (PHS), which is also undesired. Here, we report several new loci that modulate barley seed dormancy and PHS. Using genome-wide association mapping of 184 spring barley genotypes, we identified four new, highly significant associations on chromosomes 1H, 3H, and 5H previously not associated with barley seed dormancy or PHS. A total of 71 responsible genes were found mostly related to flowering time and hormone signalling. A homolog of the well-known *Arabidopsis Delay of Germination 1* (*DOG1*) gene was annotated on the barley chromosome 3H. Unexpectedly, *DOG1* appears to play only a minor role in barley seed dormancy. However, the gibberellin oxidase gene *HvGA20ox1* contributed to dormancy alleviation, and another seven important loci changed significantly during after-ripening. Furthermore, nitric oxide release correlated negatively with dormancy and shared 27 associations. Origin and growth environment affected seed dormancy and PHS more than did agronomic traits. Days to anthesis and maturity were shorter when seeds were produced under drier conditions, seeds were less dormant, and PHS increased, with a heritability of 0.57–0.80. The results are expected to be useful for crop improvement.

KEYWORDS
after-ripening, *DOG1*, germination, GWAS, *Hordeum vulgare*

1 | INTRODUCTION

Seed dormancy is the inability of a viable seed to complete germination under favourable conditions, enabling spreading of seedling establishment in time and space (Baskin & Baskin, 2004; Finch-Savage & Footitt, 2017; Finch-Savage & Leubner-Metzger, 2006). Crops were subjected to human selection pressure, and dormancy was selected against, facilitating uniform and immediate field establishment after sowing. As a trade-off, preharvest sprouting (PHS), that is, germination on the mother plant, can occur in crops, which is an agronomically and industrially undesired trait that compromises yield, nutrition, and processing quality (Rodriguez, Barrero, Corbineau, Gubler, & Benech-Arnold, 2015). Barley (*Hordeum sativum* L.) is the world's fourth most produced crop (www.fao.org/faostat). With a large haploid genome of 5.1 gigabases on seven chromosomes, barley is a diploid model plant for genetics and genomics of the Triticeae tribe (Mascher et al., 2017). Barley is used as staple food in various countries, but mainly for fodder and for beer and whiskey production. Malting relies on nondormant seeds that germinate readily, whereas PHS renders barley seeds unsuitable for malting. The underlying molecular mechanisms of dormancy are best understood in the model plant *Arabidopsis* (Shu et al., 2015). However, dormancy mechanisms can vary greatly in different plant taxa; the monocot barley is taxonomically distant from the eudicot *Arabidopsis* and produces caryopses with endospermic seeds of distinctly different morphology to the Brassicaceae seeds (note: caryopses are fruits in which the pericarp is fused with the seed coat but are hereafter referred to as "seeds," for simplicity).

Dormancy and germination are controlled by hormones, reactive oxygen species (ROS), reactive nitrogen species, and antioxidants. In addition, small noncoding ribonucleic acids are involved in the intricate hormone signalling, targeting mRNAs encoding transcription factors (Liu & El-Kassaby, 2017). During *Arabidopsis* seed maturation, abscisic acid (ABA) accumulation mediated by auxin signalling induces seed dormancy (Liu et al., 2013). In dormant seeds, ROS (Bailly, El-Maarouf-Bouteau, & Corbineau, 2008) such as hydrogen peroxide (H_2O_2), which is involved in regulating ABA catabolism, play important roles. Elevated H_2O_2 concentrations stimulate ABA degradation (Ishibashi et al., 2017) and protein oxidation (Oracz et al., 2007) and trigger dormancy release during after-ripening (Rodriguez et al., 2015). Gibberellins, brassinosteroids, ethylene, and cytokinins break dormancy or promote germination (Shu, Liu, Xie, & He, 2016).

Nitric oxide (NO) is also involved in signalling during seed development and counteracts dormancy maintenance (Arc, Galland, Godin, Cueff, & Rajjou, 2013). In aqueous environments, the biological half-life of NO is in the order of seconds, depending on concentration and cellular redox environment (Schmidt, Desch, Klatt, Kukovetz, & Mayer, 1997). It diffuses through cell membranes and acts upstream on ABA biosynthesis (Arc, Galland, et al., 2013). In ABA signalling, a basic leucine zipper transcriptional factor, namely, *ABA-insensitive 5* (*ABI5*), is a target of NO, which destabilizes the group VII ethylene response factor and suppresses ABA-responsive gene expression (Gibbs et al., 2014, 2015). Mediated by ubiquitin RING-type E3 ligase KEEP ON GOING (Stone, Williams, Farmer, Vierstra, & Callis, 2006), NO directly affects *ABI5* stability by *S*-nitrosylation that promotes germination (Albertos et al., 2015; Gibbs et al., 2014).

Environmental factors in the maternal environment, such as temperature, nitrogen availability, and light intensity, affect seed dormancy (He et al., 2016; Rodriguez, Margineda, Gonzalez-Martin, Insausti, & Benech-Arnold, 2001; Springthorpe & Penfield, 2015), and the *Arabidopsis* seed transcriptome is particularly temperature sensitive. In a pioneering study on *Arabidopsis* seed dormancy, a quantitative trait locus (QTL) was identified, termed *Delay of Germination 1* (*DOG1*; Alonso-Blanco, Bentsink, Hanhart, Vries, & Koornneef, 2003), and homologous Triticeae *DOG1-like* genes were identified, but not annotated, in barley and wheat (Ashikawa, Mori, Nakamura, & Abe, 2014). *DOG1* expression increased when seeds were matured at low temperature, which was associated with higher dormancy levels (Chiang et al., 2011). Low temperature during *Arabidopsis* seed set also led to an activation of *Flowering Locus T* (*FT*), with downstream effects on proanthocyanidin and procyanidins, affecting seed coat metabolites and deepening dormancy.

In barley, two major QTLs for seed dormancy, termed SD1 and SD2, located on chromosome 5H, were identified by linkage mapping (Gong, Li, Zhou, Bonnardeaux, & Yan, 2014; Li et al., 2003). However, no genome-wide association study (GWAS) was conducted to reveal the genetic background of barley seed dormancy. GWAS is considered as one of the most important tools that helps in understanding phenotypic and genetic variation, based on plant performance in different environments and geographic origin, also providing insights into the genetic basis of adaptation. The aim of this paper was to investigate how the genetic background in conjunction with environmental conditions during seed development is linked to PHS, dormancy and dormancy alleviation, and extracellularly released gaseous NO, a key mediating molecule for cell–cell communication and cellular signalling (Domingos, Prado, Wong, Gehring, & Feijo, 2015). To address genotypic effects, GWAS was conducted using 184 globally distributed two- and six-row spring barley genotypes, termed "EcoSeed" panel, selected according to dormancy status (from nondormant to highly dormant at maturity). To quantify whether seed germination-related traits are influenced by genetics, maternal growth environment, agronomic parameters, and/or their interaction, seeds were multiplied in the field under two contrasting environmental conditions, in a relatively "dry" growing season with 200-mm rainfall (2013) and a relatively "wet" growing season with 400-mm rainfall (2014). Based on 4,343 genetic markers, the genetic dissection of the studied traits showed that a plethora of mechanisms represented by 376 single marker–trait associations (MTAs) are involved in alleviation or maintenance of seed dormancy and PHS. In addition to known loci on chromosome 5H, 18 highly significant loci for dormancy maintenance and 33 for NO release were discovered, most of them on 2H and 3H.

2 | MATERIALS AND METHODS

2.1 | Genotyping and population structure

The EcoSeed panel comprised 116 two-row and 68 six-row barley (*Hordeum vulgare* L.) genotypes from 23 countries. Genotypes included improved cultivars (105), breeding lines (14), and landraces (65). At the three-leaf stage, leaf material of 10 seedlings each was pooled and genotyped (TraitGenetics GmbH, Gatersleben, Germany) using Illumina HD 9K chip. A set of 7,864 single-nucleotide polymorphism (SNP) markers was obtained. By applying a minor allele frequency of >10% and removal of monomorphic markers, 5,156 markers were identified. Of these, 4,343 were genetically mapped on the seven barley chromosomes with an average distance of 0.23 and a total length of 988.5 cM. Chromosomal positions that physically anchored were obtained from Comadran et al. (2012) as published by Mascher et al. (2017).

The population structure was analysed by the STRUCTURE 2.3.4 software (Pritchard, Stephens, & Donnelly, 2000) using the full marker set of 5,156 markers. By applying the admixture model, a burn-in of 10,000 iterations and a 10,000 Markov chain Monte Carlo duration, *K* values were calculated for the range from 1 to 10 using 100 replicates each. The likely number of present subpopulations (Q) was estimated by a phylogenetic tree based on the SNP marker and unweighted pair group method with arithmetic average using PHYLIP 3.695 packages dnadist, neighbour (Felsenstein, 2009). Phylogenetic tree was drawn using The Tree Drawing Tool "FigTree" Version 1.4.2 (http://tree.bio.ed.ac.uk/).

Genome-wide linkage disequilibrium (LD) was calculated through estimating the squared allele frequency correlations (r_{LD}^2) between the marker pairs using GenStat 17 (VSN International Ltd, UK). LD values (r_{LD}^2) were plotted against the genetic distance (cM) to estimate intrachromosomal LD decay. We considered r_{LD}^2 = 0.20 as a strict threshold based on the LOESS curve giving a confidence interval for the average LD decay of ±1 cM.

2.2 | Multiplication and phenotyping

All genotypes were grown in parallel in three pots each in the greenhouse and in two rows in the field at the IPK estates in 2013. In 2014, higher seed availability enabled multiplication in a randomized block design of 1 m × 1 m field subplots in two replicates (Field Plots 1 and 2). Greenhouse and field panels were evaluated for days to anthesis (DTA) and days to maturity, when half of the row/plot began to pollinate (Zadoks Growth Stage 65) and spikes were fully mature (Zadoks Growth Stage 92), respectively. When plant growth terminated after anthesis, plant height (PH) was measured. In 2014, lodging was evaluated between 1 (no lodging) and 9 (complete lodging). After harvest, key agronomic traits such as thousand seed weight (TSW) and seeds per spike (SpSs) were measured on three single spikes. Seeds were pooled, 100 g per replicate graded (<2.2, >2.2, >2.5, and > 2.8 mm) using Sortimat K3 (Pfeuffer GmbH, Germany), and expressed as percentages. Finally, all traits were related to dormancy traits.

In field material from both years, PHS was assessed using five freshly harvested mature spikes per genotype laid on wet sand for 14 days at a day/night cycle of 16/8 hr and 25/20°C. Then, PHS was scored for each spike on a scale from 1 to 7 (where 1 indicated *no sprouting* and 7 indicated *100% sprouting*). The same procedure was repeated 14 days after maturity.

Primary dormancy, referred to as "dormancy," was assayed from a sample of 120 freshly harvested seeds from mature spikes, at 0 weeks after harvest (in barley, secondary dormancy can be induced by high temperatures such as 30°C, but this was not the subject of this work as secondary dormancy has hardly any agricultural or industrial relevance for barley production or use in brewing) Sixty seeds each were placed on moistened filter paper under a 12-hr photoperiod and held either for 7 days at 20°C ($TG20_{0w}$) or for 14 days at 10°C ($TG10_{0w}$). Total germination (TG) was assessed as radicle emergence of at least 2 mm. The remaining nongerminated seeds at each temperature (D10 at 10°C and D20 at 20°C) were considered dormant and used to calculate the dormancy index (DI_{0w} = [2 × D10 + D20]/3). In 2014, higher yields allowed the investigation of dormancy release. Seeds harvested at full maturity were kept either in cold storage (CS; at −18°C, 10% moisture content) to retain dormancy or at ambient storage (AS; at 20°C, 50% relative humidity) to support after-ripening processes. Thus, 6 and 24 weeks after harvest, 60 seeds from both storage conditions were germinated at 20°C for 7 days, and DI was calculated as DI = (2 × AS + CS)/3. The following traits were evaluated: germination after CS for 6 ($TG20_{6w}$[CS]) and 24 weeks ($TG20_{24w}$[CS]), germination after AS for 6 ($TG20_{6w}$[AS]) and 24 weeks ($TG20_{24w}$[AS]), and DI after 6 (DI_{6w}) and 24 weeks (DI_{24w}).

Phenotypic data were analysed using residual maximum likelihood implemented in GenStat 17 software (VSN International Ltd, UK). Variance components were calculated by fitting a mixed linear model to multienvironment data. The broad-sense heritability (H^2) values were estimated for all traits considering the percentages of genotypic variance over the total phenotypic variance including genotype (G) by environment (E) interaction (G × E) and error variance components. The best linear unbiased estimates (BLUEs) of all the genotypes were derived by assuming fixed genotypic and random environment effects and taking into account the G × E variance. Analysis of variance was applied to genotypes, environments, treatments, and the least significance differences at $P < 0.05$ used for discrimination. If the normality test failed, the Kruskal–Wallis analysis of variance on ranks followed by the Dunn analysis was applied. Spearman correlation at $P < 0.05$ was used to find relationships between dormancy and field traits (GenStat 17, VSN International Ltd, UK).

2.3 | Genome-wide association study

For association analysis, a mixed linear model implemented in GenStat 17 (VSN International Ltd, UK) was used to calculate single MTA between 4,343 genetically mapped SNP markers and the estimated phenotypic data (BLUEs). The population structure was corrected by kinship, and associations were regarded as significant when threshold exceeded $-\log(P_M) \geq 4$. For further validation and detection of candidate genes, the false discovery rate (FDR) was used to exclude false-positive associations at $P < 0.01$. Positions of highly significant associated markers were anchored to the barley physical map based on Barke × Morex recombinant inbred line POPSEQ population (Mascher et al., 2017). Mapping of available sequence DNA information for the candidate genes from other species to the barley genome was performed using the IPK barley BLAST server, Gatersleben (http://webblast.ipk-gatersleben.de/barley_ibsc/). An important QTL was defined when the highly associated SNPs colocalized physically within chromosomal segments (LD interval) including known germination-related genes.

2.4 | NO measurements

Release of NO was estimated via its main autoxidation product dinitrogen trioxide (N_2O_3; Y. Liu, Buerk, Barbee, & Jaron, 2016; Planchet & Kaiser, 2006). This assay is suitable for quantitative assessment of this unstable, volatile compound in large samples such as whole seeds and for applications requiring differential quantification of many different samples. NO release from germinating seeds was determined in 184 barley genotypes by immerging dry mature seeds in 4-amino-5-methylamino-2′,7′-difluorescein (DAF-FM; D1821, Sigma), a nonpermeable cell membrane probe, for 6 hr. Five barley seeds per replicate were immersed in 800 μl of 5 μM DAF-FM in 10 mM Tris–HCl at pH 7.4 and 25°C. One hundred microlitres of supernatant was subsequently collected in microplates (96 wells) and used to measure fluorescence intensity with the Typhoon™ FLA 9500 laser scanner (GE Healthcare; excitation, 488 nm; emission, 518 nm). For calibration, a standard curve obtained from NO released by sodium nitroprusside was used. For a final volume of 200 μl, SNP (0, 10, 25, 50, 75, and 100 μM) was mixed with 5 μM DAF-FM in 10 mM Tris–HCl at pH 7.4 and incubated at 25°C (Figure S1). The experiments were conducted in darkness to avoid degradation of the light-sensitive DAF-FM (Sechet et al., 2015). The emission of NO by sodium nitroprusside is optimal under light, but NO is also produced from sodium nitroprusside in the dark (Shishido & de Oliveira, 2001).

ImageQuant (V.8.1.0.0) analysis software was used for data processing. Data were adjusted using univariate method using R package "outliers." Data are means ± *SD* of five biological replicates per genotype grown in Field Plot 1 and Field Plot 2, respectively. Means ± *SD* of Field Plot 1 were used for correlations with agronomic and dormancy data assessed in Field Plot 1. Means ± *SD* of both field plots, Field Plots 1 and 2, were used for GWAS.

3 | RESULTS

3.1 | The population structure of the EcoSeed panel

The 184 EcoSeed genotypes clustered into four majors Q groups representing row type and breeding status (i.e., cultivar or landrace; Figure 1, Table S1). The six-row genotypes clustered into Ethiopian landraces (Q1) and cultivars (Q3) from other countries. The two-row genotypes clustered into German cultivars (Q2) and cultivars from other countries (Q4). Accordingly, the phylogenetic tree shows a close phylogenetic relationship between Ethiopian genotypes and between some two-row cultivars (e.g., HOR 2082, HOR 2112, and HOR 2171). The average LD of 4,343 mapped markers decayed rapidly within 0.5 cM (r_{LD}^2 = 0.2; Figure S2). The majority of locus pairs in complete LD were within a genetic distance of <1.0 cM and were used for comparative analysis.

3.2 | Plant and seed phenotyping

Average agronomic performance was strongly affected by the growth environment (Figure S3). In 2013, anthesis was delayed under greenhouse conditions (DTA, 70.9 days; $P < 0.001$) and plants were larger (PH, 103.9 cm) than in field conditions (DTA, 64.0 days; PH, 98.2 cm). More SpSs ($P < 0.001$) with a higher TSW ($P < 0.001$) were produced under field conditions in both years. In 2014, it rained twice as much as in 2013, and plant performance in the field differed significantly (Figure S3B). Most genotypes required less time to anthesis

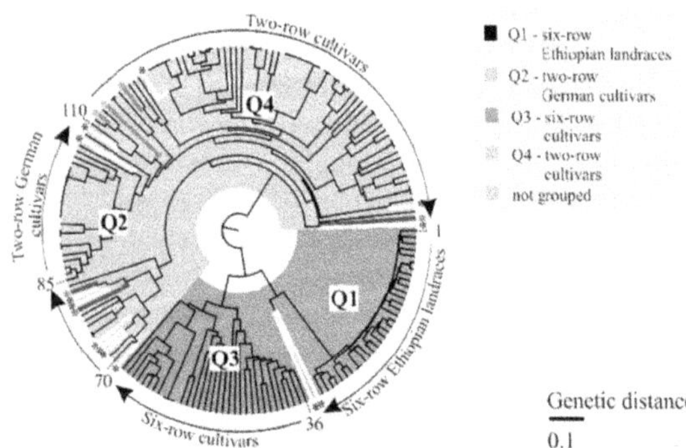

FIGURE 1 Phylogenetic tree and population structure of the EcoSeed panel. The 184 EcoSeed genotypes clustered into four Q groups. Genotypes marked with black asterisks were not assigned to any group; genotypes with red asterisks correspond to the red group. Tree numbers and details are given in Table S1

($P < 0.001$) and maturity ($P < 0.001$) but developed more SpSs ($P < 0.001$) in 2013, whereas TSW was not affected. Across all three environments (2013 and 2014 field multiplications and 2013 greenhouse multiplication), the relationship ($r = -0.31$, $P < 0.001$) between DTA and SpSs was consistent, which was confirmed by a broad-sense heritability (H^2) of 0.89 and 0.91, respectively (Table S2). In summary, strong variations in growth conditions affected plant performance, whereby DTA, PH, and SpSs expressed high heritability, indicating a strong genetic control.

All seeds were harvested at full maturity and used for dormancy, PHS, and after-ripening experiments (Figures 2 and 3, Figures S4 and S5, Table S3). Freshly harvested seeds of most genotypes showed reduced germination and PHS, compared with seeds after-ripened for 14 days in the field or seeds stratified at 10°C. In 2014, PHS was significantly lower ($P < 0.001$) than in 2013 (Table S2). When germinated at 20°C ($TG20_{0w}$), TG ranged between 0% and 100%; but when germinated at 10°C ($TG10_{0w}$), the seeds of over 120 genotypes had more than 80% germination, indicative of thermodormancy. Hardly any thermodormancy was found in most of the six-row genotypes from Ethiopia (ETH, Q1) and Sweden (SWE), and this was very useful for the subsequent quantitative genetic analysis of dormancy (Figure S4).

In most genotypes, seed dormancy was released by storage under ambient conditions (Figure S5D and F). After 6 weeks in AS ($TG20_{6w}$[AS]), the seeds of 18 six-row landraces from Q3 showed between 20% and 90% TG; but after 24 weeks ($TG20_{24w}$[AS]), seeds of almost all genotypes (181) showed TG > 90%. The DI considers the relationship between $TG10_{0w}$ and $TG20_{0w}$, and between TG at ambient (AS) and CS, expressed as DI_{0w}, DI_{6w}, and DI_{24w}. Dormancy was maintained under CS (Figure S5C and E). Across all genotypes, the DI changed from 26.7 at maturity to 14.9 after 6 weeks and 13.7 after 24 weeks of storage (Figure 2d–f), indicative of dormancy release by after-ripening.

Lack of dormancy (high TG at 20°C) and PHS at maturity (termed PHS1) were positively correlated with NO release (Figure S6). Non-dormant Ethiopian genotypes (Q1, black) had significantly higher NO release than the more dormant Q3 (red) cluster assessed in two field plots with 184 genotypes each (Figure 3c, Figure S7, Tables S3 and S4). However, genotypes from Q2 and Q4 showed a wider distribution of PHS1, with correlation coefficients for PHS1 and NO of $r = 0.30$ ($P < 0.001$) and for $TG20_{0w}$ and NO of $r = 0.44$ ($P < 0.001$; Figures S6 and S7, Table S4).

Agronomic traits correlated weakly with PHS and dormancy (Figure S6); DTA correlated positively with PHS, and with $TG10_{0w}$ and $TG20_{0w}$; and PH correlated with $TG10_{0w}$ and $TG20_{6w}$(AS). However, strong correlations existed between the dormancy traits PHS, $TG10_{0w}$ and $TG20_{0w}$, and dormancy release after 6 and 24 weeks of AS ($TG20_{6w}$[AS], $TG20_{24w}$[AS]), with high consistency across all genotypes investigated, indicative of a strong underlying genetic control.

3.3 | Marker–trait associations

GWAS of eight measured dormancy traits revealed 376 single MTAs with $-\log(P_M) \geq 4$, which were distributed among all chromosomes

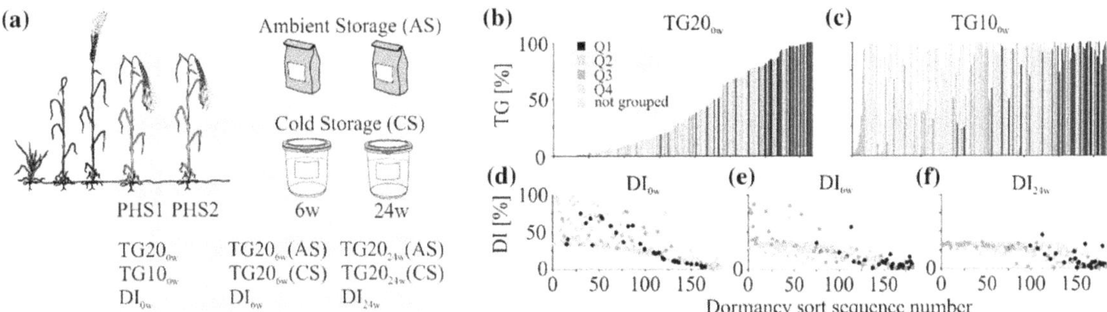

FIGURE 2 Dormancy evaluation and distribution of dormancy at full maturity and release after cold stratification and after-ripening in relation to population structure. (a) Experimental design: Dormancy was assessed by testing total germination (TG) of fully mature seeds produced in 2013 and 2014 at 20°C ($TG20_{0w}$) and 10°C ($TG10_{0w}$). Preharvest sprouting (PHS) was evaluated in sand at maturity (PHS1) and 14 days after maturity (PHS2). The effects of dry after-ripening are shown after 6 and 24 weeks in ambient conditions at 20°C and 50% relative humidity (AS) or cold storage at −18°C at 10% moisture content (CS). (b) TG at 20°C ($TG20_{0w}$) of fully mature seeds from 2014 showed various degrees of primary dormancy, which was mostly released after (c) stratification at 10°C ($TG10_{0w}$). (d) More dormant seeds had higher dormancy indexes at maturity (DI_{0w}). During after-ripening, DI decreased during storage of (e) 6 weeks (DI_{6w}) and (f) 24 weeks (DI_{24w}). The dormancy sort sequence number corresponds to barley genotypes sorted by $TG20_{0w}$ (Table S1)

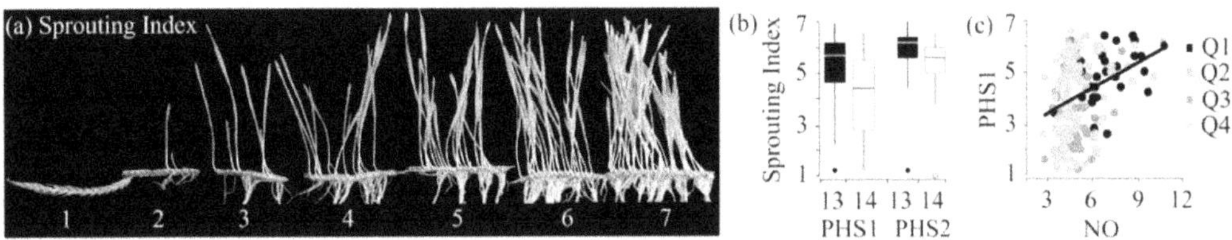

FIGURE 3 Preharvest sprouting (PHS) was higher in seeds grown under drier conditions, and PHS at maturity (PHS1) correlated positively with seed nitric oxide (NO) release. (a) Spikes that sprouted on moist sand are graded from 1 (no sprouting) to 7 (complete sprouting). (b) Box plots show differences between PHS1 and PHS 14 days after maturity (PHS2) and the years 2013 (13, black) and 2014 (14, white). (c) NO release correlated with PHS1 (r = 0.30), measured in seeds multiplied in 2014

and are indicative of a quantitatively inherited trait (Figure 4). The highest $-\log(P_M)$ for $TG20_{0w}$ was found on chromosomes 1H (SCRI_RS_60145 at 5 cM) and 3H (12_30927 at 104.3 cM and SCRI_RS_189710 at 135.6). For PHS1, the most significant marker was SCRI_RS_105688 on 5H at 169.38 cM. For traits that maintain dormancy, PHS1, $TG20_{0w}$, 6-week CS ($TG20_{6w}$[CS]) and 24-week CS ($TG20_{24w}$[CS]), 201 associations were detected, predominantly on chromosomes 2H (40), 3H (35), 4H (34), and 5H (36); but interestingly, the allelic effects of these markers were more positive and, therefore, related to dormancy release (147 of 201 MTAs; Table S5). Conversely, for traits that release dormancy, PHS1, $TG10_{0w}$ and $TG20_{6w}$(AS) and $TG20_{24w}$(AS), 175 associations were found on chromosomes 2H (45), 5H (31), and 7H (30), and the allelic effects of most of these markers were related to dormancy maintenance (163 of 175 MTAs). Therefore, it appears that these dormancy markers prepare the seeds for the reversal of their current dormancy status.

One hundred forty-eight single MTAs were found for NO release, of which 52 associations were highly significant at $-\log(P_M) > 6$; 140 markers contributed a positive allelic effect to NO release (Figure 4). Most associations were located on chromosomes 5H (55) and 2H (36) and colocated with *HvGA20ox3*, *HvCYP734A7*, *SED4*, *DSG1*, and *LOX-L2* genes, which are involved in hormone signalling and germination (Table S6). Sixty-nine markers were shared with dormancy traits (Figure S8), whereby most markers were shared with TG_{24w}(AS) (28), $TG20_{0w}$ (27), and DI_{0w} (20). The high number of shared associations was not as strongly reflected by the correlation coefficients of the phenotypes. However, the allelic effects of NO release were highly significantly correlated with the allelic effects of dormancy traits, most relevant between NO and $TG20_{0w}$ (r = 0.76, $P < 0.001$, Figure S9). Hence, it is reasonable to assume that a close genetic relationship between NO release and dormancy status exists.

PHS, germination, and NO release were evaluated during after-ripening. Interestingly, the significance levels ($-\log(P_M)$) of the markers changed during after-ripening (grey arrows in Figure 4). The most significant markers for PHS1 and for DI_{6w} were located on 5H, at 169.4 cM (SCRI_RS_105688) and 168.9 cM (11_20402), respectively, and closely located to the *HvGA20ox1* gene (Table S6). Here, the $-\log(P_M)$ for DI changed from 3.5 to 3.0 and 6.5 within 0, 6, and 24 weeks, respectively, and contributed to dormancy release with an allelic effect of up to −6.5% (a lower DI represents less dormant seeds). This is in agreement with the MTA for NO release at this locus and supports after-ripening. The most significant MTA for dormancy release was found on 4H (12_31139) and maintains dormancy by an allelic effect of up to DI = +10.3. The decreasing $-\log(P_M)$ for DI from

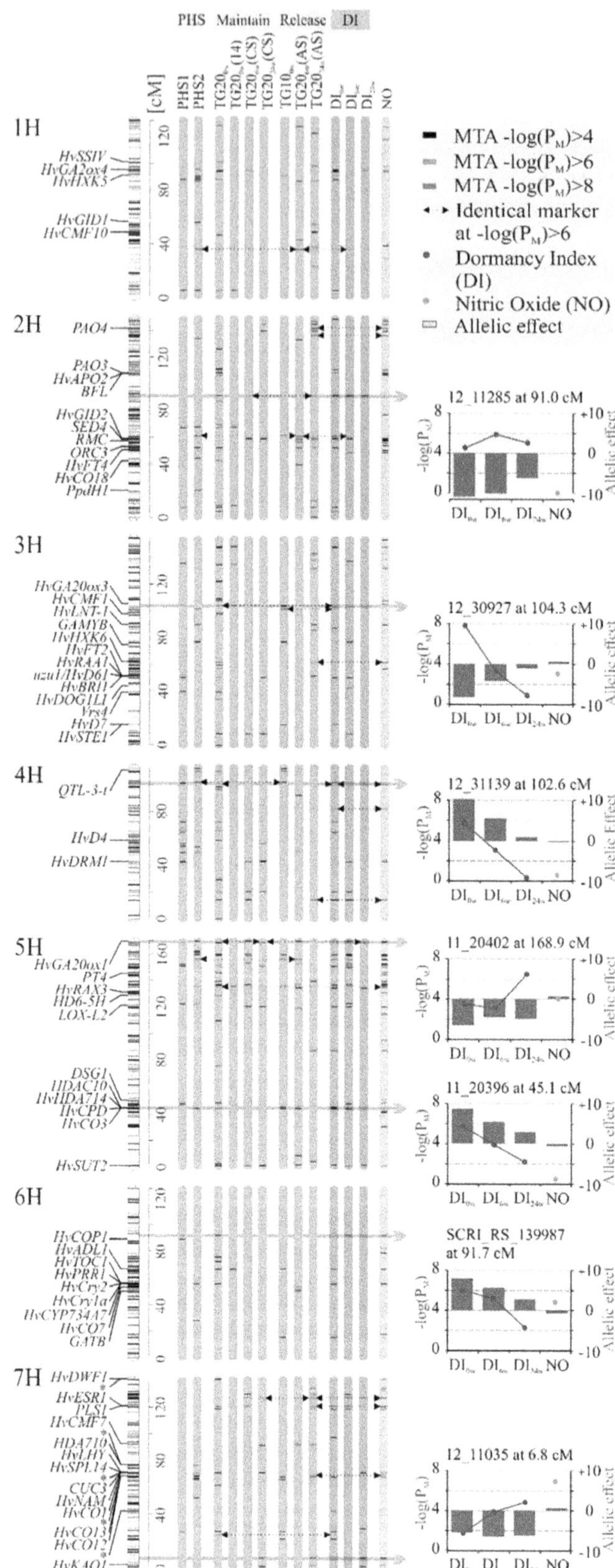

FIGURE 4 Marker–trait associations (MTAs) for preharvest sprouting (PHS), seed dormancy, and nitric oxide (NO) release. Chromosomal positions of 4,343 markers and linked genes for 376 MTAs are shown. Grey arrows: markers of which $-\log(P_M)$ for DI (green) and for NO release (pink) and the allelic effect (grey bars) changed upon storage. Abbreviations for traits (top line): PHS1 and PHS2, PHS at maturity and 14 days after maturity, respectively; TG10 or TG20 denote total germination at 10 or 20°C; subscript numbers give storage times in weeks (e.g., $TG10_{6w}$), and storage conditions (AS and CS for ambient and cold storage) are in brackets. Dormancy index was calculated at maturity (DI_{0w}) after 6 weeks (DI_{6w}) and 24 weeks of after-ripening (DI_{24w}). *Genes are listed in Table S6

5.8 and 3.1 to 0.3 within 0, 6, and 24 weeks is in agreement with dormancy release by after-ripening. In this case, NO release was not significantly associated with this locus.

Approximately 2,000 associations were detected for agronomical traits, but only few markers were shared with dormancy traits (Table S5 and Figure S8). This high number of associations is most likely due to the dominant effects of row type (two-row barley seeds are generally larger than seeds of six-row barleys). Therefore, a large number of overlaps were found for DTA (251), TSW (369), and SpSs (1074). Across all agronomic traits, most significant markers were found on 2H (412) and 7H (441) in close vicinity to the genes *PpD-H1* (2H, 19.9 cM) and *Vrs1* (2H, 79.5 cM, Table S4). In accordance with the phenotypic data, DTA shared only seven of 251 loci with PHS1 and 15 loci with DI_{OW}. Similarly, TSW shared only four of 369 markers with PHS1, which indicates that the loci responsible for agronomic traits do not influence dormancy status.

4 | DISCUSSION

The EcoSeed spring barley panel comprises 184 genotypes with dormancy levels at maturity ranging from nondormant to dormant. Most genotypes expressed nondeep physiological dormancy (Baskin & Baskin, 2004), which was broken by cold stratification or after-ripening in dry storage. Dormancy was negatively correlated with PHS. Almost all of the six-row Ethiopian landraces used were nondormant at maturity and released significantly more NO, which was useful for quantitative mapping of loci involved in the regulation of dormancy.

The GWAS of dormancy traits revealed several associated genes related to heading, widely discussed as flowering time in *Arabidopsis* (Alqudah & Schnurbusch, 2017), for example, *HvFT2, HvFT4, GAMYB, PpdH1, Vrn-H3, HvCMF1, HvCMF1, HvCMF10, HvCO1, HvCO3, HvCO7, HvCO8, HvCO12, HvCO13,* and *HvCO18*. *Arabidopsis* seeds produced by plants that flowered below 14°C were dormant, whereas those from plants flowering above 15°C were nondormant, which indicates that dormancy is partly controlled via flowering time genes (Springthorpe & Penfield, 2015). The *Arabidopsis FT* gene is expressed in the silique phloem, where temperature signals are transduced, and controls seed dormancy through inhibition of proanthocyanidin synthesis in fruits, resulting in altered seed coat tannin content (Chen et al., 2014). The synthesis of proanthocyanidin is additionally controlled, among others, by the MYB TF (Debeaujon, Lepiniec, Pourcel, & Routaboul, 2007; Nesi, Jond, Debeaujon, Caboche, & Lepiniec, 2001; Verdier et al., 2012). However, unlike in *Arabidopsis* (Chen et al., 2014), we did not find that late flowering induces higher dormancy in barley. Barley displays a coat-imposed dormancy based on the characteristics of hull, lemma, and palea (Benech-Arnold, Giallorenzi, Frank, & Rodriguez, 1999). In the Poaceae, testa, pericarp, and hulls accumulate various phenolic compounds (Rodriguez et al., 2015), some of which have strong inhibitory effects on germination (Krogmeier & Bremner, 1989). We identified the *Hvmyb10* gene, which in *Arabidopsis* encodes a R2R3 MYB domain protein that is involved in anthocyanin and proanthocyanidin syntheses, and higher levels lead to dormancy maintenance (Himi et al., 2012). In barley, an anthocyanin regulatory protein is located on 2H close to the dormancy-associated marker 12_11285 (91.0 cM, Figure 4) and was highly significant from maturity throughout after-ripening; and, against our expectations, the allelic effect indicated a role in dormancy release rather than maintenance.

The *Arabidopsis DOG1* gene is seed-specifically expressed (Bentsink, Jowett, Hanhart, & Koornneef, 2006); it is essential for dormancy induction (Dekkers & Bentsink, 2015) and contributes to natural variation in seed dormancy (Alonso-Blanco et al., 2009; Chiang et al., 2011). The barley homolog of *DOG1*, *HvDOG1L1*, was sequenced previously (Ashikawa et al., 2014), and we annotated it in the centromere region of 3H (44.3 cM). This was significantly ($-\log(P_M) > 4.0$) associated only with DI_{6w}, which contributed a positive effect of DI = +8.4, indicative of dormancy maintenance. Therefore, *DOG1* homologs appear to play minor roles in barley.

Our GWAS revealed that barley seed dormancy is intricately regulated, as shown by 201 single MTAs that maintain dormancy and 175 MTAs that release dormancy, and colocalize with 71 known genes. By linkage mapping analysis, four were validated as major dormancy loci on 4H, 5H, and 7H (Han et al., 1996). The QTLs in the centromere (SD1) and in the telomeric region of the long arm (SD2) of 5H are considered as the most important dormancy QTLs (Lohwasser, Arif, & Börner, 2013; Romagosa, Han, Clancy, & Ullrich, 1999; Sato, Matsumoto, Ooe, & Takeda, 2009). In agreement with Prada et al. (2004), our finding shows that SD1 has an effect on dormancy maintenance (11_20396; positive effect of DI = +8.8), whereas breaking of dormancy is mainly regulated by SD2 (11_20402, negative effect of DI = −6.6). The gene *MKK3* (mitogen-activated protein kinase kinase 3) is directly located at this telomeric region in barley. *MKK3* is a key signalling module for plant growth and development processes (Xu & Zhang, 2015) and was identified as a strong candidate for barley seed dormancy (Nakamura et al., 2016) and for PHS in an orthologue region of the wheat chromosome 4A (Torada et al., 2016).

We found a significant MTA in the SD2 region, colocated with the *gibberellin (GA) 20-oxidase* (*HvGA20ox1*) gene, which is also involved in seed dormancy release in barley. GA oxidases (GAoxs) are required for GA synthesis, whereby GA2 oxidase and GA20 oxidases play key roles in the regulation of bioactive GA levels and are differentially controlled by developmental and environmental cues (Huang, Wang, Ge, & Rao, 2015). Furthermore, the *HvGA2ox4* and *HvGA20ox3* genes were significantly associated with dormancy, located on 1H and 3H, respectively. *HvGA20ox* genes are best known for their effects on PH; mutations are found in dwarf and semidwarf phenotypes (Xu et al., 2017), and *HvGA20ox* genes apparently also strongly influence dormancy release. A strong relationship between PH and dormancy was not found. However, Hoang et al. (2013) demonstrated that the transcript abundance of *HvGA20ox1* remained low when secondary dormancy was induced but increased during stratification. In agreement with this finding, it was proposed that H_2O_2 induces *HvGA20ox1* expression (Bahin et al., 2011) and that H_2O_2 release may be triggered by *Polyamine Oxidase (PAO) 3* and *PAO4* genes, known to be key elements for the oxidative burst (Yoda, Hiroi, & Sano, 2006). Hence, ROS may affect GA signalling and control the dormancy status via GAoxs.

NO was highly abundant in nondormant genotypes and shared 69 loci with dormancy traits, indicative of a key role in dormancy alleviation. NO is generated from nitrate after imbibition by either nitrite reductase (Ma, Marsolais, Bykova, & Igamberdievi, 2016) or nonenzymatically in the aleuron cells (Bethke et al., 2007). NO is also stimulating both ethylene synthesis and signalling pathway (Arc, Sechet, Corbineau, Rajjou, & Marion-Poll, 2013). In *Arabidopsis* seeds, NO promotes ABA catabolism (Liu et al., 2009) and is required for the transcription of *GA3ox1* and *GA3ox2*, biosynthetic enzymes required for GA production. In our GWAS, NO release was associated with *HvGA2ox4*, *HvGA20ox3*, and *HvGA20ox1*, which play, together with cytochrome P450 (*HvCYP734A7*), essential roles in GA synthesis. The modulation of the hormonal balance by NO in combination with its involvement in oxidative processes, shown by its association with the *LOX-L2* gene and other genes, for example, *Enhancer of Shoot Regeneration* (*HvESR1*), suggests that NO is involved in molecular switchboards that control dormancy alleviation and seedling development.

In conclusion, we identified new loci for barley dormancy on 1H (5 cM) and 3H (104.3 and 135.6 cM) and for PHS on 5H (169.4 cM) and some of the responsible genes (e.g., *HvGA2ox4*, *HvGA20ox1*, *LOX-L2*, *HvCYP734A7*, and *HvDOG1LI*). We also annotated the *DOG1* homolog on the barley chromosome 3H (*HvDOG1L1*), which, unlike in *Arabidopsis*, does not seem to play a major role in barley dormancy. On the other hand, the *HvGA20ox1* gene clearly contributed to dormancy alleviation. We also showed for seven important loci on chromosomes 2H (91.0 cM), 3H (104.3 cM), 4H (102.6 cM), 5H (45.1 and 168.9 cM), 6H (91.7 cM), and 7H (6.8 cM) that the significance for dormancy and NO changes upon dormancy release by after-ripening and provided evidence for a pivotal role of NO signalling and genes related to flowering time in the regulation of dormancy. In summary, we present novel loci and genes that orchestrate dormancy alleviation. This is expected to be useful for further elucidating oxidative and hormone signalling pathways of NO synthesis and release modulating crop dormancy in the field, and for detailed functional analysis of contrasting genotypes grown under different environmental conditions.

ACKNOWLEDGEMENTS

We gratefully acknowledge support by Stefanie Thumm, Annett Marlow, Peter Schreiber, Jürgen Marlow, Michael Grau, and Karin Lipfert for excellent technical support and Erwann Arc and Gerald Kastberger for critical comments on the draft manuscript. Financial support was provided by the EU (FP7 Grant 311840 "EcoSeed" (Impacts of Envirnmental Conditions on Seed Quality)) and by Saclay Plant Sciences (SPS) LabEx supporting technical platforms within the Jean-Pierre Bourgin Institute (ANR-10-LABX-0040-SPS).

ORCID

Manuela Nagel https://orcid.org/0000-0003-0396-0333
Ahmad M. Alqudah https://orcid.org/0000-0002-0436-9724
Loïc Rajjou https://orcid.org/0000-0001-9739-1041
Sibylle Pistrick https://orcid.org/0000-0002-0325-0609
Andreas Börner https://orcid.org/0000-0003-3301-9026
Ilse Kranner https://orcid.org/0000-0003-4959-9109

REFERENCES

Albertos, P., Romero-Puertas, M. C., Tatematsu, K., Mateos, I., Sánchez-Vicente, I., Nambara, E., & Lorenzo, O. (2015). S-nitrosylation triggers ABI5 degradation to promote seed germination and seedling growth. *Nature Communications*, *6*, 8669. https://doi.org/10.1038/ncomms9669

Alonso-Blanco, C., Aarts, M. G. M., Bentsink, L., Keurentjes, J. J. B., Reymond, M., Vreugdenhil, D., & Koornneef, M. (2009). What has natural variation taught us about plant development, physiology, and adaptation? *Plant Cell*, *21*, 1877–1896. https://doi.org/10.1105/tpc.109.068114

Alonso-Blanco, C., Bentsink, L., Hanhart, C. J., Vries, H. B. E., & Koornneef, M. (2003). Analysis of natural allelic variation at seed dormancy loci of *Arabidopsis thaliana*. *Genetics*, *164*, 711–729.

Alqudah, A. M., & Schnurbusch, T. (2017). Heading date is not flowering time in spring barley. *Frontiers in Plant Science*, *8*. https://doi.org/10.3389/fpls.2017.00896

Arc, E., Galland, M., Godin, B., Cueff, G., & Rajjou, L. (2013). Nitric oxide implication in the control of seed dormancy and germination. *Frontiers in Plant Science*, *4*, 346.

Arc, E., Sechet, J., Corbineau, F., Rajjou, L., & Marion-Poll, A. (2013). ABA crosstalk with ethylene and nitric oxide in seed dormancy and germination. *Frontiers in Plant Science*, *4*, 63. https://doi.org/10.3389/fpls.2013.00063

Ashikawa, I., Mori, M., Nakamura, S., & Abe, F. (2014). A transgenic approach to controlling wheat seed dormancy level by using Triticeae *DOG1*-like genes. *Transgenic Research*, *23*, 621–629. https://doi.org/10.1007/s11248-014-9800-5

Bahin, E., Bailly, C., Sotta, B., Kranner, I., Corbineau, F., & Leymarie, J. (2011). Crosstalk between reactive oxygen species and hormonal signalling pathways regulates grain dormancy in barley. *Plant Cell and Environment*, *34*, 980–993. https://doi.org/10.1111/j.1365-3040.2011.02298.x

Bailly, C., El-Maarouf-Bouteau, H., & Corbineau, F. (2008). From intracellular signaling networks to cell death: The dual role of reactive oxygen species in seed physiology. *Comptes Rendus Biologies*, *331*, 806–814. https://doi.org/10.1016/j.crvi.2008.07.022

Baskin, J. M., & Baskin, C. C. (2004). A classification system for seed dormancy. *Seed Science Research*, *14*, 1–16.

Benech-Arnold, R. L., Giallorenzi, M. C., Frank, J., & Rodriguez, V. (1999). Termination of hull-imposed dormancy in developing barley grains is correlated with changes in embryonic ABA levels and sensitivity. *Seed Science Research*, *9*, 39–47. https://doi.org/10.1017/S0960258599000045

Bentsink, L., Jowett, J., Hanhart, C. J., & Koornneef, M. (2006). Cloning of *DOG1*, a quantitative trait locus controlling seed dormancy in *Arabidopsis*. *Proceedings of the National Academy of Sciences of the United States of America*, *103*, 17042–17047. https://doi.org/10.1073/pnas.0607877103

Bethke, P. C., Libourel, I. G. L., Aoyama, N., Chung, Y.-Y., Still, D. W., & Jones, R. L. (2007). The *Arabidopsis* aleurone layer responds to nitric oxide, gibberellin, and abscisic acid and is sufficient and necessary for seed dormancy. *Plant Physiology*, *143*, 1173–1188. https://doi.org/10.1104/pp.106.093435

Chen, M., MacGregor, D. R., Dave, A., Florance, H., Moore, K., Paszkiewicz, K., ... Penfield, S. (2014). Maternal temperature history activates *Flowering Locus T* in fruits to control progeny dormancy according to time of year. *Proceedings of the National Academy of Sciences of the United States of America*, *111*, 18787–18792. https://doi.org/10.1073/pnas.1412274111

Chiang, G. C. K., Bartsch, M., Barua, D., Nakabayashi, K., Debieu, M., Kronholm, I., ... de Meaux, J. (2011). *DOG1* expression is predicted by the seed-maturation environment and contributes to geographical variation in germination in *Arabidopsis thaliana*. *Molecular Ecology*, *20*, 3336–3349. https://doi.org/10.1111/j.1365-294X.2011.05181.x

Comadran, J., Kilian, B., Russell, J., Ramsay, L., Stein, N., Ganal, M., ... Waugh, R. (2012). Natural variation in a homolog of *Antirrhinum CENTRORADIALIS* contributed to spring growth habit and environmental adaptation in cultivated barley. *Nature Genetics*, *44*, 1388–1392. https://doi.org/10.1038/ng.2447

Debeaujon, I., Lepiniec, L., Pourcel, L., & Routaboul, J. M. (2007). Seed coat development and dormancy. In K. Bradford, & H. Nonogaki (Eds.), *Annual plant reviews, seed development, dormancy and germination* (pp. 25–49). Wiley-Blackwell. https://doi.org/10.1002/9780470988848.ch2

Dekkers, B. J. W., & Bentsink, L. (2015). Regulation of seed dormancy by abscisic acid and *DELAY OF GERMINATION 1*. *Seed Science Research*, *25*, 82–98. https://doi.org/10.1017/S0960258514000415

Domingos, P., Prado, A. M., Wong, A., Gehring, C., & Feijo, J. A. (2015). Nitric oxide: A multitasked signaling gas in plants. *Molecular Plant*, *8*, 506–520. https://doi.org/10.1016/j.molp.2014.12.010

Felsenstein, J. (2009) PHYLIP 3.69 (Phylogeny Inference Package), Washington.

Finch-Savage, W. E., & Footitt, S. (2017). Seed dormancy cycling and the regulation of dormancy mechanisms to time germination in variable field environments. *Journal of Experimental Botany*, *68*, 843–856. https://doi.org/10.1093/jxb/erw477

Finch-Savage, W. E., & Leubner-Metzger, G. (2006). Seed dormancy and the control of germination. *New Phytologist*, *171*, 501–523. https://doi.org/10.1111/j.1469-8137.2006.01787.x

Gibbs, D. J., Conde, J. V., Berckhan, S., Prasad, G., Mendiondo, G. M., & Holdsworth, M. J. (2015). Group VII ethylene response factors coordinate oxygen and nitric oxide signal transduction and stress responses in plants. *Plant Physiology*, *169*, 23–31. https://doi.org/10.1104/pp.15.00338

Gibbs, D. J., Md Isa, N., Movahedi, M., Lozano-Juste, J., Mendiondo Guillermina, M., Berckhan, S., ... Holdsworth Michael, J. (2014). Nitric oxide sensing in plants is mediated by proteolytic control of group VII ERF transcription factors. *Molecular Cell*, *53*, 369–379. https://doi.org/10.1016/j.molcel.2013.12.020

Gong, X., Li, C. D., Zhou, M. X., Bonnardeaux, Y., & Yan, G. J. (2014). Seed dormancy in barley is dictated by genetics, environments and their interactions. *Euphytica*, *197*, 355–368. https://doi.org/10.1007/s10681-014-1072-x

Han, F., Ullrich, S. E., Clancy, J. A., Jitkov, V., Kilian, A., & Romagosa, I. (1996). Verification of barley seed dormancy loci via linked molecular markers. *Theoretical and Applied Genetics*, *92*, 87–91. https://doi.org/10.1007/BF00222956

He, H. Z., Willems, L. A. J., Batushansky, A., Fait, A., Hanson, J., Nijveen, H., ... Bentsink, L. (2016). Effects of parental temperature and nitrate on seed performance are reflected by partly overlapping genetic and metabolic pathways. *Plant and Cell Physiology*, *57*, 473–487. https://doi.org/10.1093/pcp/pcv207

Himi, E., Yamashita, Y., Haruyama, N., Yanagisawa, T., Maekawa, M., & Taketa, S. (2012). *Ant28* gene for proanthocyanidin synthesis encoding the R2R3 MYB domain protein (Hvmyb10) highly affects grain dormancy in barley. *Euphytica*, *188*, 141–151. https://doi.org/10.1007/s10681-011-0552-5

Hoang, H. H., Sotta, B., Gendreau, E., Bailly, C., Leymarie, J., & Corbineau, F. (2013). Water content: a key factor of the induction of secondary dormancy in barley grains as related to ABA metabolism. *Physiologia Plantarum*, *148*, 284–296. https://doi.org/10.1111/j.1399-3054.2012.01710.x

Huang, Y., Wang, X., Ge, S., & Rao, G.-Y. (2015). Divergence and adaptive evolution of the gibberellin oxidase genes in plants. *BMC Evolutionary Biology*, *15*, 207. https://doi.org/10.1186/s12862-015-0490-2

Ishibashi, Y., Aoki, N., Kasa, S., Sakamoto, M., Kai, K., Tomokiyo, R., ... Iwaya-Inoue, M. (2017). The interrelationship between abscisic acid and reactive oxygen species plays a key role in barley seed dormancy and germination. *Frontiers in Plant Science*, *8*. https://doi.org/10.3389/fpls.2017.00275

Krogmeier, M. J., & Bremner, J. M. (1989). Effects of phenolic-acids on seed-germination and seedling growth in soil. *Biology and Fertility of Soils*, *8*, 116–122.

Li, C. D., Tarr, A., Lance, R. C. M., Harasymow, S., Uhlmann, J., Westcot, S., ... Appelsa, R. (2003). A major QTL controlling seed dormancy and preharvest sprouting/grain alpha-amylase in two-rowed barley (*Hordeum vulgare* L.). *Australian Journal of Agricultural Research*, *54*, 1303–1313. https://doi.org/10.1071/AR02210

Liu, X. D., Zhang, H., Zhao, Y., Feng, Z. Y., Li, Q., Yang, H. Q., ... He, Z. H. (2013). Auxin controls seed dormancy through stimulation of abscisic acid signaling by inducing ARF-mediated ABI3 activation in Arabidopsis. *Proceedings of the National Academy of Sciences of the United States of America*, *110*, 15485–15490. https://doi.org/10.1073/pnas.1304651110

Liu, Y., Buerk, D. G., Barbee, K. A., & Jaron, D. (2016). A mathematical model for the role of N_2O_3 in enhancing nitric oxide bioavailability following nitrite infusion. *Nitric Oxide: Biology and Chemistry*, *60*, 1–9. https://doi.org/10.1016/j.niox.2016.08.003

Liu, Y., & El-Kassaby, Y. A. (2017). Regulatory crosstalk between microRNAs and hormone signalling cascades controls the variation on seed dormancy phenotype at *Arabidopsis thaliana* seed set. *Plant Cell Reports*, *36*, 705–717. https://doi.org/10.1007/s00299-017-2111-6

Liu, Y., Shi, L., Ye, N., Liu, R., Jia, W., & Zhang, J. (2009). Nitric oxide-induced rapid decrease of abscisic acid concentration is required in breaking seed dormancy in Arabidopsis. *New Phytologist*, *183*, 1030–1042. https://doi.org/10.1111/j.1469-8137.2009.02899.x

Lohwasser, U., Arif, M. A. R., & Börner, A. (2013). Discovery of loci determining pre-harvest sprouting and dormancy in wheat and barley applying segregation and association mapping. *Biologia Plantarum*, *57*, 663–674. https://doi.org/10.1007/s10535-013-0332-2

Ma, Z., Marsolais, F., Bykova, N. V., & Igamberdievi, A. U. (2016). Nitric oxide and reactive oxygen species mediate metabolic changes in barley seed embryo during germination. *Frontiers in Plant Science*, *7*. https://doi.org/10.3389/fpls.2016.00138

Mascher, M., Gundlach, H., Himmelbach, A., Beier, S., Twardziok, S. O., Wicker, T., ... Stein, N. (2017). A chromosome conformation capture ordered sequence of the barley genome. *Nature*, *544*, 426–433.

Nakamura, S., Pourkheirandish, M., Morishige, H., Kubo, Y., Nakamura, M., Ichimura, K., ... Komatsuda, T. (2016). Mitogen-activated protein kinase kinase 3 regulates seed dormancy in barley. *Current Biology*, *26*, 775–781. https://doi.org/10.1016/j.cub.2016.01.024

Nesi, N., Jond, C., Debeaujon, I., Caboche, M., & Lepiniec, L. (2001). The *Arabidopsis TT2* gene encodes an R2R3 MYB domain protein that acts as a key determinant for proanthocyanidin accumulation in developing seed. *Plant Cell*, *13*, 2099–2114.

Oracz, K., El-Maarouf-Bouteau, H., Farrant, J. M., Cooper, K., Belghazi, M., Job, C., ... Bailly, C. (2007). ROS production and protein oxidation as a novel mechanism for seed dormancy alleviation. *The Plant Journal*, *50*, 452–465. https://doi.org/10.1111/j.1365-313X.2007.03063.x

Planchet, E., & Kaiser, W. M. (2006). Nitric oxide (NO) detection by DAF fluorescence and chemiluminescence: a comparison using abiotic and biotic NO sources. *Journal of Experimental Botany*, *57*, 3043–3055. https://doi.org/10.1093/jxb/erl070

Prada, D., Ullrich, S. E., Molina-Cano, J. L., Cistue, L., Clancy, J. A., & Romagosa, I. (2004). Genetic control of dormancy in a Triumph/Morex cross in barley. *Theoretical and Applied Genetics*, *109*, 62–70. https://doi.org/10.1007/s00122-004-1608-x

Pritchard, J. K., Stephens, M., & Donnelly, P. J. (2000). Inference of population structure using multilocus genotype data. *Genetics*, *155*, 945–959.

Rodriguez, M. V., Barrero, J. M., Corbineau, F., Gubler, F., & Benech-Arnold, R. L. (2015). Dormancy in cereals (not too much, not so little): About the mechanisms behind this trait. *Seed Science Research*, *25*, 99–119. https://doi.org/10.1017/S0960258515000021

Rodriguez, M. V., Margineda, M., Gonzalez-Martin, J. F., Insausti, P., & Benech-Arnold, R. L. (2001). Predicting preharvest sprouting susceptibility in barley: A model based on temperature during grain filling. *Agronomy Journal*, *93*, 1071–1079. https://doi.org/10.2134/agronj2001.9351071x

Romagosa, I., Han, F., Clancy, J. A., & Ullrich, S. E. (1999). Individual locus effects on dormancy during seed development and after ripening in barley. *Crop Science*, *39*, 74–79.

Sato, K., Matsumoto, T., Ooe, N., & Takeda, K. (2009). Genetic analysis of seed dormancy QTL in barley. *Breeding Science*, *59*, 645–650. https://doi.org/10.1270/jsbbs.59.645

Schmidt, K., Desch, W., Klatt, P., Kukovetz, W. R., & Mayer, B. (1997). Release of nitric oxide from donors with known half-life: A mathematical model for calculating nitric oxide concentrations in aerobic solutions. *Naunyn-Schmiedebergs Archives of Pharmacology*, *355*, 457–462. https://doi.org/10.1007/PL00004969

Sechet, J., Roux, C., Plessis, A., Effroy, D., Frey, A., Perreau, F., ... Marion-Poll, A. (2015). The ABA-deficiency suppressor locus HAS2 encodes the PPR protein LOI1/MEF11 involved in mitochondrial RNA editing. *Molecular Plant*, *8*, 644–656. https://doi.org/10.1016/j.molp.2014.12.005

Shishido, S. M., & de Oliveira, M. G. (2001). Photosensitivity of aqueous sodium nitroprusside solutions: Nitric oxide release versus cyanide toxicity. *Progress in Reaction Kinetics and Mechanism*, *26*, 239–261. https://doi.org/10.3184/007967401103165271

Shu, K., Liu, X. D., Xie, Q., & He, Z. H. (2016). Two faces of one seed: Hormonal regulation of dormancy and germination. *Molecular Plant*, *9*, 34–45. https://doi.org/10.1016/j.molp.2015.08.010

Shu, K., Meng, Y. J., Shuai, H. W., Liu, W. G., Du, J. B., Liu, J., & Yang, W. Y. (2015). Dormancy and germination: How does the crop seed decide? *Plant Biology*, *17*, 1104–1112. https://doi.org/10.1111/plb.12356

Springthorpe, V., & Penfield, S. (2015). Flowering time and seed dormancy control use external coincidence to generate life history strategy. *eLife*, *4*, e05557. https://doi.org/10.7554/eLife.05557

Stone, S. L., Williams, L. A., Farmer, L. M., Vierstra, R. D., & Callis, J. (2006). KEEP ON GOING, a RING E3 ligase essential for *Arabidopsis* growth and development, is involved in abscisic acid signaling. *Plant Cell*, *18*, 3415–3428. https://doi.org/10.1105/tpc.106.046532

Torada, A., Koike, M., Ogawa, T., Takenouchi, Y., Tadamura, K., Wu, J. Z., ... Ogihara, Y. (2016). A causal gene for seed dormancy on wheat chromosome 4A encodes a MAP kinase kinase. *Current Biology*, *26*, 782–787. https://doi.org/10.1016/j.cub.2016.01.063

Verdier, J., Zhao, J., Torres-Jerez, I., Ge, S., Liu, C., He, X., ... Udvardi, M. K. (2012). MtPAR MYB transcription factor acts as an on switch for proanthocyanidin biosynthesis in *Medicago truncatula*. *Proceedings of the National Academy of Sciences of the United States of America*, *109*, 1766–1771. https://doi.org/10.1073/pnas.1120916109

Xu, J., & Zhang, S. (2015). Mitogen-activated protein kinase cascades in signaling plant growth and development. *Trends in Plant Science*, *20*, 56–64. https://doi.org/10.1016/j.tplants.2014.10.001

Xu, Y., Jia, Q., Zhou, G., Zhang, X.-Q., Angessa, T., Broughton, S., ... Li, C. (2017). Characterization of the *sdw1* semi-dwarf gene in barley. *BMC Plant Biology*, *17*, 11. https://doi.org/10.1186/s12870-016-0964-4

Yoda, H., Hiroi, Y., & Sano, H. (2006). Polyamine oxidase is one of the key elements for oxidative burst to induce programmed cell death in tobacco cultured cells. *Plant Physiology*, *142*, 193–206. https://doi.org/10.1104/pp.106.080515

SUPPORTING INFORMATION

Additional supporting information may be found online in the Supporting Information section at the end of the article.

How to cite this article: Nagel M, Alqudah AM, Bailly M, et al. Novel loci and a role for nitric oxide for seed dormancy and preharvest sprouting in barley. *Plant Cell Environ*. 2019;42:1318–1327. https://doi.org/10.1111/pce.13483

4 EPILOG

4.1 Seed deterioration processes are determined by the storage conditions

Seed storability distinguishes between species stored under comparable temperatures and seed MC. Under ambient storage conditions at 20 °C and 50 % RH, the time required to reduce germination to 50 % (P50) was 8.6 years across 18 selected crop species. Seeds of barley, Brassica (*Brassica oleracea* L., a progenitor of oilseed rape, C genome) and wheat had an average half-life of 9.2, 7.3 and 7.2 years, respectively (Nagel and Börner, 2010) **[PAPER 1]**. When seeds were stored at -18 °C and between 3.5 % and 9.5 % MC (dependent on the species) at the NPGS in Fort Collins, USA, P50 increased to 84, 54 and 25 years for seeds of barley, wheat and oilseed rape, respectively. Across 276 species, the median P50 was 54 years. Extrapolated estimates of P50 indicated that seeds of some species can exceed more than 200 years. Based on preceding germination tests of 1,727 accessions, P50 value of lentil showed that seeds may even survive more than 365 years under cold storage conditions (Walters et al., 2005). The large P50 seems to be achievable since a date palm (*Phoenix dactylifera* L.) seed germinated after 2,000 years of storage (Sallon et al., 2008).

Seed longevity is a function of the deterioration speed (σ, slope) determined by the seed MC, storage temperature, species specific parameters and the initial seed quality (K_i) (Roberts, 1973). According to 'Harrington's rule of thumbs', seed's shelf life is doubled when either MC decreases by 1 % or temperature drops by 5 °C (Harrington, 1963). Therefore, when seed MC and storage temperature are increased, the storage behaviour of seeds can be analysed within a reasonable time frame. Thereby, the P50 of wheat seeds can be decreased to 23 days if seeds are exposed to CD at 45 °C and 60 % RH. Seeds aged under AA at 45 °C and free available water (>95 % RH) show a further reduction of P50 to around 2 days (Figure 10) (Kew, 2018). However, comparisons of germination results of 89 lettuce genotypes stored in conventional storage (9 °C and 30 % RH) and under CD (25 °C and 75 % RH) revealed that correlations are rather low $R^2<0.1$ (Schwember and Bradford, 2010). Also differences in marker-trait associations (MTAs) between artificial ageing and long-term storage in wheat (Rehman Arif et al., 2017) **[PAPER 12]**, barley (Nagel et al., 2015) and oilseed rape (Nagel et al., 2011) **[PAPER 8]** indicate that different seed deterioration mechanisms are responsible for the seed viability loss.

Water activity determines physical and chemical reactions in seeds. During seed maturation, when seed dry to about 0.06–0.12 g H_2O g^{-1} DW at 20 °C the seed cytoplasm vitrifies (Buitink et al., 1999; Ballesteros and Walters, 2011) and metabolic activities cease (Fernandez-Marin et al., 2013). In the glassy state, molecular motions are restricted to the short side chains and biochemical processes are likely between α- and β-relaxation. When pea (*Pisum sativum* L.) seeds are exposed to 0.15 g H_2O g^{-1} DW and more than 35 °C, the cytoplasm undergoes α- and β-relaxation and molecular motions occur also in the main side chains (Ballesteros and Walters, 2011). The cytoplasm of the seed will pass T_g and becomes rubbery or liquid (Buitink and Leprince, 2004). However, these conditions reflect a complete different environment in comparison to the glassy state (Ballesteros and Walters, 2011). By comparison of seeds stored under CS, AS, CD at 14 % MC (CD14) and 18 % MC (CD18) at 45 °C, differences in the contents of GSSG, tocochromanols, the appearance organic radicals and the pH could be shown in wheat (Nagel et al., 2019b) **[PAPER 4]** and barley (Nagel et al., 2015; Roach et al., 2018) **[PAPER 3]**. With increasing WC, the water acts as a solvent and lead to increased oxidation rates and the occurrence of non-enzymatic browning (Figure 14). Furthermore, enzymatic activity becomes more likely (Labuza et al., 1971). Therefore, the differences in the response of the antioxidants, organic radicals and pH are assumed to be a consequence of the increased water activity.

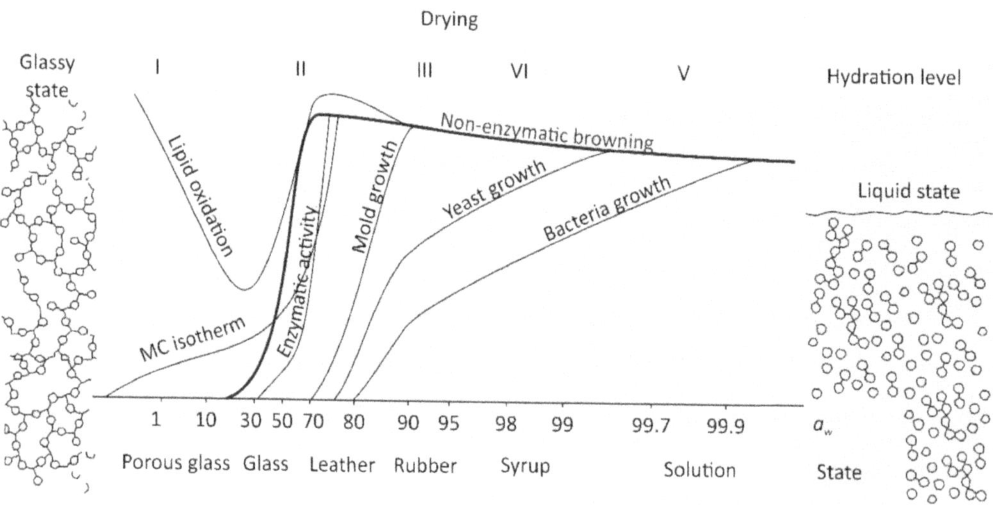

Figure 14. Relationship between reaction rate, moisture content and water activities applied to seeds. a_w, water activity Based on Labuza et al. (1971).

An increase of storage temperature and MC applied to seeds during CD or AA rises the water activity and may facilitate the resumption of the seed metabolism in combination with detrimental presumably oxidative processes. Interestingly, an increase of seed MC from 0.04 to 0.6 g H_2O g^{-1} DW at 20 °C led to the production of ATP and improved gradually the adenylate pool of sunflower (*Helianthus annuus* L.) seeds (Kibinza et al., 2006). These enzymatic and metabolic activities facilitate the production ROS required for energy transfer and signalling. The combination of high MC and high temperatures lead to mitochondrial dysfunctions and a depletion of the adenylate pool which is halved when seed viability is lost after 7 days (El-Maarouf-Bouteau et al., 2011). However, tocochromanols responsible for scavenging of ROS were not affected or tended to be produced in barley seeds under CD14 or CD18 (Nagel et al., 2015; Roach et al., 2018) **[PAPER 3]**. In parallel, the amount of GSSG dropped in wheat and barley seeds during viability loss suggesting that GR is actively recycling GSSG into GSH (Roach et al., 2018; Nagel et al., 2019b) **[PAPER 3, 4]**. In accordance, sunflower and wheat seeds had an increased or constantly high activity level of GR, CAT, GPX in conjunction with increasing amounts of tocochromanols during ageing at 40 °C and 75 % RH (Lehner et al., 2008; Morscher et al., 2015). Under extreme conditions, such as 45 °C and 100 % RH, H_2O_2 and GR level increase dramatically while SOD and CAT, responsible either for the catalyses $O_2^{\bullet-}$ in H_2O_2 and H_2O_2 into H_2O and O_2, respectively, reduce (Lehner et al., 2008). Furthermore, neither seed leachates, pH nor the content of malondialdehyde (MDA) and organic radicals changed (Bailly et al., 1996; Roach et al., 2018; Nagel et al., 2019b) **[PAPER 3, 4]** assuming that membranes remain intact and proton pumps buffer the pH under these conditions. However, other mechanisms must be responsible for the rapid viability loss. Volatile organic compounds emitted by seeds exposed to CD at 35 °C and 75 % RH reflected that fermentation-type reactions, with methanol and ethanol as predominant compounds, are mainly responsible for viability reductions (Mira et al., 2016). Alcoholic fermentation occurs under oxygen-limited conditions. Thereby, pyruvate from glycolysis is converted to acetaldehyde by pyruvate decarboxylase. The final conversion is achieved by alcohol dehydrogenase which converts acetaldehyde to ethanol (Kreuzwieser et al., 1999). Volatiles from alcoholic fermentation, Maillard reactions and lipid peroxidation were also found in pea seeds aged under 50 °C and 60 % RH and represent an additional evidence that enzymes are highly active under these conditions. However, Maillard products are formed non-enzymatically and may modify enzyme and protein activities during seed deterioration (Wettlaufer and Leopold, 1991; Colville et al., 2012). Modified proteins may be responsible for the depletion in reduction of ethanol (Colville et al., 2012) and of glucose-6-phosphatedehydrogenase, which provides NADPH for reductive processes such as required by GR during CD (Morscher et al., 2015). In conclusion, it still remains speculative which detailed processes lead to viability loss during rising temperatures and seed MCs. Though, high water activity facilitates increased antioxidative activities, they are not sufficient to avoid detrimental processes, such as presumably alcoholic fermentation, lipid peroxidation and Maillard reactions.

Under long-term seed storage conditions, at temperatures below 20 °C and 0.1 g H_2O g^{-1} DW, cytoplasm of the seeds is assumed to be glassy and metabolic activities absent. Molecules in the glass are not completely restricted in their movements. Reactions take place at very slow rates and may lead to lipid oxidation and seed deterioration (Buitink and Leprince, 2008). Under these conditions, the accumulation of organic radicals correlated negatively with seed viability and may trigger further oxidative processes. Therefore, during long-term storage under AS and CS, tocochromanols are oxidized and their amount is reduced in barley seeds. In both ,wheat and barley seeds, GSH functioned as LMW antioxidant and was converted to GSSG which, consequently, increased (Roach et al., 2018; Nagel et al., 2019b) **[PAPER 3, 4]**. The adenylate pool hardly changed and DNA alteration were not detectable in sunflower seeds stored under dry conditions, at 35 °C and 0.04 g H_2O g^{-1} DW (Kibinza et al., 2006; El-Maarouf-Bouteau et al., 2011). RNA degrades over long-term storage which is marked by a shift to smaller RNA fragments (Fleming et al., 2017). In seeds of 266 wheat and 170 barley genotypes stored at 0/20 °C and 6±2 % MC, the major metabolites were glycerol and other products of lipid degradation, i.e. glycerol phosphates and galactose, which correlated significantly negative with seed viability. Thereby, the levels of fully acylated non-oxidized storage and structural lipids decreased, whereas oxidized variants and hydrolysis products accumulated during seed viability loss (Riewe et al., 2018) **[PAPER 6]**. Interestingly, Oenel et al. (2017) found the accumulation of oxygenated triacylglycerols also in Arabidopsis seeds stored under CD at 42 °C and 85 % RH. Non-enzymatic oxidation of storage lipids was the major cause for viability loss upon storage under CS for 9 and 17 years. The appearance of oxidized fatty acids indicated that lipoxygenases (LOXs) were involved in the first oxygenation step of PUFAs. In accordance, the higher emissions of pentane and hexanal are considered to be the by-products from the peroxidation of PUFAs found in dry caraway (*Carum carvi* L.) seeds (Mira et al., 2016). Furthermore, the occurrence of by-products of the Maillard reactions were found in 1,500 years old barley seeds excavated in an Egyptian archaeological site (Evershed et al., 1997). Strelec et al. (2008) could not find significant accumulations in Maillard products in wheat seeds stored under AS at 20 °C and 55 % RH. However, the reaction rate is assumed to be very slow. Concluding, non-enzymatic reactions are most likely to occur under dry storage of seeds, among are lipid oxidation and Maillard reactions. However, products of volatile measurements and lipid degradation indicate that also enzymatic reactions, i.e. lipid peroxidation and hydrolysis, coincide with viability loss. It might be hypothesized that these reactions occur in 'hydrated pockets', seed areas where water is loosely bound and available for reactions.

Elevated O_2 levels may accelerate oxidative reactions and, hence, seed deterioration. The exposure of seeds to an increased pressure of pure O_2 (18 MPa) using EPPO at 20 °C and 40 % RH revealed a half-life of 38.5±21.8 days for 94 lines of the 'Oregon Wolfe Barley' (OWB) mapping population (Nagel et al., 2016) **[PAPER 7]**. In contrast, seeds subjected to CD at 45 °C and 75 % RH and elevated O_2 (75 %) showed hardly any changes in comparisons to seeds subjected to CD under normoxia (21 % O_2) (Roach et al., 2018) **[PAPER 3]**. Thereby, the relationship between O_2, seed MC, atmospheric pressure and storage temperature is hardly understood. Ellis and Hong (2007) postulated that oxygen is relatively more deleterious to seeds at lower MC. The effect might be explained by the diffusion rate of gas which is 10,000 fold slower through a liquid medium than through air. Thereby, the gas exchange capability depends on a combination of the structural arrangement of cells, the existence of distinct gas diffusion barriers and the MC (Borisjuk and Rolletschek, 2009). Dry seeds exposed to EPPO showed a drastic viability loss in parallel with a drop of tocopherols which overlap with symptoms of long-term stored seeds (Groot et al., 2012; Nagel et al., 2016) **[PAPER 7]**. In hydrated seeds such as developing maize seeds or in seeds exposed to CD, the internal oxygen level drops within few µm of the pericarp (Rolletschek et al., 2005) indicating that the high MC prevents O_2 diffusions and may damage only structures at the periphery of the seeds. An increase of atmospheric pressure increases also the water activity (Sun, 2002; Chaplin, 2018). Therefore, an elevated oxygen atmosphere and/or elevated atmospheric pressure is more harmful in seeds of low moisture content than in hydrated seeds and supports the idea that seed long-term storage can be extended by storage under hermetic anoxic conditions.

4.2 Differences in seed longevity are the consequence of compositional variations

Seed longevity varies greatly between species and genotypes. Across 18 selected crop species, the half-life (P50) ranges between 2.0 years (chive, *Allium schoenoprasum* L.) and 14.9 years (cucumber; *Cucumis sativus* L.) under AS (Nagel and Börner, 2010) **[PAPER 1].** When seeds were stored hermetically at -18 °C, P50 values increased to 24 years for chive and 87 years for cucumber. However, in the dataset of 276 species, there was hardly any relationships between P50, seed composition, initial viability and country of origin (Walters et al., 2005). In addition, the comparability between longevity studies is rather challenging (Walters et al., 2005; Nagel and Börner, 2010) **[PAPER 1]**. Great variation in seed longevity was found between genotypes of the same species. Seed viability ranged widely for genotypes of barley, wheat, common bean (*Phaseolus vulgaris* L.) after comparable storage intervals. Smaller variations were found for genotypes of pea, common vetch (*Vicia sativa* L.) and lettuce (Nagel and Börner, 2010) **[PAPER 1]**. For seeds of 42 oilseed rape genotypes stored under 7 °C and 6 % MC, the P50 was calculated to be between 15.2 and 50.7 years (Nagel et al., 2018) **[PAPER 2]** confirming the hypothesis that also oilseed rape genotypes show a variable seed longevity (Walters et al., 2005). Roberts (1973) assumed that each single seeds differs in the ability to survive periods of prolonged storage. Marginal differences in seed position on the parental plant, degree of maturity, post-harvest processing and storage conditions contribute to variations in seed viability (Hay et al., 1997; Kameswara Rao et al., 2017). Nevertheless, the relative longevity ranking of oat ≥ barley > wheat > rye > rice (Table 3) is mostly consistent (Priestley et al., 1985; Walters et al., 2005; Nagel and Börner, 2010) **[PAPER 1]**. Concluding, seed longevity can be roughly estimated for a species. However, the deterioration rate of a single seed depends on various other factors, most likely a combination of genetic, i.e. composition and structure, and environmental effects.

The chemical composition of the species affects deterioration processes. Wheat and barley have endospermic seeds composed of a large, death endosperm (>75 %) and a small embryo (<2 %) at the periphery (Table 3), both, protected by the pericarp (Evers et al., 1999; Xiong et al., 2013). QTL mapping has shown that the *nud* gene controlling the development of the hull affects barley seed vigour (Nagel et al., 2016) **[PAPER 7]**. The hull is an important layer protecting seeds against mechanical and oxidative damages and may buffer variations in seed MC. Therefore, the storability has shown to be improved when seeds and hull/husks were stored together (Haferkamp et al., 1953). Seeds of wheat and barley distinguish by the chemical composition. Wheat seeds contain a considerable amount of gluten proteins important for bread making quality (Shewry, 2009). In contrast, barley seeds contain high amounts of β–glucan, a hemicellulose found in endosperm cell walls and responsible for hard, steely seeds. Interestingly, a high β-glucan content is associated with a lower water uptake (Holopainen et al., 2014) and strong antioxidant capacity (Kofuji et al., 2012). Wheat cell walls are rich in the non-starch polysaccharide arabinoxylan (1.6 % to 2.7 % DW) but the content does not exceed the β-glucan content in barley (Table 3). Arabinoxylan show antioxidant properties due to the presences of phenols in their molecular structure (Bader Ul Ain et al., 2018). In contrast, oilseed rape is an endospermic seed compost of a large embryo (>60 %) containing significant amounts of lipids (41 % to 48 % DW) (Hu et al., 2013; Nagel et al., 2018) **[PAPER 2]**. Due to the higher affinity of extracted oil to peroxidation (Crapiste et al., 1999), lipids are assumed to be the major cause for the low and variable longevity of oilseed rape (P50=25 years in CS). However, seeds of Brassicaceae accessions can show between 91 % and 100 % germination after storage between 0.3 % to 3.0 % seed MC and between -5 and -10 °C for 40 years (Perez-Garcia et al., 2007). The contrasting results might be resolved by a detailed analysis of the crystallization behaviour of lipids under low temperatures and seed MC. Triacylglycerols of intermediate seeds, such as *Cuphea carthagenensis* Jacq., crystallize at 6 °C (Crane et al., 2006). For comparison, the melting point for neutral lipids in maize seeds is at -25 °C (Buitink et al., 1996). A hydration of seeds containing crystallized triacylglycerols is lethal. Therefore, the viability can be only reinstated when seeds are heated up to 45 °C before imbibition (Crane et al., 2006). If the lipids of oilseed rape crystallize above -18 °C, it may explain the variable storability of the species. However, oilseed rape seeds are well protected by a combination of various PUFAs and lipophilic antioxidants such as tocopherols, I.e. 863 $\mu g\ g^{-1}$ oil (Leckband et al., 2002; Nagel et al., 2018) **[PAPER 2]** indicating that the affinity of lipids to oxidative processes should be compensated by cellular protection mechanisms. In conclusion, the component affecting most the relative longevity ranking of barley > wheat > oilseed rape (Priestley et al., 1985; Walters et al., 2005; Nagel and Börner, 2010) **[PAPER 1]** remains uncovered. The differences in seed composition and structure has strong impact on the water uptake, seed MC, anti- and prooxidant capacity. It could be hypothesized that the high amounts of starch and proteins in wheat and barley seed lead to Maillard reactions whereby oilseed rape seeds may deal with lipid degradation as main cause of viability loss.

Table 3. Seed longevity, structural and compositional characteristics of the three species studied. P50, half-viability period; DW, dry weight; UKN, unknown.

	Wheat (*Triticum aestivum* L.)	Barley (*Hordeum vulgare* L.)	Oilseed rape (*Brassica napus* L.)
Relative longevity*[1]	Medium	Medium long	Variable
P50 cold storage [years]	54*[1]	84*[1]	25*[1]
P50 ambient storage [years]	7.2*[2]-7.6*[8]	7.2*[8]-9.2*[2]	7.3*[2] (*B. oleracea*) – 13.9*[8]
Thousand Seed weight [mg]	27-50*[3]	32 - 36*[3]	3.0 - 5.7*[10]
Proportion pericarp & testa [%]	8.5*[3]	2.9 - 3.3*[3]	12.5 - 23.7*[11]
Proportion embryo/ cotyledon [%]	1.3*[3]	1.7 - 1.9*[3]	59.5 - 75.3*[11]
Proportion hull [%]	-*[3]	13*[3]	-
Proportion aleurone [%]	6.7*[3]	4.8 - 5.5*[3]	UKN
Proportion endo-sperm [%]	82.0*[3]	76.3 - 87.0*[3]	UKN
Proportion scutel-lum [%]	1.5*[3]	1.3 - 1.5*[3]	UKN
Starch [% DW]	82.0*[7]	61.2 - 71.8*[4]	UKN
Protein [% DW]	14.0*[7]	11.2 - 12.5*[4]	19.1 - 27.0*[10]
Lipids [% DW]	2.0*[7]	2.9 - 3.6 %*[4]	41.1 - 48.2*[10]
β-glucan [% DW]	0.57*[5]	2.9 - 5.7*[4]	UKN
Tocopherols		3.6 $\mu g\ g^{-1}$ DW *[12]	579 - 863 $\mu g\ g^{-1}$ Oil*[9]
α-tocopherol	12.2 - 61.0 $\mu g\ g^{-1}$ DW*[6]	2.8 $\mu g\ g^{-1}$ DW *[12]	208-329 $\mu g\ g^{-1}$ Oil*[9]

*[1] Walters et al. (2005), *[2] Nagel and Börner (2010), *[3] Evers et al. (1999), *[4]Holopainen et al. (2014), *[6] Ohm et al. (2016)*[6]Whent et al. (2012), *[7] Earle and Jones (1962) and Jones and Earle (1966), *[8] Priestley et al. (1985), *[9] Leckband et al. (2002), *[10] Nagel et al. (2018), *[11] Hu et al. (2013), *[12] Roach et al. (2018)

Variations in chemical components and allocation in the seed body determines ageing processes. A comparison of seed germination between two wheat genotypes stored, both, under AS and CS identified a storage tolerant (TRI 23248) and storage sensitive (TRI 10230) type. Seeds of the sensitive genotype revealed lowest seed viability under AS and highest levels of lipid peroxidation and membrane damages. The relative abundance of the transcripts of CAT1, CAT2, APX1, APX2, DHAR1 and DHAR2 was significant lower with downstream effects on the contents of ASA and DHA, which were also lower in comparison to the storage tolerant type after 24 hours of imbibition. The proteomic approach showed that proteins of seeds of the sensitive type were associated to defences, energy and metabolism. Interestingly, the abundance of globulins, the major storage protein family in wheat, decreased (Chen et al., 2018) **[PAPER 5]** and support the important role of storage proteins for seed longevity (Nguyen et al., 2015). In oilseed rape, significant relationships were found between oil content (r=0.48), α-linolenic acid (18:3) (r=0.42) and glucosinolates (r=-0.43). Interestingly, a combination of stearic (18:0), linoleic (18:2), arachidic (20:0), eicosadienoic (20:2) and erucic acid (22:1) predicted best seed viability after storage at 7 °C for 31 years (Nagel et al., 2018) **[PAPER 2]** and indicate that some fatty acids are more involved in seed deterioration or protection mechanisms than others. Woodfield et al. (2017) revealed a 'heterogeneous landscape' in the distribution of molecular species and identified that linoleic (18:2) and α-linolenic (18:3) acids are accumulated in seed coat. The outer cotyledon was also more enriched in linoleic (18:2) and α-linolenic (18:3) fatty acids and both, seed coat and outer cotyledon, might be the first targets of external oxidative stress. The inner cotyledon accumulated higher amounts of stearic (18:0) and oleic (18:1) fatty acids and the embryonic axis was more enriched with palmitic acid (16:0). A higher amount of PUFAs associated with a lower degree of saturation was demonstrated as a plant response to adapt to heat stress. Lipid saturation determines the membrane fluidity and is assumed to be associated to a higher plasticity and ability to deal with oxidative and environmental stresses (Spicher et al., 2016). Membranes and lipids composed of saturated fatty acids are less prone for oxidative stress (Reis and Spickett, 2012). The specific composition may be associated also to specific functions during imbibition. Thereby, the water allocates through a small section of the seed coat close to the hilum and hydrates first the testa from inside (Figure 15). The extension of the testa is not uniform and might predetermined by the localized variation of plasticity, hence, by the chemical structure. Simultaneously, water is moved through the micropyle into the radicle (Munz et al., 2017). The higher degree of saturation of lipids in the inner part of the seed may indicate that these parts are essential for seed development and require a lower affinity to oxidative stress. The extremely 'heterogenous landscape' of components and functions must be regulated by a complex molecular network. The control of the molecular mobility and fluid motion of protons via cellular constituents (Leubner-Metzger, 2005) is required to regulate seed dormancy, germination and storability.

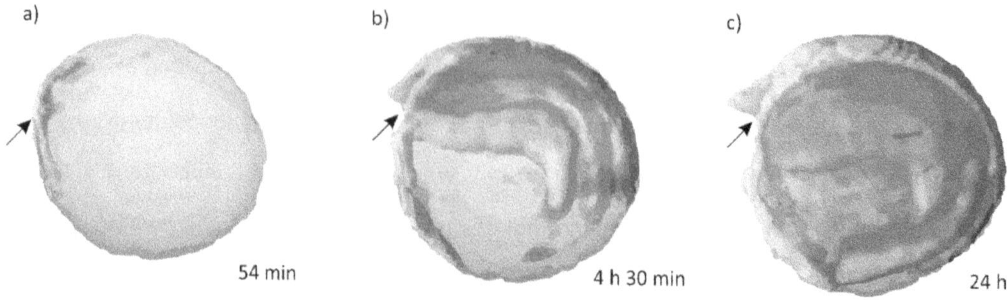

Figure 15. Water uptake of germinating oilseed rape seeds at 20 °C. The blue zone indicates hydrated tissue. The black arrows indicate the site of protrusion of the radicle. Based on Munz et al. (2017).

4.3 Maternal growth environment influences seed longevity and seed dormancy

The environmental conditions during seed development on the mother plant in conjunction with the genetic background are assumed to be the major determinants of seed longevity for species and genotypes. The reproduction of 28 winter oilseed rape varieties in different locations has shown that growth environment affects the initial seed quality. The mean seed germination differed significantly and ranged between 97.2 % and 98.8 % for six locations. In addition, each location distinguished significantly in the acquisition of secondary seed dormancy, thousand seed weight, oil and protein content. The differences in the initial quality are assumed to cause variations in seed germination to between 35.2 % and 74.4 % after CD (Schatzki et al., 2013) **[PAPER 9]**. In a barley population of 184 genotypes, seed dormancy and initial seed germination varied strongly in two years of contrasting precipitation and temperature. In the drier year, days to anthesis and maturity were shorter, seeds were less dormant but pre-harvest sprouting (PHS) increased (Nagel et al., 2019a) **[PAPER 13]**. Gualano and Benech-Arnold (2009) found a positive relationship between the thermal environment during seed filling and the initial seed quality of barley. Thereby, the timing seems to play an important role. At early developmental stages, wheat seeds are sensitive to high temperatures and the subsequent seed longevity is reduced. Later exposure to high temperature improves seed longevity but decreases yield parameters and bread making quality (Nasehzadeh and Ellis, 2017). A similar response can be found for oilseed rape which produces higher seed germination but lower yield under higher growth temperatures (Awan et al., 2018). Interestingly, the reverse is true for the Australian species *Plantago cunninghamii* Decne. Seeds from cool growth conditions had higher seed quality and survived longer after CD (Kochanek et al., 2011). In conclusion, seed quality and longevity is affected throughout the seed development; and species-specific optimum growth conditions lead to high seed quality and yield. At later stages of seed development, wheat, barley and oilseed rape can produce high seed germination under higher temperatures but at the cost of seed yield. Although crops had been domesticated up to 10,000 years ago, the maintenance of seed germination is prioritized for most species. Under adverse environmental conditions, seed dormancy is developed to ensure the survival of the species.

Growth conditions during seed development modulate gene expressions. Many QTLs, MTAs or genes detected in longevity studies indicate that differences in germination after storage is the consequence of phenotypic variations to adverse environmental conditions during seed development. In oilseed rape, QTLs found in the double-haploid (DH) mapping population 'YE2' overlap with orthologous loci in the Arabidopsis genome (Nagel et al., 2011) **[PAPER 8]** which are involved in the ABA biosynthesis (Thorlby et al., 1999; Clerkx et al., 2004). The alteration of ABA content and sensitivity has been found as a major mechanism to respond to environmental stress including large effects on seed vigour (Awan et al., 2018). In barley, seed germination and dormancy after harvest were associated with genes controlling flowering time, i.e. *HvFT2, HvFT4, PpdH1, Vrn-H3* (Nagel et al., 2019a) **[PAPER 13]**. The Arabidopsis *FLOWERING TIME* (*FT)* gene controls seed dormancy via proanthocyanidin synthesis which results in an altered seed coat tannin content (Chen et al., 2014). On chromosome 2H, the genomic region around the *ZEOCRITON1* (*Zeo1*) gene was highly associated with seed germination after CD and EPPO (Nagel et al., 2016) **[PAPER 7]**. *Zeo1* determines plant height, spike compactness and seed width (Franckowiak et al., 1997) via a polymorphisms in the binding site of the barley ortholog of *APELATA2 (AP2)*-like TF, termed *HvAP2*. Here, genetic variations can lead to different developmental windows with consequences on the shape of the barley inflorescence (Houston et al., 2013). To prove the concept that plant height may affect longevity, seeds of near isogenic lines carrying different *REDUCED HEIGHT* (*Rht*) dwarfing alleles were exposed to CD for one and two days. Shoot length was significantly affected by the *Rht* alleles across all four tested varieties. However, a significant reduction of shoot length in combination with a signification delay in germination time was only found for the allele *Rht-B1c+-D1b*, most pronounced in the 'Bersee' background. *Rht-D1* encodes DELLA proteins which react in response to endogenous GA (Pearce et al., 2011). Therefore, we assume that different regulation of the GA metabolisms in the different varieties may modulate the seed germination more than the plant height per se (Nagel et al., 2013) **[PAPER 11]**. However, GWAS analysis of seed germination after long-term storage, AA and CD revealed loci associated with cellulose synthase, chaperonins and dehydrin/LEA group 2 like proteins (Rehman Arif et al., 2017) **[PAPER 12]** which are involved during maturation drying (Leprince et al., 2017). Plants and seeds have an enormous plasticity to respond to different environmental stresses. He et al. (2016b) investigated systematically the seed performance after low temperature, low nitrate, low and high light intensity. Interestingly, high light intensity during seed maturation increased galactinol content, an RFO accumulating during maturation drying, and displaying a high correlation with seed longevity. At low nitrate concentration, the correlation between galactinol and seed longevity was lower and negative. In conclusion, a complex network of gene, TFs and proteins coordinate the response mechanisms to different environmental conditions. RFOs, LEA proteins and APETALA2/ethylene-responsive element (ERE)-binding proteins (AP2/EREBP) TFs can be associated with seed longevity. However, a universal genetic marker providing reliable information on the initial seed quality could not be developed so far.

4.4 Molecular markers for seed deterioration at different water contents

Seed longevity is a quantitatively inherited trait and QTLs stable across different growth and storage environments are rare. Seed storability was studied first in the progeny of the Arabidopsis parents 'Cape Verde Islands' and 'Ler' that distinguished in the content of soluble oligosaccharides (Bentsink et al., 2000). The so-called linkage mapping approach is based on the co-segregation of alleles of mapped marker loci and phenotypic traits which allow the identification of linked markers, so-called QTLs (Mackay and Powell, 2007). Bentsink et al. (2000) identified two QTLs for seed germination after AS for 4 years and four QTLs for seed germination after CD at 40 °C and 85 % RH. However, these did not overlap with the major QTLs for stachyose, raffinose and sucrose at the same locus. In rice, seed germination after CD at 40 °C and 95 % RH revealed three major loci on chromosomes Os9, Os11 and Os12 in the 'ZYQ8 x JX17' population (Zeng et al., 2006) and another tree loci on chromosomes Os1, Os3 and Os9 in the 'Asominori x IR24' population (Xue et al., 2008). The exposition of seeds CD using 30 °C and 15 % seed MC for 2 months led to a strong segregation in germination between the parents 'Nipponbare' and 'Kasalath'. Here, QTLs were found on chromosomes Os2, Os4 and Os9 (Miura et al., 2002). The non-reproducibility of seed longevity QTLs in different mapping populations and between different ageing approaches was also shown in barley. By germinating seeds of three different biparental mapping populations after CD at 44 °C and 18 %

seed MC and AA at 43 °C and >95 % RH, four different loci differing between the populations were found on chromosomes 2H, 5H (two) and 7H. The use of expressed sequence tag (EST) markers enabled the identification of molecular regulators, i.e. dehydration responsive element binding proteins (DREB), AP2/EREBP and the *nud* gene (Nagel et al., 2009). This gene, an ethylene response factor (ERF) family TF gene, controls via lipid biosynthesis the adhesion between lemma and palea and, hence, naked/ hulled caryopsis (Taketa et al., 2008). Naked seeds exposed to CD loose viability faster than hulled ones (Nagel et al., 2009) assuming that either the modulated lipid composition, the physical protection or antioxidative capacity of the hull affect the seed deterioration behaviour. The *Nud* gene is a homolog of the Arabidopsis *WIN1/SHN1* TF gene (Taketa et al., 2008) but the naked/ hulled seed feature is a special case for barley. By combining linkage mapping, proteomic and transcriptomic and CD at 60 °C and 17 % RH, Wozny et al. (2018) identified the UDP-glycosyltransferase MLOC_11661.1 as candidate gene on chromosome 2H and as possible downstream target gene the NADP-dependent malic enzyme (NADP-ME) MLOC_35785.1 as specific regulators for seed longevity. NADPH, the product of NADP-ME, provides reducing equivalents for the oxidation-reduction involved in the protection against ROS. Arabidopsis mutants of *nadp-me1* had an extreme short-lived phenotype which could be rescued by the original promotors. However, it can be assumed that the appearance of QTLs and the regulation of genes may differ between the ageing approaches, i.e. CD and EPPO. In the OWB mapping population, one major loci was found on 5H for seed germination after CD, whereas three QTLs on chromosomes 3H, 4H and 5H were found for seeds exposed to EPPO (Nagel et al., 2016) **[PAPER 7]**. In oilseed rape, minor changes in the ageing treatments resulted in completely different QTLs for seed longevity (Nagel et al., 2011) **[PAPER 8]**. In general, one may speculate that the restricted number of meiotic events captured in biparental mapping population reduce the genetic resolution. Hence, QTLs for seed longevity cover a wide area which may include thousands of genes. However, different loci found for different ageing approaches point towards the various detrimental processes appearing under different storage/ageing conditions.

GWAS studies have higher genetic resolution and are, in dependency on the markers, more informative but QTLs for seed longevity remain unstable. GWAS is based on the principle of linkage disequilibrium (LD) at the population level. LD is the non-random association between alleles at different loci which is created by events of mutations, drift, and selection during evolution (Mackay and Powell, 2007). To detect an adequate linkage between the phenotype and causal loci, a higher number of individuals and genetic markers is required. In humans, to decipher exceptional longevities, 2,070 individuals were analysed for longevity-associated variants (Sebastiani et al., 2017). Although, the results are quite different from those revealed for longevity in dry seeds, it demonstrates, that the approach is valid to identify genetic loci for seed longevity. The first GWAS study on seed longevity has been conducted by Rehman Arif et al. (2012) after exposition of seeds of 183 genotypes to AA at 43 °C and >95 % RH and CD at 43 °C and 18 % seed MC. In total, 101 MTAs were identified for seed germination after CS, 74 MTAs for seed germination after AA and 97 MTAs for seed germination after CD. By comparing MTAs of seeds stored under CS for 40 years, the data show that only few loci were associated for more than one ageing/ storage treatment (Rehman Arif et al., 2017) **[PAPER 12]**. Interestingly, a number of MTAs for AA and CD overlapped with loci found for seed longevity in barley (Nagel et al., 2015). Here, the Diversity Array Technology (DArT) markers could be blasted and showed sequence homology, among others, to RNAseH, which is involved DNA in replication, repair and transcription. In addition, again *AP2/ERF*, *DREB* and *ERE* were identified for seeds aged under AA at 43 °C and >95 % RH and under CD at 45 °C and 60 % RH. In rice, the comparison between seeds exposed to CD at 45 °C and 14 % MC for 14 days and seeds stored at 25 °C and 14 % MC for two years revealed 17 MTAs (Li et al., 2014). Three markers were associated with both traits and had been found in Miura et al. (2002) and Sasaki et al. (2005) before. However, the available GWAS studies on seed longevity do lack either on information of the markers or the usage of long-term stored seeds. Meanwhile, 'omic' approaches and mutant research identified possible markers for seed longevity. In wheat and rice, RNAi mutants with lower levels of lipoxygenase (LOX) transcripts showed significant higher germination after AA at 40 °C and 85 % RH (Huang et al., 2014; Dong et al., 2015). In accordance, mutations in *VTE1* and *VTE2* led to a tocopherol deficiency including increased levels of lipid hydroperoxides and hydroxy fatty acids and reduced germination (Sattler et al., 2004). In tomato, QTL mapping revealed a co-locating QTL for galactinol content and seed longevity. By knocking out the key enzymes of the galactinol pathway, i.e. *GALACTINOL SYTHASE1*, *RAFFINOSE SYNTHASE*, the galactinol level was at minimum and, in comparison to the wildtype, seeds had a significant reduced germination after CD at 40 °C and 85 % RH (de Souza Vidigal et al., 2016). However, Leprince et al. (2017) showed that correlations between seed longevity and galactinol are not valid for all species and needs further confirmation. Unfortunately, most, if not all, marker for seed longevity were confirmed by artificial ageing treatments. Therefore, the upcoming manuscript will make use of the GWAS approach and study the genetic background of germination and metabolomic data after storage in CS for 40 years; presented by Riewe et al. (2018) **[PAPER 6]**. Then, identified genetic marker for seed longevity are planned to be evaluated in long-term stored material.

Seed longevity and seed dormancy are considered as distinct traits (Bewley, 1997) which are evaluated via seed germination. Dormancy is a state of metabolic arrest that ensures the survival of the organism. Most organisms, i.e. *Drosophila* spp., use this mechanism to extend their lifespan (Nguyen et al., 2012). If central genes are knocked down, i.e. *ABI5* important for the acquisition of desiccation tolerance, seed dormancy and longevity are impaired as consequence of the modulation of the RFO metabolism and LEA proteins (Zinsmeister et al., 2016). Over thousands of years, in crops, it has been selected against seed dormancy, whereas gene banks selected accidentally for seed longevity. Therefore, it might be hypothesized that a negative correlation could be found for seedbank accessions. By studying six recombinant inbred line populations, Nguyen et al. (2012) revealed that the loci *Germination Ability After Storage1* (*GAAS1*) to *GAAS5* co-located with seed dormancy loci *DOG*. Thereby, a deep seed dormancy was correlated with low seed longevity. However, studies on wild population do not show this clear pattern as the initial quality is strongly affected by the environmental conditions (Mondoni et al., 2018). In wheat, seed longevity of seeds exposed to CD at 45 °C and 60 % RH was compared with pre-harvest sprouting assessed through the falling number. Thereby, the major QTL for seed longevity on chromosome 1A was not co-located with the four QTLs identified for pre-harvest sprouting (Börner et al., 2018) **[PAPER 10]**. By using the power of GWAS analysis, Rehman Arif et al. (2017) **[PAPER 12]** identified 78 MTAs for seed dormancy and 110 MTAs for pre-harvest sprouting. Thereby, 11 MTAs for seed dormancy and 19 MTAs for pre-harvest sprouting co-located with MTAs found for germination after CS for 40 years or the initial germination indicating again the necessity to test long-term stored seed material. In barley, the relationship between seed longevity and dormancy needs to be elucidated. So far, after-ripening experiments indicated that seeds of low dormancy started to loose viability quickly during after-ripening assuming a positive relation between seed dormancy and longevity in barley (Nagel et al., 2019a) **[PAPER 13]**. The nexus between seed longevity and dormancy remains controversial and might be distinct between wild plants and crops or even between crop species. Again, a realistic relationship between seed dormancy and seed longevity can only be drawn when seeds are stored under dry conditions.

4.5 The implications for gene bank management

In the 20th century, gene banks were established to prevent the irretrievable loss of plant genetic resources. Many gene banks were founded in the 1970ies. Assuming an median half-life of crop species of 53 years (Walters et al., 2005), the germination of many accessions might have depleted by 50 % by now, if seeds were stored under CS and were not reproduced in the meantime.

Coincidently, with the setup of seedbanks, Roberts (1972) started his pioneering work on seed storage and assumed that seed longevity is dependent on genotype, maternal growth and storage conditions. The introduction of AA and CD tests to the ISTA rules (Hampton and TeKrony, 1995) and their ease of handling let to a vast amount of seed longevity studies. Unfortunately, only few comparisons have been done between conventional ambient or long-term storage conditions. Based on different approaches, Schwember and Bradford (2010), Ballesteros and Walters (2011), Roach et al. (2018) **[PAPER 3]** and Nagel et al. (2019b) **[PAPER 4]** showed, seed deterioration mechanisms are dependent on water activity; and dry long-term storage does not reflect the same environment as artificial ageing conditions at elevated temperatures and seed MCs or O_2. Interestingly, the liability of accelerated storage for the shelf-life estimation of foods, pharmaceuticals, cosmetics and other products is also controversially discussed. Thereby, the microbial growth is an unpredictable parameter which leads to deviations between the observation and extrapolation. Corradini and Peleg (2007) suggested to use at least two different mathematical models for reliable predictions. Therefore, Walters et al. (2005) applied the Avrami kinetics to predict the seeds half-life under CS. A comparable study has been initiated in Gatersleben. Here, a combination of seed viability equation and machine learning is used to extrapolate the P50 for a range of crops stored under CS.

The maintenance of plant genetic resource is extremely challenging. Maternal growth and storage conditions can vary between different years and accessions and affect strongly the initial seed viability and, consequently, the storability. Therefore, reproduction should meet the conditions which plants require in their natural environment. This aspect is rather challenging for most gene banks. Nevertheless, the NPGS holds a system of different stations in different climatic zones. Therefore, it might be worth to consider a similar setup for European gene banks.

Scientific studies should focus on the specific requirements of species during growth, maturation, drying and storage. Especially, more attention should be laid on the optimization of the seed MC and the O_2 concentration. Groot et al. (2014) showed clearly that a complete lack of oxygen prolongs the storage period. It is assumed that the date palm germinated after 2,000 years (Sallon et al., 2008) was kept under anoxia. Hypoxic conditions might be also achieved by plasma treatments. So far, cold plasma has been used as alternative for seed coating and improved seed vigour in the field (Jiang et al., 2014). However, a special plasma coating of seeds for long-term storage has been discussed but is not investigated, so far. Equivalent, cryopreservation of seeds is actively done at NPGS but only few data exist about the reliability of the storage approach. Critical remarks indicate that cryostorage might be not suitable for seeds (Buitink and Leprince, 2004; Walters et al., 2004; Walters, 2014) but further studies must elaborated the potential of liquid nitrogen for long-term storage of seeds. An old idea, but recently investigated, includes the addition of antioxidants during seed germination. Deng et al. (2017) could show that not ASC or GSH but melatonin improved germination results of aged seeds. Melatonin induces indirectly CAT and SOD activities and prevents lipid peroxidation. There is still potential to improve and prolong the storability of gene bank accessions. Therefore, studies must focus on new technologies but should also elucidate simple storage kinetics for each species separately.

The maintenance of genetic diversity can only be guaranteed by a sophisticated viability monitoring. So far, germination tests are done manually and about 10,000 accessions can be tested every year at IPK. Based on 150,000 accessions, each accessions can be re-tested all 15 years. For most accessions, the viability monitoring might be appropriate but germination of some accessions may deplete quickly. Therefore, non-destructive and rapid viability estimations would be helpful. Most methods, such as electrical conductivity (ISTA, 2018) or infrared thermography (Kranner et al., 2010a) lack on the appropriate adaptation to various species. Accordingly, the diversity of biochemical mechanisms during seed deterioration challenges the development of genetic markers for seed viability. A promising method promoted by the breeding industry involves high-throughput phenotyping. The use of achieved innovations in image analysis might be promising in future. However, a combination of image analysis and seed morphology characterization would support the detailed evaluation of the genetic diversity stored in the gene bank vaults.

5 SUMMARY AND CONCLUSIONS

Seed deterioration of wheat, barley and oilseed rape, some of the major crops worldwide, is important for seed industry, agriculture and gene bank management. The loss of seed viability has been investigated over time applying seed MCs between 5 % and >25 %, storage temperatures between 0 °C and 45 °C and a modified atmosphere of increased O_2 concentration (75 %) or increased atmospheric pressure (18 MPa).

In both, wheat and barley, the elevation of seed MCs and storage temperatures resulted in a gradual change of biochemical mechanisms during seed deterioration. Under dry seed storage conditions, GSH and tocochromanols functioned as LMW antioxidants and depleted. GSH was partly converted to GSSG, which increased. Lipids were oxidised or hydrolysed, the pH decreased, whereas organic radicals accumulated over time and correlated negatively with seed viability. Under high seed MCs (>13 %) and storage temperatures (>40 °C), membrane damages, changes of pH or organic radicals were absent. Furthermore, GSH and GSSG depleted whereas tocochromanols remained stable or increased indicating that enzymatic activities are reinstated but repair processes are outweighed by other detrimental reactions. Therefore, seeds stored under high MCs and temperatures are exposed to a different environment than seeds exposed to lower MCs and temperatures where water activity and deterioration rate is reduced and the cytoplasm is assumed to be glassy.

The genotype determines the seed structure and composition. Environmental stresses during seed development modulate compounds and the acquisition of desiccation tolerance, seed longevity and dormancy. In barley, seed development under high temperatures and low rainfall facilitated high seed vigour and reduced seed dormancy whereby this response turned, when seeds were produced under colder and wetter conditions. Similarly, nutritional supply and growth conditions at different locations facilitate different degrees of secondary dormancy and seed storability in oilseed rape. Genotypic variations between oilseed rape accessions can vary strongly in the amount and composition of oil, glucosinolates and fatty acids. Oilseed rape seeds rich in α-linolenic acid (18:3) and or in a combination of stearic (18:0), linoleic (18:2), arachidic (20:0), eicosadienoic (20:2) and erucic acid (22:1) had highest viabilities. In wheat and barley, seeds producing high amounts of glycerol and products of lipid degradation were more affected in seed germination after 40 years of storage. The compounds are heterogeneously distributed in seeds. Therefore, biochemical processes possibly distinguish in dependency on the seed compartment (embryo, endosperm, seed coat).

Linkage and association mapping approaches revealed that associations between seed longevity and genetic loci can be various. QTLs varied with ageing treatment, genetic constitution of the panel and maternal growth conditions. Most QTLs overlapped with QTLs under control conditions and verify the major impact of growth environment on seed longevity. Few genes have been frequently identified, i.e. *DREB* and *APETALA2*. Their impacts require verification in future. To gain specific knowledge on seed deterioration mechanisms and their genetic control in seedbank storage, the studies must be carried out on seeds stored under dry conditions. Otherwise, the results are only valid for those conditions for which seeds are tested for.

Seedbanks are essential to preserve the genetic resources and provide the basis for our crop production in future. The maintenance of seed viability is the key to preserve the genetic integrity and the genotype of those plants which farmers and breeders have been selected for centuries. However, species and genotypes differ in their requirements on storage conditions and the ability to deteriorate. Aspects about the function of water, temperature, O_2, the physical state of the cytoplasm and biochemical reactions in seeds or even compartments exposed to different conditions are hardly understood and optimized storage procedures for each species are desirable. Techniques, such as seed cryopreservation or plasma coating, may potentially prolong the storage period. However, a comprehensive understanding of seed deterioration under dry condition is required first, to gain success with other approaches.

Universal tools for rapid and non-destructive viability monitoring remain illusory. The diversity of damaging chemical reactions compromises the development of universal genetic and biochemical makers. Though, the redox potential $E_{GSSG/2GSH}$ represents a reliable viability marker for wheat and barley seeds, its measurement is complex and not adapted to many species. Therefore, the phenotyping industry may fill the gap and provide destructive, but high-throughput approaches for rapid evaluation of germination and other morphological traits. However, all approaches are finally challenged by the high and fascinating plant genetic diversity maintained in a gene bank.

6 REFERENCES OF PROLOG AND EPILOG

Adl, S.M., Simpson, A.G.B., Farmer, M.A., Andersen, R.A., Anderson, O.R., Barta, J.R., Bowser, S.S., Brugerolle, G.U.Y., Fensome, R.A., Fredericq, S., James, T.Y., Karpov, S., Kugrens, P., Krug, J., Lane, C.E., Lewis, L.A., Lodge, J., Lynn, D.H., Mann, D.G., Mccourt, R.M., Mendoza, L., Moestrup, Ø., Mozley-Standridge, S.E., Nerad, T.A., Shearer, C.A., Smirnov, A.V., Spiegel, F.W., and Taylor, M.F.J.R. (2005). The new higher level classification of eukaryotes with emphasis on the taxonomy of protists. *Journal of Eukaryotic Microbiology* 52, 399-451.https://doi.org/10.1111/j.1550-7408.2005.00053.x

Alercia, A., López, F.M., Sackville Hamilton, N.R., and Marsella, M. (2018). *Object Identifiers for food crops - Descriptors and guidelines of the Global Information System.* Rome.

Almoguera, C., Personat, J.M., Prieto-Dapena, P., and Jordano, J. (2015). Heat shock transcription factors involved in seed desiccation tolerance and longevity retard vegetative senescence in transgenic tobacco. *Planta* 242, 461-475.https://doi.org/10.1007/s00425-015-2336-y

Alscher, R.G., Erturk, N., and Heath, L.S. (2002). Role of superoxide dismutases (SODs) in controlling oxidative stress in plants. *Journal of Experimental Botany* 53, 1331-1341.https://doi.org/10.1093/jexbot/53.372.1331

Angell, C.A. (2008). Insights into phases of liquid water from study of its unusual glass-forming properties. *Science* 319, 582-587.https://doi.org/10.1126/science.1131939

Angelovici, R., Galili, G., Fernie, A.R., and Fait, A. (2010). Seed desiccation: A bridge between maturation and germination. *Trends in Plant Science* 15, 211-218.http://dx.doi.org/10.1016/j.tplants.2010.01.003

Apel, K., and Hirt, H. (2004). Reactive oxygen species: Metabolism, oxidative stress, and signal transduction. *Annual Review of Plant Biology* 55, 373-399.https://doi.org/10.1146/annurev.arplant.55.031903.141701

Awan, S., Footitt, S., and Finch-Savage, W.E. (2018). Interaction of maternal environment and allelic differences in seed vigour genes determines seed performance in *Brassica oleracea*. *The Plant Journal* 94, 1098-1108.https://doi.org/10.1111/tpj.13922

Ayala, A., Muñoz, M.F., and Argüelles, S. (2014). Lipid peroxidation: Production, metabolism, and signaling mechanisms of malondialdehyde and 4-hydroxy-2-nonenal. *Oxidative Medicine and Cellular Longevity* 2014, 31.https://doi.org/10.1155/2014/360438

Bader Ul Ain, H., Saeed, F., Ahmad, N., Imran, A., Niaz, B., Afzaal, M., Imran, M., Tufail, T., and Javed, A. (2018). Functional and health-endorsing properties of wheat and barley cell wall's non-starch polysaccharides. *International Journal of Food Properties* 21, 1463-1480.https://doi.org/10.1080/10942912.2018.1489837

Bailly, C., Benamar, A., Corbineau, F., and Come, D. (1996). Changes in malondialdehyde content and in superoxide dismutase, catalase and glutathione reductase activities in sunflower seeds as related to deterioration during accelerated aging. *Physiologia Plantarum* 97, 104-110.https://doi.org/10.1111/j.1399-3054.1996.tb00485.x

Baldwin, R.L. (2014). Dynamic hydration shell restores Kauzmann's 1959 explanation of how the hydrophobic factor drives protein folding. *Proceedings of the National Academy of Sciences of the United States of America* 111, 13052-13056.https://doi.org/10.1073/pnas.1414556111

Ball, P. (2008). Water as an active constituent in cell biology. *Chemical Reviews* 108, 74-108.https://doi.org/10.1021/cr068037a

Ball, P. (2017). Water is an active matrix of life for cell and molecular biology. *Proceedings of the National Academy of Sciences of the United States of America* 114, 13327-13335.https://doi.org/10.1073/pnas.1703781114

Ballesteros, D., and Walters, C. (2011). Detailed characterization of mechanical properties and molecular mobility within dry seed glasses: Relevance to the physiology of dry biological systems. *The Plant Journal* 68, 607-619.https://doi.org/10.1111/j.1365-313X.2011.04711.x

Bancroft, I., Morgan, C., Fraser, F., Higgins, J., Wells, R., Clissold, L., Baker, D., Long, Y., Meng, J.L., Wang, X.W., Liu, S.Y., and Trick, M. (2011). Dissecting the genome of the polyploid crop oilseed rape by transcriptome sequencing. *Nature Biotechnology* 29, 762-U128.https://doi.org/10.1038/nbt.1926

Bartels-Rausch, T., Bergeron, V., Cartwright, J.H.E., Escribano, R., Finney, J.L., Grothe, H., Gutiérrez, P.J., Haapala, J., Kuhs, W.F., Pettersson, J.B.C., Price, S.D., Sainz-Díaz, C.I., Stokes, D.J., Strazzulla, G., Thomson, E.S., Trinks, H., and Uras-Aytemiz, N. (2012). Ice structures, patterns, and processes: A view across the icefields. *Reviews of Modern Physics* 84, 885-944.https://doi.org/10.1103/RevModPhys.84.885

Baskin, J.M., and Baskin, C.C. (2004). A classification system for seed dormancy. *Seed Science Research* 14, 1-16.https://doi.org/10.1079/SSR2003150

Baud, S., Boutin, J.P., Miquel, M., Lepiniec, L., and Rochat, C. (2002). An integrated overview of seed development in *Arabidopsis thaliana* ecotype WS. *Plant Physiology and Biochemistry* 40, 151-160.https://doi.org/10.1016/S0981-9428(01)01350-X

Bayer, P.E., Hurgobin, B., Golicz, A.A., Chan, C.K.K., Yuan, Y., Lee, H., Renton, M., Meng, J., Li, R., Long, Y., Zou, J., Bancroft, I., Chalhoub, B., King, G.J., Batley, J., and Edwards, D. (2017). Assembly and comparison of two closely related *Brassica napus* genomes. *Plant Biotechnology Journal* 15, 1602-1610.https://doi.org/10.1111/pbi.12742

Bela, K., Horváth, E., Gallé, Á., Szabados, L., Tari, I., and Csiszár, J. (2015). Plant glutathione peroxidases: Emerging role of the antioxidant enzymes in plant development and stress responses. *Journal of Plant Physiology* 176, 192-201.https://doi.org/10.1016/j.jplph.2014.12.014

Bentsink, L., Alonso-Blanco, C., Vreugdenhil, D., Tesnier, K., Groot, S.P.C., and Koornneef, M. (2000). Genetic analysis of seed-soluble oligosaccharides in relation to seed storability of *Arabidopsis*. *Plant Physiology* 124, 1595-1604.https://doi.org/10.1104/pp.124.4.1595

Bewley, J.D. (1997). Seed germination and dormancy. *Plant Cell* 9, 1055-1066

Bewley, J.D., Bradford, K.J., Hilhorst, H.W.M., and Nonogaki, H. (2013). *Seeds: Physiology of development, germination and dormancy, 3rd edition.* New York: Springer.

Borisjuk, L., and Rolletschek, H. (2009). The oxygen status of the developing seed. *New Phytologist* 182, 17-30.https://doi.org/10.1111/j.1469-8137.2008.02752.x

Borisjuk, L., Walenta, S., Rolletschek, H., Mueller-Klieser, W., Wobus, U., and Weber, H. (2002). Spatial analysis of plant metabolism: Sucrose imaging within *Vicia faba* cotyledons reveals specific developmental patterns. *The Plant Journal* 29, 521-530.https://doi.org/10.1046/j.1365-313x.2002.01222.x

Börner, A., Nagel, M., Agacka-Mołdoch, M., Gierke, P.U., Oberforster, M., Albrecht, T., and Mohler, V. (2018). QTL analysis of falling number and seed longevity in wheat (*Triticum aestivum* L.). *Journal of Applied Genetics* 59, 35-42.https://doi.org/10.1007/s13353-017-0422-5

Buitink, J., Hemminga, M.A., and Hoekstra, F.A. (1999). Characterization of molecular mobility in seed tissues: An electron paramagnetic resonance spin probe study. *Biophysical Journal* 76, 3315-3322.https://doi.org/10.1016/S0006-3495(99)77484-9

Buitink, J., and Leprince, O. (2004). Glass formation in plant anhydrobiotes: Survival in the dry state. *Cryobiology* 48, 215-228.https://doi.org/10.1016/j.cryobiol.2004.02.011

Buitink, J., and Leprince, O. (2008). Intracellular glasses and seed survival in the dry state. *Comptes Rendus Biologies* 331, 788-795.https://doi.org/10.1016/j.crvi.2008.08.002

Buitink, J., Leprince, O., Hemminga, M.A., and Hoekstra, F.A. (2000). Molecular mobility in the cytoplasm: An approach to describe and predict lifespan of dry germplasm. *Proceedings of the National Academy of Sciences of the United States of America* 97, 2385-2390.https://doi.org/10.1073/pnas.040554797

Buitink, J., Waltersvertucci, C., Hoekstra, F.A., and Leprince, O. (1996). Calorimetric properties of dehydrating pollen - Analysis of a desiccation-tolerant and an intolerant species. *Plant Physiology* 111, 235-242.https://doi.org/10.1104/pp.111.1.235

Cairns, N.G., Pasternak, M., Wachter, A., Cobbett, C.S., and Meyer, A.J. (2006). Maturation of Arabidopsis seeds is dependent on glutathione biosynthesis within the embryo. *Plant Physiology* 141, 446-455.https://doi.org/10.1104/pp.106.077982

Candat, A., Paszkiewicz, G., Neveu, M., Gautier, R., Logan, D.C., Avelange-Macherel, M.-H., and Macherel, D. (2014). The ubiquitous distribution of late embryogenesis abundant proteins across cell compartments in *Arabidopsis* offers tailored protection against abiotic stress. *The Plant Cell Online*.https://doi.org/10.1105/tpc.114.127316

Cbd (2018a). *The Convention on Biological Diversity* [Online]. The Secretariat of the Convention on Biological Diversity. Available: https://www.cbd.int/convention/ [Accessed 5th October 2018].

Cbd (2018b). *Global Strategy for Plant Conservation 2011-2020: The targets 2011-2020* [Online]. The Secretariat of the Convention on Biological Diversity. Available: https://www.cbd.int/gspc/targets.shtml [Accessed 5th October 2018].

Chalhoub, B., Denoeud, F., Liu, S., Parkin, I.a.P., Tang, H., Wang, X., Chiquet, I., Belcram, H., Tong, C., Samans, B., Corréa, M., Da Silva, C., Just, J., Falentin, C., Koh, C.S., Le Clainche, I., Bernard, M., Bento, P., Noel, B., Labadie, K., Alberti, A., Charles, M., Arnaud, D., Guo, H., Daviaud, C., Alamery, S., Jabbari, K., Zhao, M., Edger, P.P., Chelaifa, H., Tack, D., Lassalle, G., Mestiri, I., Schnel, N., Le Paslier, M.-C., Fan, G., Renault, V., Bayer, P.E., Golicz, A.A., Manoli, S., Lee, T.-H., Thi, V.H.D., Chalabi, S., Hu, Q., Fan, C., Tollenaere, R., Lu, Y., Battail, C., Shen, J., Sidebottom, C.H.D., Wang, X., Canaguier, A., Chauveau, A., Bérard, A., Deniot, G., Guan, M., Liu, Z., Sun, F., Lim, Y.P., Lyons, E., Town, C.D., Bancroft, I., Wang, X., Meng, J., Ma, J., Pires, J.C., King, G.J., Brunel, D., Delourme, R., Renard, M., Aury, J.-M., Adams, K.L., Batley, J., Snowdon, R.J., Tost, J., Edwards, D., Zhou, Y., Hua, W., Sharpe, A.G., Paterson, A.H., Guan, C., and Wincker, P. (2014). Early allopolyploid evolution in the post-Neolithic *Brassica napus* oilseed genome. *Science* 345, 950-953.https://doi.org/10.1126/science.1253435

Champion, D., Le Meste, M., and Simatos, D. (2000). Towards an improved understanding of glass transition and relaxations in foods: molecular mobility in the glass transition range. *Trends in Food Science & Technology* 11, 41-55.https://doi.org/10.1016/S0924-2244(00)00047-9

Chang, B.S., Beauvais, R.M., Dong, A., and Carpenter, J.F. (1996). Physical factors affecting the storage stability of freeze-dried Interleukin-1 receptor antagonist: Glass transition and protein conformation. *Archives of Biochemistry and Biophysics* 331, 249-258.https://doi.org/10.1006/abbi.1996.0305

Chaplin, M.F. (2011). "The water molecule, liquid water, hydrogen bonds, and water networks," in *Water. The forgotten biological molecule,* eds. D. Le Bihan & H. Fukuyama. (Singapore: Pan Standford Publishing), 3-19.

Chaplin, M.F. (2018). *Water structure and science* [Online]. UK. Available: http://www1.lsbu.ac.uk/water/water_structure_science.html [Accessed 18th September 2018].

Chen, D.F., Li, Y.L., Fang, T., Shi, X.L., and Chen, X.W. (2016). Specific roles of tocopherols and tocotrienols in seed longevity and germination tolerance to abiotic stress in transgenic rice. *Plant Science* 244, 31-39.https://doi.org/10.1016/j.plantsci.2015.12.005

Chen, M., Ko, H.Y., Remsing, R.C., Calegari Andrade, M.F., Santra, B., Sun, Z., Selloni, A., Car, R., Klein, M.L., Perdew, J.P., and Wu, X. (2017). *Ab initio* theory and modeling of water. *Proceedings of the National Academy of Sciences of the United States of America* 114, 10846-10851.https://doi.org/10.1073/pnas.1712499114

Chen, M., Macgregor, D.R., Dave, A., Florance, H., Moore, K., Paszkiewicz, K., Smirnoff, N., Graham, I.A., and Penfield, S. (2014). Maternal temperature history activates *Flowering Locus T* in fruits to control progeny dormancy according to time of year. *Proceedings of the National Academy of Sciences of the United States of America* 111, 18787-18792.https://doi.org/10.1073/pnas.1412274111

Chen, X., Yin, G., Börner, A., Xin, X., He, J., Nagel, M., Liu, X., and Lu, X. (2018). Comparative physiology and proteomics of two wheat genotypes differing in seed storage tolerance. *Plant Physiology and Biochemistry* 130, 455-463.https://doi.org/10.1016/j.plaphy.2018.07.022

Choudhury, F.K., Rivero, R.M., Blumwald, E., and Mittler, R. (2017). Reactive oxygen species, abiotic stress and stress combination. *The Plant Journal* 90, 856-867.https://doi.org/10.1111/tpj.13299

Clerkx, E.J.M., Blankestijn-De Vries, H., Ruys, G.J., Groot, S.P.C., and Koornneef, M. (2004). Genetic differences in seed longevity of various Arabidopsis mutants. *Physiologia Plantarum* 121, 448-461.https://doi.org/10.1111/j.1399-3054.2004.00339.x

Colville, L., Bradley, E.L., Lloyd, A.S., Pritchard, H.W., Castle, L., and Kranner, I. (2012). Volatile fingerprints of seeds of four species indicate the involvement of alcoholic fermentation, lipid peroxidation, and Maillard reactions in seed deterioration during ageing and desiccation stress. *Journal of Experimental Botany* 63, 6519-6530.https://doi.org/10.1093/Jxb/Ers307

Colville, L., Saez, C.M.B., Lewis, G.P., and Kranner, I. (2015). The distribution of glutathione and homoglutathione in leaf, root and seed tissue of 73 species across the three sub-families of the Leguminosae. *Phytochemistry* 115, 175-183.https://doi.org/10.1016/j.phytochem.2015.01.011

Corradini, M.G., and Peleg, M. (2007). Shelf-life estimation from accelerated storage data. *Trends in Food Science & Technology* 18, 37-47.https://doi.org/10.1016/j.tifs.2006.07.011

Costa, M.C.D., Cooper, K., Hilhorst, H.W.M., and Farranta, J.M. (2017). Orthodox seeds and resurrection plants: Two of a kind? *Plant Physiology* 175, 589-599.https://doi.org/10.1104/pp.17.00760

Costa, M.C.D., Farrant, J.M., Oliver, M.J., Ligterink, W., Buitink, J., and Hilhorst, H.M.W. (2016). Key genes involved in desiccation tolerance and dormancy across life forms. *Plant Science* 251, 162-168.https://doi.org/10.1016/j.plantsci.2016.02.001

Costa, M.C.D., Righetti, K., Nijveen, H., Yazdanpanah, F., Ligterink, W., Buitink, J., and Hilhorst, H.W.M. (2015). A gene co-expression network predicts functional genes controlling the re-establishment of desiccation tolerance in germinated *Arabidopsis thaliana* seeds. *Planta* 242, 435-449.https://doi.org/10.1007/s00425-015-2283-7

Couto, N., Wood, J., and Barber, J. (2016). The role of glutathione reductase and related enzymes on cellular redox homoeostasis network. *Free Radical Biology and Medicine* 95, 27-42.https://doi.org/10.1016/j.freeradbiomed.2016.02.028

Crane, J., Kovach, D., Gardner, C., and Walters, C. (2006). Triacylglycerol phase and 'intermediate' seed storage physiology: a study of Cuphea carthagenensis. *Planta* 223, 1081-1089.https://doi.org/10.1007/s00425-005-0157-0

Crapiste, G.H., Brevedan, M.I.V., and Carelli, A.A. (1999). Oxidation of sunflower oil during storage. *Journal of the American Oil Chemists' Society* 76, 1437-1443.https://doi.org/10.1007/s11746-999-0181-5

Crop trust (2015). "Sowing the seed for tomorrow", (ed.) G.C.D.T.A. Icarda. (Bonn, Germany).

Curaba, J., Moritz, T., Blervaque, R., Parcy, F., Raz, V., Herzog, M., and Vachon, G. (2004). *AtGA3ox2*, a key gene responsible for bioactive gibberellin biosynthesis, is regulated during embryogenesis by *LEAFY COTYLEDON2* and *FUSCA3* in Arabidopsis. *Plant Physiology* 136, 3660-3669.https://doi.org/10.1104/pp.104.047266

Damania, A.B. (2008). History, achievements, and current status of genetic resources conservation. *Agronomy Journal* 100, S27-S39.https://doi.org/10.2134/agronj2005.0239cs

De Carvalho, M.a.a.P., Bebeli, P.J., Bettencourt, E., Costa, G., Dias, S., Dos Santos, T.M.M., and Slaski, J.J. (2013). Cereal landraces genetic resources in worldwide GeneBanks. A review. *Agronomy for Sustainable Development* 33, 177-203.https://doi.org/10.1007/s13593-012-0090-0

De Gara, L., De Pinto, M.C., and Arrigoni, O. (1997). Ascorbate synthesis and ascorbate peroxidase activity during the early stage of wheat germination. *Physiologia Plantarum* 100, 894-900.https://doi.org/10.1111/j.1399-3054.1997.tb00015.x

De Gara, L., De Pinto, M.C., Moliterni, V.M.C., and D'egidio, M.G. (2003). Redox regulation and storage processes during maturation in kernels of *Triticum durum*. *Journal of Experimental Botany* 54, 249-258.https://doi.org/10.1093/Jxb/Erg021

De Souza Vidigal, D., Willems, L., Van Arkel, J., Dekkers, B.J.W., Hilhorst, H.W.M., and Bentsink, L. (2016). Galactinol as marker for seed longevity. *Plant Science* 246, 112-118.https://doi.org/10.1016/j.plantsci.2016.02.015

Dekkers, B.J.W., He, H., Hanson, J., Willems, L.a.J., Jamar, D.C.L., Cueff, G., Rajjou, L., Hilhorst, H.W.M., and Bentsink, L. (2016). The Arabidopsis *DELAY OF GERMINATION 1* gene affects *ABSCISIC ACID INSENSITIVE 5* (*ABI5*) expression and genetically interacts with *ABI3* during Arabidopsis seed development. *The Plant Journal* 85, 451-465.https://doi.org/10.1111/tpj.13118

Delmas, F., Sankaranarayanan, S., Deb, S., Widdup, E., Bournonville, C., Bollier, N., Northey, J.G.B., Mccourt, P., and Samuel, M.A. (2013). *ABI3* controls embryo degreening through Mendel's I locus. *Proceedings of the National Academy of Sciences of the United States of America* 110, E3888-E3894.https://doi.org/10.1073/pnas.1308114110

Delwiche, Charles f., and Cooper, Endymion d. (2015). The evolutionary origin of a terrestrial flora. *Current Biology* 25, R899-R910.https://doi.org/10.1016/j.cub.2015.08.029

Demir, I., Mavi, K., Kenanoglu, B.B., and Matthews, S. (2008). Prediction of germination and vigour in naturally aged commercially available seed lots of cabbage (*Brassica oleracea* var. *capitata*) using the bulk conductivity method. *Seed Science and Technology* 36, 509-523.https://doi.org/10.15258/sst.2008.36.3.01

Deng, B.L., Yang, K.J., Zhang, Y.F., and Li, Z.T. (2017). Can antioxidant's reactive oxygen species (ROS) scavenging capacity contribute to aged seed recovery? Contrasting effect of melatonin, ascorbate and glutathione on germination ability of aged maize seeds. *Free Radical Research* 51, 765-771.https://doi.org/10.1080/10715762.2017.1375099

Dixon, D.P., Davis, B.G., and Edwards, R. (2002). Functional divergence in the glutathione transferase superfamily in plants: Identification of two classes with putative functions in redox homeostasis in *Arabidopsis thaliana*. *Journal of Biological Chemistry* 277, 30859-30869.https://doi.org/10.1074/jbc.M202919200

Dixon, D.P., Steel, P.G., and Edwards, R. (2011). Roles for glutathione transferases in antioxidant recycling. *Plant Signaling & Behavior* 6, 1223-1227.https://doi.org/10.4161/psb.6.8.16253

Dong, Z.Y., Feng, B., Liang, H., Rong, C.W., Zhang, K.P., Cao, X.M., Qin, H.J., Liu, X., Wang, T., and Wang, D.W. (2015). Grain-specific reduction in lipoxygenase activity improves flour color quality and seed longevity in common wheat. *Molecular Breeding* 35.https://doi.org/10.1007/s11032-015-0347-9

Dormann, P. (2007). Functional diversity of tocochromanols in plants. *Planta* 225, 269-276.https://doi.org/10.1007/s00425-006-0438-2

Earle, F.R., and Jones, Q. (1962). Analyses of seed samples from 113 plant families. *Economic Botany* 16, 221-250

Edwards, E.A., Rawsthorne, S., and Mullineaux, P.M. (1990). Subcellular distribution of multiple forms of glutathione reductase in leaves of pea (*Pisum sativum* L.). *Planta* 180, 278-284.https://doi.org/10.1007/bf00194008

El-Maarouf-Bouteau, H., Mazuy, C., Corbineau, F., and Bailly, C. (2011). DNA alteration and programmed cell death during ageing of sunflower seed. *Journal of Experimental Botany* 62, 5003-5011.https://doi.org/10.1093/jxb/err198

Ellis, R.H., and Hong, T.D. (2007). Seed longevity - moisture content relationships in hermetic and open storage. *Seed Science and Technology* 35, 423-431.https://doi.org/10.15258/sst.2007.35.2.17

Ellis, R.H., and Roberts, E.H. (1980). Improved equations for the prediction of seed longevity. *Annals of Botany* 45, 13-30.https://doi.org/10.1093/oxfordjournals.aob.a085797

Evers, A.D., Blakeney, A.B., and O´Brien, L. (1999). Cereal structure and composition. *Australian Journal of Agricultural Research* 50, 629-650.https://doi.org/10.1071/AR98158

Evershed, R.P., Bland, H.A., Bergen, P.F.V., Carter, J.F., Horton, M.C., and Rowleyconwy, P.A. (1997). Volatile compounds in archaeological plant remains and the Maillard reaction during decay of organic matter. *Science* 278, 432-433.https://doi.org/10.1126/science.278.5337.432

Falk, J., Krahnstover, A., Van Der Kooij, T.a.W., Schlensog, M., and Krupinska, K. (2004). Tocopherol and tocotrienol accumulation during development of caryopses from barley (*Hordeum vulgare* L.). *Phytochemistry* 65, 2977-2985.https://doi.org/10.1016/j.phytochem.2004.08.047

Falk, J., and Munné-Bosch, S. (2010). Tocochromanol functions in plants: antioxidation and beyond. *Journal of Experimental Botany* 61, 1549-1566.https://doi.org/10.1093/jxb/erq030

Fan, Z. (2017). Barley starch: Composition, structure, properties, and modifications. *Comprehensive Reviews in Food Science and Food Safety* 16, 558-579.https://doi.org/10.1111/1541-4337.12265

Fao (2010). "The second report on the state of the world's plant genetic resources for food and agriculture". (Rome).

Fao (2014). *Genebank standards for plant genetic resources for food and agriculture.* Food and Agriculture Organization of the United Nations, Rome.

Fernandez Marin, B., Kranner, I., San Sebastian, M., Artetxe, U., Laza, J.M., Vilas, J.L., Pritchard, H.W., Nadajaran, J., Miguez, F., Becerril, J.M., and Garcia-Plazaola, J.I. (2013). Evidence for the absence of enzymatic reactions in the glassy state. A case study of xanthophyll cycle pigments in the desiccation-tolerant moss *Syntrichia ruralis*. *Journal of Experimental Botany* 64, 3033-3043.https://doi.org/10.1093/jxb/ert145

Finch-Savage, W.E., and Leubner-Metzger, G. (2006). Seed dormancy and the control of germination. *New Phytologist* 171, 501-523.https://doi.org/10.1111/j.1469-8137.2006.01787.x

Fleming, M.B., Richards, C.M., and Walters, C. (2017). Decline in RNA integrity of dry-stored soybean seeds correlates with loss of germination potential. *Journal of Experimental Botany* 68, 2219-2230.https://doi.org/10.1093/jxb/erx100

Fox, G.P. (2010). "Chemical composition in barley grains and malt quality," in *Genetics and Improvement of Barley Malt Quality,* eds. G. Zhang & C. Li. (Berlin, Heidelberg: Springer Berlin Heidelberg), 63-98.

Foyer, C.H., and Noctor, G. (2005). Redox homeostasis and antioxidant signaling: A metabolic interface between stress perception and physiological responses. *Plant Cell* 17, 1866-1875.https://doi.org/10.1105/tpc.105.033589

Foyer, C.H., and Noctor, G. (2011). Ascorbate and glutathione: The heart of the redox hub. *Plant Physiology* 155, 2-18.https://doi.org/10.1104/pp.110.167569

Franckowiak, D., Lundqvist, U., and Konishi, T. (1997). New and revised names for barley genes. *Barley Genetic Newsletter* 26, 22-516

Gaff, D.F., and Oliver, M. (2013). The evolution of desiccation tolerance in angiosperm plants: A rare yet common phenomenon. *Functional Plant Biology* 40, 315-328.http://dx.doi.org/10.1071/FP12321

Gepts, P. (2006). Plant genetic resources conservation and utilization: The accomplishments and future of a societal insurance policy. *Crop Science* 46, 2278-2292.https://doi.org/10.2135/cropsci2006.03.0169gas

Gerna, D., Roach, T., Stoggl, W., Wagner, J., Vaccino, P., Limonta, M., and Kranner, I. (2017). Changes in low-molecular-weight thiol-disulphide redox couples are part of bread wheat seed germination and early seedling growth. *Free Radical Research* 51, 568-581.https://doi.org/10.1080/10715762.2017.1338344

Gill, S.S., Anjum, N.A., Hasanuzzaman, M., Gill, R., Trivedi, D.K., Ahmad, I., Pereira, E., and Tuteja, N. (2013). Glutathione and glutathione reductase: A boon in disguise for plant abiotic stress defense operations. *Plant Physiology and Biochemistry* 70, 204-212.https://doi.org/10.1016/j.plaphy.2013.05.032

Gill, S.S., and Tuteja, N. (2010). Reactive oxygen species and antioxidant machinery in abiotic stress tolerance in crop plants. *Plant Physiology and Biochemistry* 48, 909-930.https://doi.org/10.1016/j.plaphy.2010.08.016

Grando, S., and Gormez Macpherson, H. (Year). "Food barley: Importance, uses and local knowledge", in: *Proceedings of the International Workshop on Food Barley Improvement, 14th-17th January 2002*).

Groot, S.P.C., De Groot, L., Kodde, J., and Van Treuren, R. (2014). Prolonging the longevity of ex situ conserved seeds by storage under anoxia. *Plant Genetic Resources: Characterization and Utilization* 13, 18-26.https://doi.org/10.1017/S1479262114000586

Groot, S.P.C., Surki, A.A., De Vos, R.C.H., and Kodde, J. (2012). Seed storage at elevated partial pressure of oxygen, a fast method for analysing seed ageing under dry conditions. *Annals of Botany* 110, 1149-1159.https://doi.org/10.1093/aob/mcs198

Gualano, N.A., and Benech-Arnold, R.L. (2009). Predicting pre-harvest sprouting susceptibility in barley: Looking for "sensitivity windows" to temperature throughout grain filling in various commercial cultivars. *Field Crops Research* 114, 35-44.https://doi.org/10.1016/j.fcr.2009.06.016

Gutierrez, L., Van Wuytswinkel, O., Castelain, M., and Bellini, C. (2007). Combined networks regulating seed maturation. *Trends in Plant Science* 12, 294-300.https://doi.org/10.1016/j.tplants.2007.06.003

Haferkamp, M.E., Smith, L., and Nilan, R.A. (1953). Studies on aged seeds I. Relation of age of seed to germination and longevity. *Agronomy Journal* 45, 434-437

Halewood, M., Chiurugwi, T., Hamilton, R.S., Kurtz, B., Marden, E., Welch, E., Michiels, F., Mozafari, J., Sabran, M., Patron, N., Kersey, P., Bastow, R., Dorius, S., Dias, S., Mccouch, S., and Powell, W. (2018). Plant genetic resources for food and agriculture: opportunities and challenges emerging from the science and information technology revolution. *New Phytologist* 217, 1407-1419.https://doi.org/10.1111/nph.14993

Hampton, J.G., and Tekrony, D.M. (1995). *Handbook of vigour test methods.* Zürich: International Seed Testing Association.

Harrington, J.F. (1963). "Practical instructions and advice on seed storage", in: *Proceedings of the International Seed Testing Association*).

Hay, F.R., Adams, J., Manger, K., and Probert, R. (2008). The use of non-saturated lithium chloride solutions for experimental control of seed water content. *Seed Science and Technology* 36, 737-746.https://doi.org/10.15258/sst.2008.36.3.23

Hay, F.R., Probert, R.J., and Smith, R.D. (1997). The effect of maturity on the moisture relations of seed longevity in foxglove (*Digitalis purpurea* L.). *Seed Science Research* 7, 341-349.https://doi.org/10.1017/S0960258500003743

Hay, F.R., Timple, S., and Van Duijn, B. (2015). Can chlorophyll fluorescence be used to determine the optimal time to harvest rice seeds for long-term genebank storage? *Seed Science Research* 25, 321-334.https://doi.org/10.1017/S0960258515000082

He, G., Guan, C.-N., Chen, Q.-X., Gou, X.-J., Liu, W., Zeng, Q.-Y., and Lan, T. (2016a). Genome-wide analysis of the glutathione S-transferase gene family in *Capsella rubella*: Identification, expression, and biochemical functions. *Frontiers in Plant Science* 7.https://doi.org/10.3389/fpls.2016.01325

He, H.Z., Willems, L.a.J., Batushansky, A., Fait, A., Hanson, J., Nijveen, H., Hilhorst, H.W.M., and Bentsink, L. (2016b). Effects of parental temperature and nitrate on seed performance are reflected by partly overlapping genetic and metabolic pathways. *Plant and Cell Physiology* 57, 473-487.https://doi.org/10.1093/pcp/pcv207

Heck, D.E., Shakarjian, M., Kim, H.D., Laskin, J.D., and Vetrano, A.M. (2010). Mechanisms of oxidant generation by catalase. *Annals of the New York Academy of Sciences* 1203, 120-125.https://doi.org/10.1111/j.1749-6632.2010.05603.x

Herschbach, C., and Rennenberg, H. (1994). Influence of glutathione (GSH) on net uptake of sulphate and sulphate transport in tobacco plants. *Journal of Experimental Botany* 45, 1069-1076.https://doi.org/10.1093/jxb/45.8.1069

Hilhorst, H.W.M., Costa, M.C.D., and Farrant, J.M. (2018). A footprint of plant desiccation tolerance. Does it exist? *Molecular Plant* 11, 1003-1005.https://doi.org/10.1016/j.molp.2018.07.001

Hincha, D.K., and Thalhammer, A. (2012). LEA proteins: IDPs with versatile functions in cellular dehydration tolerance. *Biochemical Society Transactions* 40, 1000-1003.https://doi.org/10.1042/Bst20120109

Hincha, D.K., Zuther, E., Hellwege, E.M., and Heyer, A.G. (2002). Specific effects of fructo- and gluco-oligosaccharides in the preservation of liposomes during drying. *Glycobiology* 12, 103-110.https://doi.org/10.1093/glycob/12.2.103

Hincha, D.K., Zuther, E., and Heyer, A.G. (2003). The preservation of liposomes by raffinose family oligosaccharides during drying is mediated by effects on fusion and lipid phase transitions. *Biochimica et Biophysica Acta-Biomembranes* 1612, 172-177.https://doi.org/10.1016/S0005-2736(03)00116-0

Holdsworth, M.J., Bentsink, L., and Soppe, W.J.J. (2008). Molecular networks regulating Arabidopsis seed maturation, after-ripening, dormancy and germination. *New Phytologist* 179, 33-54.https://doi.org/10.1111/j.1469-8137.2008.02437.x

Holopainen, U.R.M., Pihlava, J.M., Serenius, M., Hietaniemi, V., Wilhelmson, A., Poutanen, K., and Lehtinen, P. (2014). Milling, water uptake, and modification properties of different barley (*Hordeum vulgare* L.) lots in relation to grain composition and structure. *Journal of Agricultural and Food Chemistry* 62, 8875-8882.https://doi.org/10.1021/jf500857e

Horvath, G., Wessjohann, L., Bigirimana, J., Jansen, M., Guisez, Y., Caubergs, R., and Horemans, N. (2006). Differential distribution of tocopherols and tocotrienols in photosynthetic and non-photosynthetic tissues. *Phytochemistry* 67, 1185-1195.https://doi.org/10.1016/j.phytochem.2006.04.004

Houston, K., Mckim, S.M., Comadran, J., Bonar, N., Druka, I., Uzrek, N., Cirillo, E., Guzy-Wrobelska, J., Collins, N.C., Halpin, C., Hansson, M., Dockter, C., Druka, A., and Waugh, R. (2013). Variation in the interaction between alleles of *HvAPETALA2* and microRNA172 determines the density of grains on the barley inflorescence. *Proceedings of the National Academy of Sciences of the United States of America* 110, 16675-16680.https://doi.org/10.1073/pnas.1311681110

Hu, Z.-Y., Hua, W., Zhang, L., Deng, L.-B., Wang, X.-F., Liu, G.-H., Hao, W.-J., and Wang, H.-Z. (2013). Seed structure characteristics to form ultrahigh oil content in rapeseed. *PLOS One* 8, e62099.https://doi.org/10.1371/journal.pone.0062099

Huang, J.X., Cai, M.H., Long, Q.Z., Liu, L.L., Lin, Q.Y., Jiang, L., Chen, S.H., and Wan, J.M. (2014). *OsLOX2*, a rice type I lipoxygenase, confers opposite effects on seed germination and longevity. *Transgenic Research* 23, 643-655.https://doi.org/10.1007/s11248-014-9803-2

Hummer, K.E., and Hancock, J.F. (2015). Vavilovian Centers of plant diversity: Implications and impacts. *HortScience* 50, 780-783

Icarda (2016). "Expanded crop genebank opens in Lebanon". (Lebanon).

Ipk (2018). *Scientific Report 2016/ 2017* [Online]. IPK Gatersleben. Available: http://www.ipk-gatersleben.de/fileadmin/content-ipk/content-ipk-forschung/Forschung/Download/Forschungsbericht_2016_2017.pdf [Accessed 18th September 2018].

Ista (2018). *International Rules for Seed Testing.* Bassersdorf, Switzerland: International Seed Testing Association.

Iwgsc (2018). Shifting the limits in wheat research and breeding using a fully annotated reference genome. *Science* 361.https://doi.org/10.1126/science.aar7191

Jacob, P., Hirt, H., and Bendahmane, A. (2017). The heat-shock protein/chaperone network and multiple stress resistance. *Plant Biotechnology Journal* 15, 405-414.https://doi.org/10.1111/pbi.12659

Jeevan Kumar, S.P., Rajendra Prasad, S., Banerjee, R., and Thammineni, C. (2015). Seed birth to death: Dual functions of reactive oxygen species in seed physiology. *Annals of Botany* 116, 663-668.https://doi.org/10.1093/aob/mcv098

Jiang, J.F., He, X., Li, L., Li, J.G., Shao, H.L., Xu, Q.L., Ye, R.H., and Dong, Y.H. (2014). Effect of cold plasma treatment on seed germination and growth of wheat. *Plasma Science & Technology* 16.https://doi.org/10.1088/1009-0630/16/1/12

Jones, Q., and Earle, F.R. (1966). Chemical analyses of seeds II: Oil and protein content of 759 species. *Economic Botany* 20, 127-155

Kameswara Rao, N., Dulloo, M.E., and Engels, J.M.M. (2017). A review of factors that influence the production of quality seed for long-term conservation in genebanks. *Genetic Resources and Crop Evolution* 64, 1061-1074.https://doi.org/10.1007/s10722-016-0425-9

Karmas, R., Pilar Buera, M., and Karel, M. (1992). Effect of glass transition on rates of nonenzymic browning in food systems. *Journal of Agricultural and Food Chemistry* 40, 873-879.https://doi.org/10.1021/jf00017a035

Kew, R.B.G. (2018). *Seed Information Database (SID). Version 7.1* [Online]. Available: http://data.kew.org/sid/ [Accessed 27th August 2018].

Kibinza, S., Bazin, J., Bailly, C., Farrant, J.M., Corbineau, F., and El-Maarouf-Bouteau, H. (2011). Catalase is a key enzyme in seed recovery from ageing during priming. *Plant Science* 181, 309-315.https://doi.org/10.1016/j.plantsci.2011.06.003

Kibinza, S., Vinel, D., Come, D., Bailly, C., and Corbineau, F. (2006). Sunflower seed deterioration as related to moisture content during ageing, energy metabolism and active oxygen species scavenging. *Physiologia Plantarum* 128, 496-506.https://doi.org/10.1111/j.1399-3054.2006.00771.x

Kilian, B., and Graner, A. (2012). NGS technologies for analyzing germplasm diversity in genebanks. *Briefings in Functional Genomics* 11, 38-50.https://doi.org/10.1093/Bfgp/Elr046

Kliebenstein, D.J., Monde, R.-A., and Last, R.L. (1998). Superoxide dismutase in Arabidopsis: An eclectic enzyme family with disparate regulation and protein localization. *Plant Physiology* 118, 637-650.https://doi.org/10.1104/pp.118.2.637

Koch, K. (2004). Sucrose metabolism: Regulatory mechanisms and pivotal roles in sugar sensing and plant development. *Current Opinion in Plant Biology* 7, 235-246.https://doi.org/10.1016/j.pbi.2004.03.014

Kochanek, J., Steadman, K.J., Probert, R.J., and Adkins, S.W. (2011). Parental effects modulate seed longevity: Exploring parental and offspring phenotypes to elucidate pre-zygotic environmental influences. *New Phytologist* 191, 223-233.https://doi.org/10.1111/j.1469-8137.2011.03681.x

Kofuji, K., Aoki, A., Tsubaki, K., Konishi, M., Isobe, T., and Murata, Y. (2012). Antioxidant activity of β-glucan. *ISRN Pharmaceutics* 2012, 125864.https://doi.org/10.5402/2012/125864

Kong, L.G., Guo, H.H., and Sun, M.Z. (2015). Signal transduction during wheat grain development. *Planta* 241, 789-801.https://doi.org/10.1007/s00425-015-2260-1

Kopriva, S., and Rennenberg, H. (2004). Control of sulphate assimilation and glutathione synthesis: interaction with N and C metabolism. *Journal of Experimental Botany* 55, 1831-1842.https://doi.org/10.1093/jxb/erh203

Kranner, I., Birtic, S., Anderson, K.M., and Pritchard, H.W. (2006). Glutathione half-cell reduction potential: A universal stress marker and modulator of programmed cell death? *Free Radical Biology and Medicine* 40, 2155-2165.https://doi.org/10.1016/j.freeradbiomed.2006.02.013

Kranner, I., and Grill, D. (1996). Significance of thiol-disulfide exchange in resting stages of plant development. *Botanica Acta* 109, 8-14.https://doi.org/10.1111/j.1438-8677.1996.tb00864.x

Kranner, I., Kastberger, G., Hartbauer, M., and Pritchard, H.W. (2010a). Noninvasive diagnosis of seed viability using infrared thermography. *Proceedings of the National Academy of Sciences of the United States of America* 107, 3912-3917.https://doi.org/10.1073/pnas.0914197107

Kranner, I., Minibayeva, F.V., Beckett, R.P., and Seal, C.E. (2010b). What is stress? Concepts, definitions and applications in seed science. *New Phytologist* 188, 655-673.https://doi.org/10.1111/j.1469-8137.2010.03461.x

Kreuzwieser, J., Schnitzler, J.-P., and Steinbrecher, R. (1999). Biosynthesis of organic compounds emitted by plants. *Plant Biology* 1, 149-159.https://doi.org/10.1111/j.1438-8677.1999.tb00238.x

Krieger-Liszkay, A., Fufezan, C., and Trebst, A. (2008). Singlet oxygen production in photosystem II and related protection mechanism. *Photosynthesis Research* 98, 551-564.https://doi.org/10.1007/s11120-008-9349-3

Kuznetsova, I.M., Turoverov, K.K., and Uversky, V.N. (2014). What macromolecular crowding can do to a protein. *International Journal of Molecular Sciences* 15, 23090-23140.https://doi.org/10.3390/ijms151223090

Labuza, T.P., Heidelbaugh, N.D., Silver, M., and Karel, M. (1971). Oxidation at intermediate moisture contents. *Journal of the American Oil Chemists' Society* 48, 86-90.https://doi.org/10.1007/BF02635692

Leckband, G., Frauen, M., and Friedt, W. (2002). NAPUS 2000. Rapeseed (*Brassica napus*) breeding for improved human nutrition. *Food Research International* 35, 273-278.https://doi.org/10.1016/S0963-9969(01)00196-X

Lee, Y.P., Baek, K.-H., Lee, H.-S., Kwak, S.-S., Bang, J.-W., and Kwon, S.-Y. (2010). Tobacco seeds simultaneously over-expressing Cu/Zn-superoxide dismutase and ascorbate peroxidase display enhanced seed longevity and germination rates under stress conditions. *Journal of Experimental Botany* 61, 2499-2506.10.1093/jxb/erq085

Lehner, A., Bailly, C., Fleche, B., Poels, P., Come, D., and Corbineau, F. (2006). Changes in wheat seed germination ability, soluble carbohydrate and antioxidant enzyme activities in the embryo during the desiccation phase of maturation. *Journal of Cereal Science* 43, 175-182.https://doi.org/10.1016/j.jcs.2005.07.005

Lehner, A., Mamadou, N., Poels, P., Come, D., Bailly, C., and Corbineau, F. (2008). Changes in soluble carbohydrates, lipid peroxidation and antioxidant enzyme activities in the embryo during ageing in wheat grains. *Journal of Cereal Science* 47, 555-565.https://doi.org/10.1016/j.jcs.2007.06.017

Leprince, O., Deltour, R., Thorpe, P.C., Atherton, N.M., and Hendry, G.a.F. (1990). The role of free-radicals and radical processing systems in loss of desiccation tolerance in germinating maize (*Zea mays* L.). *New Phytologist* 116, 573-580.https://doi.org/10.1111/j.1469-8137.1990.tb00541.x

Leprince, O., Pellizzaro, A., Berriri, S., and Buitink, J. (2017). Late seed maturation: Drying without dying. *Journal of Experimental Botany* 68, 827-841.https://doi.org/10.1093/jxb/erw363

Leubner-Metzger, G. (2005). β-1,3-glucanase gene expression in low-hydrated seeds as a mechanism for dormancy release during tobacco after-ripening. *The Plant Journal* 41, 133-145.https://doi.org/10.1111/j.1365-313X.2004.02284.x

Li, G., Na, Y.-W., Kwon, S.-W., and Park, Y.-J. (2014). Association analysis of seed longevity in rice under conventional and high-temperature germination conditions. *Plant Systematics and Evolution* 300, 389-402.https://doi.org/10.1007/s00606-013-0889-4

Lu, S.C. (2013). Glutathione synthesis. *Biochimica et Biophysica Acta (BBA) - General Subjects* 1830, 3143-3153.https://doi.org/10.1016/j.bbagen.2012.09.008

Lubzens, E., Cerdà, J., and Clark, M.S. (2010). "Introduction," in *Dormancy and resistance in harsh environments,* eds. E. Lubzens, J. Cerda & M. Clark. (Berlin, Heidelberg: Springer), 1-4.

Mackay, I., and Powell, W. (2007). Methods for linkage disequilibrium mapping in crops. *Trends in Plant Science* 12, 57-63.https://doi.org/10.1016/j.tplants.2006.12.001

Mascher, M., Gundlach, H., Himmelbach, A., Beier, S., Twardziok, S.O., Wicker, T., Radchuk, V., Dockter, C., Hedley, P.E., Russell, J., Bayer, M., Ramsay, L., Liu, H., Haberer, G., Zhang, X.Q., Zhang, Q.S., Barrero, R.A., Li, L., Taudien, S., Groth, M., Felder, M., Hastie, A., Simkova, H., Stankova, H., Vrana, J., Chan, S., Munoz-Amatrian, M., Ounit, R., Wanamaker, S., Bolser, D., Colmsee, C., Schmutzer, T., Aliyeva-Schnorr, L., Grasso, S., Tanskanen, J., Chailyan, A., Sampath, D., Heavens, D., Clissold, L., Cao, S.J., Chapman, B., Dai, F., Han, Y., Li, H., Li, X., Lin, C.Y., Mccooke, J.K., Tan, C., Wang, P.H., Wang, S.B., Yin, S.Y., Zhou, G.F., Poland, J.A., Bellgard, M.I., Borisjuk, L., Houben, A., Dolezel, J., Ayling, S., Lonardi, S., Kersey, P., Lagridge, P., Muehlbauer, G.J., Clark, M.D., Caccamo, M., Schulman, A.H., Mayer, K.F.X., Platzer, M., Close, T.J., Scholz, U., Hansson, M., Zhang, G.P., Braumann, I., Spannagl, M., Li, C.D., Waugh, R., and Stein, N. (2017). A chromosome conformation capture ordered sequence of the barley genome. *Nature* 544, 426-433.https://doi.org/10.1038/nature22043

Matthaus, B., Ozcan, M.M., and Al Juhaimi, F. (2016). Some rape/canola seed oils: fatty acid composition and tocopherols. *Zeitschrift für Naturforschung C-A Journal of Biosciences* 71, 73-77.https://doi.org/10.1515/znc-2016-0003

Mayer, K.F.X., Waugh, R., Langridge, P., Close, T.J., Wise, R.P., Graner, A., Matsumoto, T., Sato, K., Schulman, A., Muehlbauer, G.J., Stein, N., Ariyadasa, R., Schulte, D., Poursarebani, N., Zhou, R.N., Steuernagel, B., Mascher, M., Scholz, U., Shi, B.J., Langridge, P., Madishetty, K., Svensson, J.T., Bhat, P., Moscou, M., Resnik, J., Close, T.J., Muehlbauer, G.J., Hedley, P., Liu, H., Morris, J., Waugh, R., Frenkel, Z., Korol, A., Berges, H., Graner, A., Stein, N., Steuernagel, B., Taudien, S., Groth, M., Felder, M., Platzer, M., Brown, J.W.S., Schulman, A., Platzer, M., Fincher, G.B., Muehlbauer, G.J., Sato, K., Taudien, S., Sampath, D., Swarbreck, D., Scalabrin, S., Zuccolo, A., Vendramin, V., Morgante, M., Mayer, K.F.X., and Schulman, A. (2012). A physical, genetic and functional sequence assembly of the barley genome. *Nature* 491, 711-716.https://dx.doi.org/10.1038/nature11543

Mccouch, S.R., Mcnally, K.L., Wang, W., and Sackville Hamilton, R. (2012). Genomics of gene banks: A case study in rice. *American Journal of Botany* 99, 407-423.https://doi.org/10.3732/ajb.1100385

Mène-Saffrané, L., Dubugnon, L., Chételat, A., Stolz, S., Gouhier-Darimont, C., and Farmer, E.E. (2009). Nonenzymatic oxidation of trienoic fatty acids contributes to reactive oxygen species management in Arabidopsis. *Journal of Biological Chemistry* 284, 1702-1708.https://doi.org/10.1074/jbc.M807114200

Mène-Saffrané, L., Jones, A.D., and Dellapenna, D. (2010). Plastochromanol-8 and tocopherols are essential lipid-soluble antioxidants during seed desiccation and quiescence in Arabidopsis. *Proceedings of the National Academy of Sciences of the United States of America* 107, 17815-17820.https://doi.org/10.1073/pnas.1006971107

Miller, G., Shulaev, V., and Mittler, R. (2008). Reactive oxygen signaling and abiotic stress. *Physiologia Plantarum* 133, 481-489.https://doi.org/10.1111/j.1399-3054.2008.01090.x

Mira, S., Hill, L.M., Gonzalez-Benito, M.E., Ibanez, M.A., and Walters, C. (2016). Volatile emission in dry seeds as a way to probe chemical reactions during initial asymptomatic deterioration. *Journal of Experimental Botany* 67, 1783-1793.https://doi.org/10.1093/jxb/erv568

Mittler, R., Vanderauwera, S., Gollery, M., and Van Breusegem, F. (2004). Reactive oxygen gene network of plants. *Trends in Plant Science* 9, 490-498.https://doi.org/10.1016/j.tplants.2004.08.009

Miura, K., Lin, S.Y., Yano, M., and Nagamine, T. (2002). Mapping quantitative trait loci controlling seed longevity in rice (*Oryza sativa* L.). *Theoretical and Applied Genetics* 104, 981-986.https://doi.org/10.1007/s00122-002-0872-x

Moise, J.A., Han, S., Gudynaite-Savitch, L., Johnson, D.A., and Miki, B.L.A. (2005). Seed coats: Structure, development, composition, and biotechnology. *In Vitro Cellular & Developmental Biology-Plant* 41, 620-644.https://doi.org/10.1079/ivp2005686

Mondoni, A., Orsenigo, S., Muller, J.V., Carlsson-Graner, U., Jimenez-Alfaro, B., and Abeli, T. (2018). Seed dormancy and longevity in subarctic and alpine populations of *Silene suecica*. *Alpine Botany* 128, 71-81.https://doi.org/10.1007/s00035-017-0194-x

Moreau, R.A., Wayns, K.E., Flores, R.A., and Hicks, K.B. (2007). Tocopherols and tocotrienols in barley oil prepared from germ and other fractions from scarification and sieving of hulless barley. *Cereal Chemistry* 84, 587-592.https://doi.org/10.1094/Cchem-84-6-0587

Morscher, F., Kranner, I., Arc, E., Bailly, C., and Roach, T. (2015). Glutathione redox state, tocochromanols, fatty acids, antioxidant enzymes and protein carbonylation in sunflower seed embryos associated with after-ripening and ageing. *Annals of Botany* 116, 669-678.https://doi.org/10.1093/aob/mcv108

Mubarakshina, M.M., Ivanov, B.N., Naydov, I.A., Hillier, W., Badger, M.R., and Krieger-Liszkay, A. (2010). Production and diffusion of chloroplastic H_2O_2 and its implication to signalling. *Journal of Experimental Botany* 61, 3577-3587.https://doi.org/10.1093/jxb/erq171

Munné-Bosch, S., and Alegre, L. (2002). The function of tocopherols and tocotrienols in plants. *Critical Reviews in Plant Sciences* 21, 31-57.https://doi.org/10.1080/0735-260291044179

Munz, E., Rolletschek, H., Oeltze-Jafra, S., Fuchs, J., Guendel, A., Neuberger, T., Ortleb, S., Jakob, P.M., and Borisjuk, L. (2017). A functional imaging study of germinating oilseed rape seed. *New Phytologist* 216, 1181-1190.https://doi.org/10.1111/nph.14736

Nagel, M., Alqudah, A.M., Bailly, M., Rajjou, L., Pistrick, S., Matzig, G., Börner, A., and Kranner, I. (2019a). Novel loci and a role for nitric oxide for seed dormancy and pre-harvest sprouting in barley. *Plant Cell and Environment* 42, 1318-1327.https://doi.org/10.1111/pce.13483

Nagel, M., Behrens, A.-K., and Börner, A. (2013). Effects of *Rht* dwarfing alleles on wheat seed vigour after controlled deterioration. *Crop and Pasture Science* 64, 857-864.https://doi.org/10.1071/CP13041

Nagel, M., and Börner, A. (2010). The longevity of crop seeds stored under ambient conditions. *Seed Science Research* 20, 1-12.https://doi.org/10.1017/S0960258509990213

Nagel, M., Holstein, K., Willner, E., and Börner, A. (2018). Machine learning links seed composition, glucosinolates and viability of oilseed rape after 31 years of long-term storage. *Seed Science Research*, 1 9.https://doi.org/10.1017/S0960258518000259

Nagel, M., Kodde, J., Pistrick, S., Mascher, M., Börner, A., and Groot, S.P.C. (2016). Barley seed ageing: genetics behind the dry elevated pressure of oxygen ageing and moist controlled deterioration. *Frontiers in Plant Science* 7, 388.https://doi.org/10.3389/fpls.2016.00388

Nagel, M., Kranner, I., Neumann, K., Rolletschek, H., Seal, C.E., Colville, L., Fernandez-Marin, B., and Börner, A. (2015). Genome-wide association mapping and biochemical markers reveal that seed ageing and longevity are intricately affected by genetic background and developmental and environmental conditions in barley. *Plant Cell and Environment* 38, 1011-1022.https://doi.org/10.1111/pce.12474

Nagel, M., Rosenhauer, M., Willner, E., Snowdon, R.J., Friedt, W., and Börner, A. (2011). Seed longevity in oilseed rape (*Brassica napus* L.) - genetic variation and QTL mapping. *Plant Genetic Resources: Characterization and Utilization* 9, 260-263.https://doi.org/10.1017/S1479262111000372

Nagel, M., Seal, C.E., Colville, L., Rodenstein, A., Un, S., Richter, J., Pritchard, H.W., Börner, A., and Kranner, I. (2019b). Wheat seed ageing viewed through the cellular redox environment and changes in pH. *Free Radical Research* 53, 641-654.https://doi.org/10.1080/10715762.2019.1620226

Nagel, M., Vogel, H., Landjeva, S., Buck-Sorlin, G., Lohwasser, U., Scholz, U., and Börner, A. (2009). Seed conservation in ex situ genebanks—genetic studies on longevity in barley. *Euphytica* 170, 5-14.https://doi.org/10.1007/s10681-009-9975-7

Narayana Murthy, U.M., and Sun, W.Q. (2000). Protein modification by Amadori and Maillard reactions during seed storage: roles of sugar hydrolysis and lipid peroxidation. *Journal of Experimental Botany* 51, 1221-1228.https://doi.org/10.1093/jexbot/51.348.1221

Nasehzadeh, M., and Ellis, R.H. (2017). Wheat seed weight and quality differ temporally in sensitivity to warm or cool conditions during seed development and maturation. *Annals of Botany* 120, 479-493.https://doi.org/10.1093/aob/mcx074

Needham, P. (2008). Is water a mixure? Bridging the distinction between physical and chemical properties. *Studies in History and Philosophy of Science Part A* 39, 66-77.https://doi.org/10.1016/j.shpsa.2007.11.005

Nguyen, T.-P., Cueff, G., Hegedus, D.D., Rajjou, L., and Bentsink, L. (2015). A role for seed storage proteins in Arabidopsis seed longevity. *Journal of Experimental Botany* 66, 6399-6413.https://doi.org/10.1093/jxb/erv348

Nguyen, T.P., Keizer, P., Van Eeuwijk, F., Smeekens, S., and Bentsink, L. (2012). Natural variation for seed longevity and seed dormancy are negatively correlated in Arabidopsis. *Plant Physiology* 160, 2083-2092.https://doi.org/10.1104/pp.112.206649

Noctor, G., Mhamdi, A., Chaouch, S., Han, Y., Neukermans, J., Marquez-Garcia, B., Queval, G., and Foyer, C.H. (2012). Glutathione in plants: An integrated overview. *Plant Cell and Environment* 35, 454-484.https://doi.org/10.1111/j.1365-3040.2011.02400.x

Nunhems netherlands (2018). *First royalty payment to the treaty's benefit-sharing fund* [Online]. Available: https://subsites.wur.nl/en/show/First-Royalty-payment-to-the-Treatys-Benefit-sharing-Fund.htm [Accessed 28th September 2018].

Obendorf, R.L., and Gorecki, R.J. (2012). Soluble carbohydrates in legume seeds. *Seed Science Research* 22, 219-242.https://doi.org/10.1017/S0960258512000104

Oenel, A., Fekete, A., Krischke, M., Faul, S.C., Gresser, G., Havaux, M., Mueller, M.J., and Berger, S. (2017). Enzymatic and non-enzymatic mechanisms contribute to lipid oxidation during seed aging. *Plant and Cell Physiology* 58, 925-933.https://doi.org/10.1093/pcp/pcx036

Ohm, J.-B., Lee, C.W., and Cho, K. (2016). Germinated wheat: Phytochemical composition and mixing characteristics. *Cereal Chemistry Journal* 93, 612-617.https://doi.org/10.1094/CCHEM-01-16-0006-R

Oliver, M.J., Tuba, Z., and Mishler, B.D. (2000). The evolution of vegetative desiccation tolerance in land plants. *Plant Ecology* 151, 85-100.https://doi.org/10.1023/a:1026550808557

Pankin, A., and Von Korff, M. (2017). Co-evolution of methods and thoughts in cereal domestication studies: a tale of barley (*Hordeum vulgare*). *Current Opinion in Plant Biology* 36, 15-21.https://doi.org/10.1016/j.pbi.2016.12.001

Parzies, H.K., Spoor, W., and Ennos, R.A. (2000). Genetic diversity of barley landrace accessions (*Hordeum vulgare* ssp. *vulgare*) conserved for different lengths of time in *ex situ* gene banks. *Heredity* 84, 476-486.http://dx.doi.org/10.1046/j.1365-2540.2000.00705.x

Pasternak, M., Lim, B., Wirtz, M., Hell, R., Cobbett, C.S., and Meyer, A.J. (2008). Restricting glutathione biosynthesis to the cytosol is sufficient for normal plant development. *The Plant Journal* 53, 999-1012.https://doi.org/10.1111/j.1365-313X.2007.03389.x

Pearce, S., Saville, R., Vaughan, S.P., Chandler, P.M., Wilhelm, E.P., Sparks, C.A., Al-Kaff, N., Korolev, A., Boulton, M.I., Phillips, A.L., Hedden, P., Nicholson, P., and Thomas, S.G. (2011). Molecular characterization of *Rht-1* dwarfing genes in hexaploid wheat. *Plant Physiology* 157, 1820-1831.https://doi.org/10.1104/pp.111.183657

Perez-Garcia, F., Gonzalez-Benito, M.E., and Gomez-Campo, C. (2007). High viability recorded in ultra-dry seeds of 37 species of Brassicaceae after almost 40 years of storage. *Seed Science and Technology* 35, 143-153.https://doi.org/10.15258/sst.2007.35.1.13

Persson, E., and Halle, B. (2008). Cell water dynamics on multiple time scales. *Proceedings of the National Academy of Sciences of the United States of America* 105, 6266-6271.https://doi.org/10.1073/pnas.0709585105

Potters, G., De Gara, L., Asard, H., and Horemans, N. (2002). Ascorbate and glutathione: Guardians of the cell cycle, partners in crime? *Plant Physiology and Biochemistry* 40, 537-548.https://doi.org/10.1016/s0981-9428(02)01414-6

Priestley, D.A., Cullinan, V.I., and Wolfe, J. (1985). Differences in seed longevity at the species level. *Plant Cell and Environment* 8, 557-562.https://doi.org/10.1111/j.1365-3040.1985.tb01693.x

Probert, R., Adams, J., Coneybeer, J., Crawford, A., and Hay, F. (2007). Seed quality for conservation is critically affected by pre-storage factors. *Australian Journal of Botany* 55, 326-335.https://doi.org/10.1071/BT06046

Queval, G., Issakidis-Bourguet, E., Hoeberichts, F.A., Vandorpe, M., Gakiere, B., Vanacker, H., Miginiac-Maslow, M., Van Breusegem, F., and Noctor, G. (2007). Conditional oxidative stress responses in the *Arabidopsis* photorespiratory mutant cat2 demonstrate that redox state is a key modulator of daylength-dependent gene expression, and define photoperiod as a crucial factor in the regulation of H_2O_2-induced cell death. *The Plant Journal* 52, 640-657.https://doi.org/10.1111/j.1365-313X.2007.03263.x

Queval, G., Thominet, D., Vanacker, H., Miginiac-Maslow, M., Gakiere, B., and Noctor, G. (2009). H2O2-activated up-regulation of glutathione in Arabidopsis involves induction of genes encoding enzymes involved in cysteine synthesis in the chloroplast. *Molecular Plant* 2, 344-356.https://doi.org/10.1093/Mp/Ssp002

Rathjen, J.R., Strounina, E.V., and Mares, D.J. (2009). Water movement into dormant and non-dormant wheat (*Triticum aestivum* L.) grains. *Journal of Experimental Botany* 60, 1619-1631.https://doi.org/10.1093/jxb/erp037

Re, M.D., Gonzalez, C., Escobar, M.R., Sossi, M.L., Valle, E.M., and Boggio, S.B. (2017). Small heat shock proteins and the postharvest chilling tolerance of tomato fruit. *Physiologia Plantarum* 159, 148-160.https://doi.org/10.1111/ppl.12491

Rehman Arif, M.A., Nagel, M., Lohwasser, U., and Börner, A. (2017). Genetic architecture of seed longevity in bread wheat (*Triticum aestivum* L.). *Journal of Biosciences*, 1-9.https://doi.org/10.1007/s12038-016-9661-6

Rehman Arif, M.A., Nagel, M., Neumann, K., Kobiljski, B., Lohwasser, U., and Börner, A. (2012). Genetic studies of seed longevity in hexaploid wheat using segregation and association mapping approaches. *Euphytica* 186, 1-13.https://doi.org/10.1007/s10681-011-0471-5

Reis, A., and Spickett, C.M. (2012). Chemistry of phospholipid oxidation. *Biochimica et Biophysica Acta (BBA) - Biomembranes* 1818, 2374-2387.https://doi.org/10.1016/j.bbamem.2012.02.002

Riewe, D., Nagel, M., Altmann, T., Börner, A., and Wiebach, J. (2018). Ageing is highly associated with lipid oxidation and hydrolysis in wheat seeds. *Nature Plant* (submitted)

Righetti, K., Vu, J.L., Pelletier, S., Vu, B.L., Glaab, E., Lalanne, D., Pasha, A., Patel, R.V., Provart, N.J., Verdier, J., Leprince, O., and Buitink, J. (2015). Inference of longevity-related genes from a robust coexpression network of seed maturation identifies regulators linking seed storability to biotic defense-related pathways. *Plant Cell*.https://doi.org/10.1105/tpc.15.00632

Roach, T., Nagel, M., Börner, A., Eberle, C., and Kranner, I. (2018). Changes in tocochromanols and glutathione reveal differences in the mechanisms of seed ageing under seedbank conditions and controlled deterioration in barley. *Environmental and Experimental Botany* 156, 8-15.https://doi.org/10.1016/j.envexpbot.2018.08.027

Roberts, E.H. (1960). The viability of cereal seed in relation to temperature and moisture. *Annals of Botany* 24, 12-31.https://doi.org/10.1093/oxfordjournals.aob.a083684

Roberts, E.H. (1961). The viability of rice seed in relation to temperature, moisture content, and gaseous environment. *Annals of Botany* 25, 381-390.https://doi.org/10.1093/oxfordjournals.aob.a083759

Roberts, E.H. (1972). *Viability of seeds.* London: Chapman & Hall.

Roberts, E.H. (1973). Predicting the storage life of seeds. *Seed Science and Technology* 1, 499-514

Roberts, E.H., and Abdalla, F.H. (1968). The influence of temperature, moisture, and oxygen on period of seed viability in barley, broad beans, and peas. *Annals of Botany* 32, 97-117.https://doi.org/10.1093/oxfordjournals.aob.a084202

Rolletschek, H., Koch, K., Wobus, U., and Borisjuk, L. (2005). Positional cues for the starch/lipid balance in maize kernels and resource partitioning to the embryo. *The Plant Journal* 42, 69-83.https://doi.org/10.1111/j.1365-313X.2005.02352.x

Rolletschek, H., Weschke, W., Weber, H., Wobus, U., and Borisjuk, L. (2004). Energy state and its control on seed development: starch accumulation is associated with high ATP and steep oxygen gradients within barley grains. *Journal of Experimental Botany* 55, 1351-1359.https://doi.org/10.1093/jxb/erh130

Roos, Y. (2010). *Glass transition temperature and its relevance in food processing.*

Sallon, S., Solowey, E., Cohen, Y., Korchinsky, R., Egli, M., Woodhatch, I., Simchoni, O., and Kislev, M. (2008). Germination, genetics, and growth of an ancient date seed. *Science* 320, 1464-1464.https://doi.org/10.1126/science.1153600

Santos-Mendoza, M., Dubreucq, B., Baud, S., Parcy, F., Caboche, M., and Lepiniec, L. (2008). Deciphering gene regulatory networks that control seed development and maturation in Arabidopsis. *The Plant Journal* 54, 608-620.https://doi.org/10.1111/j.1365-313X.2008.03461.x

Sasaki, K., Fukuta, Y., and Sato, T. (2005). Mapping of quantitative trait loci controlling seed longevity of rice (*Oryza sativa* L.) after various periods of seed storage. *Plant Breeding* 124, 361-366.https://doi.org/10.1111/j.1439-0523.2005.01109.x

Sattler, S.E., Gilliland, L.U., Magallanes-Lundback, M., Pollard, M., and Dellapenna, D. (2004). Vitamin E is essential for seed longevity, and for preventing lipid peroxidation during germination. *Plant Cell* 16, 1419-1432.https://doi.org/10.1105/Tpc.021360

Schafer, F.Q., and Buettner, G.R. (2001). Redox environment of the cell as viewed through the redox state of the glutathione disulfide/glutathione couple. *Free Radical Biology and Medicine* 30, 1191-1212.https://doi.org/10.1016/S0891-5849(01)00480-4

Schatzki, J., Allam, M., Klöppel, C., Nagel, M., Börner, A., and Möllers, C. (2013). Genetic variation for secondary seed dormancy and seed longevity in a set of black-seeded European winter oilseed rape cultivars. *Plant Breeding* 132, 174-179.https://doi.org/10.1111/pbr.12023

Schmutzer, T., Samans, B., Dyrszka, E., Ulpinnis, C., Weise, S., Stengel, D., Colmsee, C., Lespinasse, D., Micic, Z., Abel, S., Duchscherer, P., Breuer, F., Abbadi, A., Leckband, G., Snowdon, R., and Scholz, U. (2015). Species-wide genome sequence and nucleotide polymorphisms from the model allopolyploid plant *Brassica napus*. *Scientific Data* 2, 150072.http://dx.doi.org/10.1038/sdata.2015.72

Schwember, A.R., and Bradford, K.J. (2010). Quantitative trait loci associated with longevity of lettuce seeds under conventional and controlled deterioration storage conditions. *Journal of Experimental Botany* 61, 4423-4436.https://doi.org/10.1093/Jxb/Erq248

Sebastiani, P., Gurinovich, A., Bae, H., Andersen, S., Malovini, A., Atzmon, G., Villa, F., Kraja, A.T., Ben-Avraham, D., Barzilai, N., Puca, A., and Perls, T.T. (2017). Four genome-wide association studies identify new extreme longevity variants. *Journals of Gerontology Series a-Biological Sciences and Medical Sciences* 72, 1453-1464.https://doi.org/10.1093/gerona/glx027

Sgsv (2018). *Svalbard Global Seed Vault* [Online]. The Crop Trust. Available: https://www.seedvault.no/ [Accessed 24th September 2018].

Shen-Miller, J., Mudgett, M.B., Schopf, J.W., Clarke, S., and Berger, R. (1995). Exceptional seed longevity and robust growth - ancient sacred lotus from China. *American Journal of Botany* 82, 1367-1380.https://doi.org/10.2307/2445863

Shewry, P.R. (2009). Wheat. *Journal of Experimental Botany* 60, 1537-1553.https://doi.org/10.1093/Jxb/Erp058

Singh, D.P., Jermakow, A.M., and Swain, S.M. (2002). Gibberellins are required for seed development and pollen tube growth in Arabidopsis. *Plant Cell* 14, 3133-3147.https://doi.org/10.1105/tpc.003046

Smirnoff, N. (2010). Tocochromanols: Rancid lipids, seed longevity, and beyond. *Proceedings of the National Academy of Sciences of the United States of America* 107, 17857-17858.https://doi.org/10.1073/pnas.1012749107

Specht, C.E., and Börner, A. (1998). An interim report on a long term storage experiment on rye (*Secale cereale* L.) under a range of temperatures and atmospheres. *Genetic Resources and Crop Evolution* 45, 483-488.https://doi.org/10.1023/A:1008660831897

Sperling, L.H. (2005). *Introduction to physical polymer science*. Hoboken, New Jersey: John Wiley & Sons.

Spicher, L., Glauser, G., and Kessler, F. (2016). Lipid antioxidant and galactolipid remodeling under temperature stress in tomato plants. *Frontiers in Plant Science* 7.https://doi.org/10.3389/fpls.2016.00167

Sreenivasulu, N., Radchuk, V., Strickert, M., Miersch, O., Weschke, W., and Wobus, U. (2006). Gene expression patterns reveal tissue-specific signaling networks controlling programmed cell death and ABA-regulated maturation in developing barley seeds. *The Plant Journal* 47, 310-327.https://doi.org/10.1111/j.1365-313X.2006.02789.x

Steiner, A.M., and Ruckenbauer, P. (1995). Germination of 110-year-old cereal and weed seeds, the Vienna Sample of 1877. Verification of effective ultra-dry storage at ambient temperature. *Seed Science Research* 5, 195-199.https://doi.org/10.1017/S0960258500002853

Strelec, I., Ugarcic-Hardi, Z., and Hlevnjak, M. (2008). Accumulation of Amadori and Maillard products in wheat seeds aged under different storage conditions. *Croatica Chemica Acta* 81, 131-137

Sun, W.Q. (2002). "Methods for the study of water relations under desiccation stress," in *Dessication and suvival in plants: Drying without dying*, eds. M. Black & H.W. Pritchard. (Wallingford, UK: CAB International).

Sun, W.Q., and Leopold, A.C. (1997). Cytoplasmic vitrification acid survival of anhydrobiotic organisms. *Comparative Biochemistry and Physiology a-Physiology* 117, 327-333.https://doi.org/10.1016/S0300-9629(96)00271-X

Swanson, J.M.J., Maupin, C.M., Chen, H., Petersen, M.K., Xu, J., Wu, Y., and Voth, G.A. (2007). Proton solvation and transport in aqueous and biomolecular systems: Insights from computer simulations. *The Journal of Physical Chemistry B* 111, 4300-4314.https://doi.org/10.1021/jp070104x

Szarka, A., Tomasskovics, B., and Bánhegyi, G. (2012). The Ascorbate-glutathione-α-tocopherol triad in abiotic stress response. *International Journal of Molecular Sciences* 13, 4458-4483.https://doi.org/10.3390/ijms13044458

Taketa, S., Amano, S., Tsujino, Y., Sato, T., Saisho, D., Kakeda, K., Nomura, M., Suzuki, T., Matsumoto, T., Sato, K., Kanamori, H., Kawasaki, S., and Takeda, K. (2008). Barley grain with adhering hulls is controlled by an ERF family transcription factor gene regulating a lipid biosynthesis pathway. *Proceedings of the National Academy of Sciences of the United States of America* 105, 4062-4067.https://doi.org/10.1073/pnas.0711034105

Telewski, F.W., and Zeevaart, J.a.D. (2002). The 120-yr period for Dr. Beal's seed viability experiment. *American Journal of Botany* 89, 1285-1288.https://doi.org/10.3732/ajb.89.8.1285

Thoke, H.S., Bagatolli, L.A., and Olsen, L.F. (2018). Effect of macromolecular crowding on the kinetics of glycolytic enzymes and the behaviour of glycolysis in yeast. *Integrative Biology*.https://doi.org/10.1039/C8IB00099A

Thorlby, G., Veale, E., Butcher, K., and Warren, G. (1999). Map positions of *SFR* genes in relation to other freezing-related genes of *Arabidopsis thaliana*. *The Plant Journal* 17, 445-452.https://doi.org/10.1046/j.1365-313X.1999.00395.x

Thornton, J.M., Powell, A.A., and Matthews, S. (1990). Investigation of the relationship between seed leachate conductivity and the germination of Brassica seed. *Annals of Applied Biology* 117, 129-135.https://doi.org/10.1111/j.1744-7348.1990.tb04201.x

Tiedemann, J., Rutten, T., Monke, G., Vorwieger, A., Rolletschek, H., Meissner, D., Milkowski, C., Petereck, S., Mock, H.P., Zank, T., and Baumlein, H. (2008). Dissection of a complex seed phenotype: Novel insights of *FUSCA3* regulated developmental processes. *Developmental Biology* 317, 1-12.https://doi.org/10.1016/j.ydbio.2008.01.034

Tolleter, D., Hincha, D.K., and Macherel, D. (2010). A mitochondrial late embryogenesis abundant protein stabilizes model membranes in the dry state. *Biochimica et Biophysica Acta (BBA) - Biomembranes* 1798, 1926-1933.http://dx.doi.org/10.1016/j.bbamem.2010.06.029

Triantaphylidès, C., and Havaux, M. (2009). Singlet oxygen in plants: production, detoxification and signaling. *Trends in Plant Science* 14, 219-228.https://doi.org/10.1016/j.tplants.2009.01.008

Verdier, J., Lalanne, D., Pelletier, S., Torres-Jerez, I., Righetti, K., Bandyopadhyay, K., Leprince, O., Chatelain, E., Vu, B.L., Gouzy, J., Gamas, P., Udvardi, M.K., and Buitink, J. (2013). A regulatory network-based approach dissects late maturation processes related to the acquisition of desiccation tolerance and longevity of *Medicago truncatula* seeds. *Plant Physiology* 163, 757-774.https://doi.org/10.1104/pp.113.222380

Vom Dorp, K., Hölzl, G., Plohmann, C., Eisenhut, M., Abraham, M., Weber, A.P.M., Hanson, A.D., and Dörmann, P. (2015). Remobilization of phytol from chlorophyll degradation is essential for tocopherol synthesis and growth of Arabidopsis. *Plant Cell* 27, 2846-2859.https://doi.org/10.1105/tpc.15.00395

Vries, J., and Archibald, J.M. (2018). Plant evolution: Landmarks on the path to terrestrial life. *New Phytologist* 217, 1428-1434.https://doi.org/10.1111/nph.14975

Walters, C. (2014). Extreme Biology: Probing Life at Low Water Contents and Temperatures. *Ii International Symposium on Plant Cryopreservation* 1039, 49-56

Walters, C. (2015). Orthodoxy, recalcitrance and in-between: Describing variation in seed storage characteristics using threshold responses to water loss. *Planta* 242, 397-406.https://doi.org/10.1007/s00425-015-2312-6

Walters, C., Ballesteros, D., and Vertucci, V.A. (2010). Structural mechanics of seed deterioration: Standing the test of time. *Plant Science* 179, 565-573.https://doi.org/10.1016/j.plantsci.2010.06.016

Walters, C., Wheeler, L., and Stanwood, P.C. (2004). Longevity of cryogenically stored seeds. *Cryobiology* 48, 229-244.DOI 10.1016/j.cryobiol.2004.01.007

Walters, C., Wheeler, L.M., and Grotenhuis, J.M. (2005). Longevity of seeds stored in a genebank: Species characteristics. *Seed Science Research* 15, 1-20.https://doi.org/10.1079/Ssr2004195

Wang, X.H., Long, Y., Wang, N., Zou, J., Ding, G.D., Broadley, M.R., White, P.J., Yuan, P., Zhang, Q.W., Luo, Z.L., Liu, P.F., Zhao, H., Zhang, Y., Cai, H.M., King, G.J., Xu, F.S., Meng, J.L., and Shi, L. (2017). Breeding histories and selection criteria for oilseed rape in Europe and China identified by genome wide pedigree dissection. *Scientific Reports* 7.https://doi.org/10.1038/s41598-017-02188-z

Waszczak, C., Carmody, M., and Kangasjarvi, J. (2018). Reactive oxygen species in plant signaling. *Annual Review of Plant Biology* 69, 209-236.https://doi.org/10.1146/annurev-arplant-042817-040322

Weber, H., Borisjuk, L., and Wobus, U. (2005). Molecular physiology of legume seed development. *Annual Review of Plant Biology* 56, 253-279.https://doi.org/10.1146/annurev.arplant.56.032604.144201

Weschke, W., Panitz, R., Gubatz, S., Wang, Q., Radchuk, R., Weber, H., and Wobus, U. (2003). The role of invertases and hexose transporters in controlling sugar ratios in maternal and filial tissues of barley caryopses during early development. *The Plant Journal* 33, 395-411.https://doi.org/10.1046/j.1365-313X.2003.01633.x

Wettlaufer, S.H., and Leopold, A.C. (1991). Relevance of Amadori and Maillard products to seed deterioration. *Plant Physiology* 97, 165-169.https://doi.org/10.1104/pp.97.1.165

Whent, M., Huang, H., Xie, Z., Lutterodt, H., Yu, L., Fuerst, E.P., Morris, C.F., Yu, L., and Luthria, D. (2012). Phytochemical composition, anti-inflammatory, and antiproliferative activity of whole wheat flour. *Journal of Agricultural and Food Chemistry* 60, 2129-2135.https://doi.org/10.1021/jf203807w

Wilkinson, M.D., Dumontier, M., Aalbersberg, I.J., Appleton, G., Axton, M., Baak, A., Blomberg, N., Boiten, J.-W., Da Silva Santos, L.B., Bourne, P.E., Bouwman, J., Brookes, A.J., Clark, T., Crosas, M., Dillo, I., Dumon, O., Edmunds, S., Evelo, C.T., Finkers, R., Gonzalez-Beltran, A., Gray, A.J.G., Groth, P., Goble, C., Grethe, J.S., Heringa, J., 'T Hoen, P.a.C., Hooft, R., Kuhn, T., Kok, R., Kok, J., Lusher, S.J., Martone, M.E., Mons, A., Packer, A.L., Persson, B., Rocca-Serra, P., Roos, M., Van Schaik, R., Sansone, S.-A., Schultes, E., Sengstag, T., Slater, T., Strawn, G., Swertz, M.A., Thompson, M., Van Der Lei, J., Van Mulligen, E., Velterop, J., Waagmeester, A., Wittenburg, P., Wolstencroft, K., Zhao, J., and Mons, B. (2016). The FAIR guiding principles for scientific data management and stewardship. *Scientific Data* 3, 160018.https://doi.org/10.1038/sdata.2016.18

Williams, R.J., and Leopold, A.C. (1989). The glassy state in corn embryos. *Plant Physiology* 89, 977-981.https://doi.org/10.1104/pp.89.3.977

Wolkers, W.F., Alberda, M., Koornneef, M., Léon-Kloosterziel, K.M., and Hoekstra, F.A. (1998). Properties of proteins and the glassy matrix in maturation-defective mutant seeds of *Arabidopsis thaliana*. *The Plant Journal* 16, 133-143.https://doi.org/10.1046/j.1365-313x.1998.00277.x

Wolkers, W.F., Oliver, A.E., Tablin, F., and Crowe, J.H. (2004). A Fourier-transform infrared spectroscopy study of sugar glasses. *Carbohydrate Research* 339, 1077-1085.https://doi.org/10.1016/j.carres.2004.01.016

Woodfield, H.K., Sturtevant, D., Borisjuk, L., Munz, E., Guschina, I.A., Chapman, K., and Harwood, J.L. (2017). Spatial and temporal mapping of key lipid species in *Brassica napus* seeds. *Plant Physiology* 173, 1998-2009.https://doi.org/10.1104/pp.16.01705

Wozny, D., Kramer, K., Finkemeier, I., Acosta Ivan, F., and Koornneef, M. (2018). Genes for seed longevity in barley identified by genomic analysis on near isogenic lines. *Plant Cell and Environment* in press.https://doi.org/10.1111/pce.13330

Xiong, F., Yu, X.R., Zhou, L., Wang, F., and Xiong, A.S. (2013). Structural and physiological characterization during wheat pericarp development. *Plant Cell Reports* 32, 1309-1320.https://doi.org/10.1007/s00299-013-1445-y

Xue, Y., Zhang, S.Q., Yao, Q.H., Peng, R.H., Xiong, A.S., Li, X., Zhu, W.M., Zhu, Y.Y., and Zha, D.S. (2008). Identification of quantitative trait loci for seed storability in rice (*Oryza sativa* L.). *Euphytica* 164, 739-744.https://doi.org/10.1007/s10681-008-9696-3

Yashina, S., Gubin, S., Maksimovich, S., Yashina, A., Gakhova, E., and Gilichinsky, D. (2012). Regeneration of whole fertile plants from 30,000-y-old fruit tissue buried in Siberian permafrost. *Proceedings of the National Academy of Sciences of the United States of America* 109, 4008-4013.https://doi.org/10.1073/pnas.1118386109

Yashina, S.G., Gubin, S.V., Shabaeva, E.V., Egorova, E.F., and Maksimovich, S.V. (2002). Viability of higher plant seeds of late pleistocene age from permafrost deposits as determined by *in vitro* culturing. *Doklady Biological Sciences* 383, 151-154.https://doi.org/10.1023/A:1015350209946

Zazula, G.D., Harington, C.R., Telka, A.M., and Brock, F. (2009). Radiocarbon dates reveal that *Lupinus arcticus* plants were grown from modern not Pleistocene seeds. *New Phytologist* 182, 788-792.https://doi.org/10.1111/j.1469-8137.2009.02818.x

Zeng, D.L., Guo, L.B., Xu, Y.B., Yasukumi, K., Zhu, L.H., and Qian, Q. (2006). QTL analysis of seed storability in rice. *Plant Breeding* 125, 57-60.https://doi.org/10.1111/j.1439-0523.2006.01169.x

Zinsmeister, J., Lalanne, D., Terrasson, E., Chatelain, E., Vandecasteele, C., Vu, B.L., Dubois-Laurent, C., Geoffriau, E., Le Signor, A., Dalmais, M., Gutbrod, K., Dormann, P., Gallardo, K., Bendahmane, A., Buitink, J., and Leprince, O. (2016). *ABI5* is a regulator of seed maturation and longevity in legumes. *Plant Cell* 28, 2735-2754.https://doi.org/10.1105/tpc.16.00470

ACKNOWLEDGEMENTS

Mit Beginn des Jahres 2012 fiel die Entscheidung in ferner Zukunft eine Habilitation zu verfassen. Nun ist es tatsächlich soweit. Die Zeilen sind nach bestmöglichen Wissen und Gewissen gefüllt und ich möchte mich für die großartige Unterstützung bedanken. Der größte Dank geht dabei an Herr PD Dr. Andreas Börner. Ohne dessen Begeisterung für die Wissenschaft, Ideenreichtum, Vertrauen, Motivation und Tatendrang wären so manche illusionären Vorhaben nicht möglich gewesen. Ein tiefer Dank gebührt ebenso Prof. Ilse Kranner für die Vielzahl an fachlichen Diskussionen und dass sie mich in vielen Situationen geerdet und motiviert hat. Gelernt habe ich vor allem drei Dinge: 1. Wir machen keine schlampigen Experimente! 2. Schreiben ist eine Kunst, die nicht jeder beherrscht! und 3. Wenn Du zu langsam bist, werde schneller! Für das Vertrauen in meine Fähigkeiten Vorlesungen zu geben, möchte ich mich bei Prof. Wolfgang Link, Prof. Rolf Rauber und besonders bei Prof. Michael Kruse bedanken, der mich großartig bei der formalen Realisierung unterstützt und mit Rat und Tat zur Seite steht.

Ein ganz großes Dankeschön geht dabei an die Menschen, die mir dauerhaft geholfen, mich beraten, motiviert und die Versuche kritisch beleuchtet haben. Ohne die technische Unterstützung von, allen voran Sibylle Pistrick, Gabriele Matzig, Stefanie Thumm, Annette Marlow, Josefine Richter and den bereits ausgeschiedenen Kolleginnen Anita Winger und Renate Voss wäre die Vielzahl an erfolgreichen Publikationen in den vergangenen 10 Jahren nicht möglich gewesen. Hinzukommt ein großes Dankeschön für die Bereitstellung an Materialien und wertvollen Erfahrungen der einzelnen Sortimentsgruppen vertreten durch Karina Krusch, Bärbel Schmidt und ihrer Nachfolgerin Britta Ruckwied, Renate Kurch, Matthias Kotter and Michael Grau, und die Ermöglichung von Gewächshaus- and Feldversuchen durch Jürgen Marlow und Peter Schreiber.

Ein besonderer Dank geht auch an Herrn Dr. Joachim Keller und meinen Kolleginnen der AG CSB, ohne deren Motivation, Unterstützung und Toleranz eine Saatgutwissenschaftlerin nie eine Habilitation unter dem Deckmantel der Krybiologie hätte schreiben können.

Zum Schluss möchte ich auch meiner Familie, meinen Freunden und meinen engsten Kollegen, Rasha, Ahmad und Natalia, ein großes Dankeschön entgegenbringen, durch die auch arbeitsreiche Wochenenden Spaß machen konnten.

CURRICULUM VITAE

Name	PD Dr. sc. agr. Manuela Nagel
Date of birth	4th October 1981
Nationality	German
Current position	Head of Cryo and Stress Biology group Leibniz Institute of Plant Genetics and Crop Plant Research (IPK), Gatersleben, Seeland, Germany
Email	nagel@ipk-gatersleben.de
Webpage	www.ipk-gatersleben.de/en/genebank/cryo-and-stress-biology/

Identification

- OrcID http://orcid.org/0000-0003-0396-0333
- ResearcherID http://www.researcherid.com/rid/T-8816-2017

Education

- PhD thesis (2011): Seed survival in genebanks –genetic and biochemical aspects of seed deterioration in barley, University of Gottingen
- MSc thesis (2007): Langlebigkeit von Saatgut unter ambienten Lagerungsbedingungen in der *ex situ* Genbank für landwirtschaftliche und gartenbauliche Kulturpflanzen in Gatersleben, University of Gottingen

Employment history

- 19 Oct 2019 to current: Lecturer at University of Hohenheim
- 01 Jan 2016 to current: Head of Cryo and Stress Biology working group, IPK Gatersleben
- 01 July 2011 to 31 Dec 2015: PostDoc project 'EcoSeed', IPK Gatersleben
- Jan 2008 to 30 June 2011: PhD project, IPK Gatersleben

Research interests

- Long-term preservation of plant genetic resources, focussing on potato, mint, barley, wheat, oilseed rape
- Survival strategies of meristematic tissues, seeds, pollen in extreme environments
- Plant stress physiology: free radicals, antioxidants and redox signalling
- Biochemical markers of cell viability, seed quality and dormancy
- Lipid peroxidation, fats and fatty acids
- Quantitative mapping